T0245287

# Buckwheat Germplasm in the World

# Buckwheat Germplasm in the World

Edited by

## Meiliang Zhou

Institute of Crop Sciences, Chinese Academy of Agricultural Sciences, Beijing, China

## Ivan Kreft

Nutrition Institute, Ljubljana, Slovenia

## Galina Suvorova

All-Russia Research Institute of Legumes and Groat Crops, Orel, Russia

## Yu Tang

Department of Tourism, Sichuan Tourism University, Chengdu, China

## Sun Hee Woo

College of Agriculture, Life and Environment Science, Chungbuk National University, Cheongju, South Korea

ACADEMIC PRESS

An imprint of Elsevier

Academic Press is an imprint of Elsevier
125 London Wall, London EC2Y 5AS, United Kingdom
525 B Street, Suite 1800, San Diego, CA 92101-4495, United States
50 Hampshire Street, 5th Floor, Cambridge, MA 02139, United States
The Boulevard, Langford Lane, Kidlington, Oxford OX5 1GB, United Kingdom

Copyright © 2018 Elsevier Inc. All rights reserved.

No part of this publication may be reproduced or transmitted in any form or by any means, electronic or
mechanical, including photocopying, recording, or any information storage and retrieval system, without
permission in writing from the publisher. Details on how to seek permission, further information about the
Publisher's permissions policies and our arrangements with organizations such as the Copyright Clearance
Center and the Copyright Licensing Agency, can be found at our website: www.elsevier.com/permissions.

This book and the individual contributions contained in it are protected under copyright by the Publisher (other
than as may be noted herein).

**Notices**
Knowledge and best practice in this field are constantly changing. As new research and experience broaden
our understanding, changes in research methods, professional practices, or medical treatment may become
necessary.

Practitioners and researchers must always rely on their own experience and knowledge in evaluating and using
any information, methods, compounds, or experiments described herein. In using such information or methods
they should be mindful of their own safety and the safety of others, including parties for whom they have a
professional responsibility.

To the fullest extent of the law, neither the Publisher nor the authors, contributors, or editors, assume any
liability for any injury and/or damage to persons or property as a matter of products liability, negligence or
otherwise, or from any use or operation of any methods, products, instructions, or ideas contained in the
material herein.

**British Library Cataloguing-in-Publication Data**
A catalogue record for this book is available from the British Library

**Library of Congress Cataloging-in-Publication Data**
A catalog record for this book is available from the Library of Congress

ISBN: 978-0-12-811006-5

For Information on all Academic Press publications
visit our website at https://www.elsevier.com/books-and-journals

Working together
to grow libraries in
developing countries

www.elsevier.com • www.bookaid.org

*Publisher*: Andre G. Wolff
*Acquisition Editor*: Nancy Maragioglio
*Editorial Project Manager*: Billie Jean Fernandez
*Production Project Manager*: Punithavathy Govindaradjane
*Cover Designer*: Miles Hitchen

Typeset by MPS Limited, Chennai, India

# Contents

**v**

# List of Contributors

**Taiji Adachi**
Institute for Plant Biotechnology R & D, Ltd., Osaka, Japan

**Anton N. Akulov**
Kazan Institute of Biochemistry and Biophysics of Kazan Science Centre of the Russian Academy of Sciences, Kazan, Russia; Kazan Federal University, Kazan, Russia

**Gerardo Baviello**
ENEA, Rome, Italy

**Alexander Betekhtin**
University of Silesia in Katowice, Katowice, Poland

**Wioletta Biel**
West Pomeranian University of Technology in Szczecin, Szczecin, Poland

**Marian Brestic**
Slovak University of Agriculture in Nitra, Nitra, Slovak Republic

**Andrea Brunori**
ENEA, Rome, Italy

**Kwang-Soo Cho**
Highland Agriculture Research Institute, National Institute of Crop Science, Rural Development Administration, Pyeongchang, South Korea

**Seong-Woo Cho**
Chonbuk National University, Jeonju, South Korea

**Jong-Soon Choi**
Chungnam National University, Daejeon, South Korea; Korea Basic Science Institute, Daejeon, South Korea

**Yu-Mi Choi**
National Academy of Agricultural Science, Suwon, South Korea

**Keun-Yook Chung**
College of Agriculture, Life and Environment Science, Chungbuk National University, Cheongju, South Korea

**Fayaz A. Dar**
University of Kashmir, Srinagar, Jammu and Kashmir, India

**Xianyu Deng**
Kunming University of Science and Technology, Kunming, Yunnan, China

**Mengqi Ding**
Sichuan Agricultural University, Yaan, China

**Aleksey N. Fesenko**
All-Russia Research Institute of Legumes and Groat Crops, Orel, Russia

**Ivan N. Fesenko**
All-Russia Research Institute of Legumes and Groat Crops, Orel, Russia

**Nikolay N. Fesenko**
All-Russia Research Institute of Legumes and Groat Crops, Orel, Russia

**Mateja Germ**
University of Ljubljana, Ljubljana, Slovenia

**Ravinder N. Gohil**
University of Jammu, Jammu, Jammu and Kashmir, India

**Elena A. Gumerova**
Kazan Institute of Biochemistry and Biophysics of Kazan Science Centre of the Russian Academy of Sciences, Kazan, Russia

**Robert Hasterok**
University of Silesia in Katowice, Katowice, Poland

**Su-Young Hong**
Highland Agriculture Research Institute, National Institute of Crop Science, Rural Development Administration, Pyeongchang, South Korea

**Yeon-Ju Jung**
National Academy of Agricultural Science, Suwon, South Korea

**Abu Hena Mostafa Kamal**
University of Texas at Arlington, Arlington, TX, United States

**Chung-Kon Kim**
National Academy of Agricultural Science, Suwon, South Korea

**Hag H. Kim**
Woosong College, Daejeon, South Korea

**Alexey Grigoryevich Klykov**
Federal State Budget Scientific Institution "Far Eastern Regional Agricultural Scientific Centre" and Federal State Budget Scientific Institution "Primorsky Scientific Research Institute of Agriculture", Russia

**Vladimir A. Koshkin**
N. I. Vavilov Institute of Plant Genetic Resources (VIR), St. Petersburg, Russia

**Ivan Kreft**
Nutrition Institute, Ljubljana, Slovenia

**Soo J. Kwon**
College of Agriculture, Life and Environment Science, Chungbuk National University, Cheongju, South Korea

**Gi-An Lee**
National Academy of Agricultural Science, Suwon, South Korea

**Jung-Ro Lee**
National Academy of Agricultural Science, Suwon, South Korea

**Moon-Soon Lee**
College of Agriculture, Life and Environment Science, Chungbuk National University, Cheongju, South Korea

**Myung-Chul Lee**
National Academy of Agricultural Science, Suwon, South Korea

**Zlata Luthar**
University of Ljubljana, Ljubljana, Slovenia

**Kyung-Ho Ma**
National Academy of Agricultural Science, Suwon, South Korea

**Bisma Malik**
University of Kashmir, Srinagar, Jammu and Kashmir, India

**Chiara Nobili**
ENEA, Rome, Italy

**Ohmi Ohnishi**
Kyoto University, Kyoto-fu, Japan

**Domenico Palumbo**
ENEA, Rome, Italy

**Chang Ha Park**
Chungnam National University, Daejeon, South Korea

**Cheol-Ho Park**
Kangwon National University, Chuncheon, South Korea

**Hong-Jae Park**
National Academy of Agricultural Science, Suwon, South Korea

**Sang Un Park**
Chungnam National University, Daejeon, South Korea

**Tanveer B. Pirzadah**
University of Kashmir, Srinagar, Jammu and Kashmir, India

**Silvia Procacci**
ENEA, Rome, Italy

**Reiaz U. Rehman**
University of Kashmir, Srinagar, Jammu and Kashmir, India

**Olga I. Romanova**
N. I. Vavilov Institute of Plant Genetic Resources (VIR), St. Petersburg, Russia

**Swapan K. Roy**
College of Agriculture, Life and Environment Science, Chungbuk National University, Cheongju, South Korea

**Chen Ruan**
Kunming University of Science and Technology, Kunming, Yunnan, China

**Natalya I. Rumyantseva**
Kazan Institute of Biochemistry and Biophysics of Kazan Science Centre of the Russian Academy of Sciences, Kazan, Russia; Kazan Federal University, Kazan, Russia

**Agnieszka Rybicka**
University of Silesia in Katowice, Katowice, Poland

**Jirong Shao**
Sichuan Agricultural University, Yaan, China

**Geeta Sharma**
University of Jammu, Jammu, Jammu and Kashmir, India

**Vida Škrabanja**
University of Ljubljana, Ljubljana, Slovenia

**Iryna Smetanska**
Hochschule Weihenstephan-Triesdorf, Weidenbach, Germany

**Jae Young Song**
National Academy of Agricultural Science, Suwon, South Korea

**Galina Suvorova**
All-Russia Research Institute of Legumes and Groat Crops, Orel, Russia

**Oksana Sytar**
Slovak University of Agriculture in Nitra, Nitra, Slovak Republic; Taras
Shevchenko National University of Kyiv, Kyiv, Ukraine

**Inayatullah Tahir**
University of Kashmir, Srinagar, Jammu and Kashmir, India

**Yixiong Tang**
Biotechnology Research Institute, Chinese Academy of Agricultural
Sciences, Beijing, China

**Yu Tang**
Sichuan Tourism University, Chengdu, China

**Lyubov K. Taranenko**
Scientific-Production enterprise Antaria, SEO, Ukraine

**Pavlo P. Taranenko**
Scientific-Production enterprise Antaria, SEO, Ukraine

**Taras P. Taranenko**
Scientific-Production enterprise Antaria, SEO, Ukraine

**György Végvári**
Kaposvár University, Kaposvár, Hungary

**Joanna Wolna**
University of Silesia in Katowice, Katowice, Poland

**Sun Hee Woo**
College of Agriculture, Life and Environment Science, Chungbuk National
University, Cheongju, South Korea

**Yanmin Wu**
Biotechnology Research Institute, Chinese Academy of Agricultural
Sciences, Beijing, China

**Oleh L. Yatsyshen**
National Scientific Center "Institute of Agriculture" of the National Academy
of Agricultural Sciences, Ukraine

**Mun-Sup Yoon**
National Academy of Agricultural Science, Suwon, South Korea

**Meiliang Zhou**
Institute of Crop Sciences, Chinese Academy of Agricultural Sciences,
Beijing, China

# Foreword by Prof. Ohnishi

In 2016 I acquired a copy of the book *Molecular Breeding and Nutritional Aspects of Buckwheat*, edited by M. Zhou, I. Kreft, S. H. Woo, N. Chrungoo, and G. Wieslander. At the same time I heard that Dr. Meiliang Zhou (China) and Prof. Ivan Kreft (Slovenia) were planning to publish another book on buckwheat, titled *Buckwheat Germplasm in the World*.

As I said in the foreword of the book *Molecular Breeding and Nutritional Aspects of Buckwheat*, we, buckwheat scientists, are slightly lagging behind the scientists involved in molecular studies of other major crops. Now I say that it is time for us to catch up with them, and by saying so I want to encourage buckwheat scientists, in particular young buckwheat scientists.

I was very glad to hear that the second book on buckwheat, *Buckwheat Germplasm in the World*, was being planned. It is my pleasure to have a chance to write a foreword for the book.

Farmers now cultivate mainly local landraces of buckwheat, even in Japan, where many registered cultivars are available. A collection of buckwheat germplasm is a set of collected landraces that have been passed down to farmers from their ancestors—from generation to generation. Buckwheat germplasm must be first collected and then preserved in appropriate institutions. By hearing the reports of participants and reading the proceedings of the 13th International Symposium on Buckwheat at Cheongju in South Korea (2016), I realized that the collection of landraces of cultivated buckwheat had been finished in many countries. In China, the collection of wild buckwheat species is now underway. Now is the time for conservations, for money- and labor-consuming operations. Afterwards, germplasm must be evaluated. After that, the germplasm which has been deemed valuable will undergo the routine of conservation and wait until it is used as breeding material.

The publication of a book on buckwheat germplasm in the world will facilitate the work of young buckwheat scientists who may collect, maintain, and evaluate buckwheat germplasm in near future. By reading the book, young buckwheat scientists may learn about the kinds of buckwheat germplasm available for breeding, and they will know which institution provides them.

Since buckwheat cultivation has been declining in many countries, the genetic germplasm of buckwheat is also on the decline. It is high time for many countries to start collecting and conserving buckwheat germplasm.

By collecting, conserving, and evaluating buckwheat germplasm, we can obtain a good buckwheat germplasm collection which will in future enable conducting appropriate molecular breeding of buckwheat, and this may in turn increase the production of buckwheat grains and lead to a greater consumption of buckwheat flour.

**Ohmi Ohnishi**
*Kyoto University, Kyoto-fu, Japan*

# Foreword by Prof. Fan

The chapters of the present book, *Buckwheat Germplasm in the World*, are written by established scientists from many countries, and within them they describe buckwheat genetic resources as the basis for buckwheat breeding. Contemporary buckwheat breeding uses diverse genetic resources, including those of the wild relatives of cultivated buckwheat, to reach the goal—namely of improving buckwheat cultivars and the technology of buckwheat cultivation. Nowadays not only a high quantity of crop yield is appreciated, but also the high quality of buckwheat yield, as used in human and animal nutrition, as well as in pharmaceutical, medical, and veterinary products based on natural buckwheat materials.

China is the global center of common and Tartary buckwheat origin, especially the high-altitude regions of southwest China, where the diverse ancient cultivars of common and Tartary buckwheat as well as the wild relatives of cultivated buckwheat are grown. The chapters of the present book describe many wild buckwheat species which could be used as a valuable genetic source for buckwheat breeding.

Buckwheat growing and utilization are described in many ancient Chinese books on medicine and agriculture. Chinese folk legends, ancient poems and songs describe the importance of buckwheat for people. We expect that this book, written and edited by an international team of established buckwheat scientists, will be a useful tool to even better understand and utilize buckwheat genetic resources for further buckwheat breeding, cultivation, and product development.

This book is highly recommended for a broad circle of scientists, professors, students, and experts working in agriculture and in agro-industry, as well as for the general public.

**Prof. Yunliu Fan**
*Chinese Academy of Agricultural Sciences, Beijing, China*

# Foreword by Prof. Park

Buckwheat germplasm is the basis of the past and present improvements of buckwheat species. The recently increasing demand for buckwheat materials in food and pharmaceutical industry requires wider and more sophisticated improvements in buckwheat. In addition to higher yields and good quality, the buckwheat improvements based on advances in scientific and agricultural disciplines (botany, taxonomy, genetics, molecular biology, biotechnology) are expected to facilitate functional foods production, medicinal use, and other industrial benefits. Buckwheat germplasm in particular plays important roles in producing, breeding, processing, and cultural application of buckwheat seeds.

Despite its pivotal role in the development of buckwheat science and buckwheat industry, buckwheat germplasm has been slighted in the scientific communication system. Few journals are devoted exclusively to buckwheat improvement. Although there are some international journals of plant breeding and genetics, topics in buckwheat germplasm have been few and diminishing. The resultant fragmentation of the buckwheat improvement literature suggests that a review publication devoted to this subject is particularly appropriate.

Thus, the goal of *Buckwheat Germplasm in the World* will be to consolidate all aspects related to buckwheat improvement. It will endeavor to emphasize the major botanical information on buckwheat species, including classification, distribution, and description, and to overview biotechnological approaches, such as interspecific hybridization, cell culture, genetic markers and molecular genetics of buckwheat species. It will contribute to the expansion of the scientific knowledge of buckwheat species and also to the development of the buckwheat industry.

**Prof. Cheol Ho Park**
*Kangwon National University, Chuncheon, South Korea*

# Preface

Three-sided buckwheat grain is an interface between past and future buckwheat generations. In the past, buckwheat plants were inter-pollinated, which generated and maintained genotype diversity. The buckwheat plants which were most suitable for survival and propagation in each actual environment were able to maintain their place in the field and in the population. Thus, for example, buckwheat populations in high altitude areas accumulated genes for the synthesis of metabolites, protecting plants against UV radiation, as well as against pests and diseases. Human-made environments were conducive to plants with higher yields and larger grain, and to those resistant to lodging and grain-shedding. Different populations and domestic varieties were developed in diverse environments and with different cultivation techniques.

Buckwheat seeds and cultivation moved eastwards from the most probable area of origin of cultivated buckwheat in the Eastern Himalayas over China to Korea and Japan. It also moved westwards to Bhutan, Nepal, North India, and probably further west. Caravans and merchants on the Silk Road apparently took buckwheat grain with them to West China and further west. The Black Sea region, once dominated by Greeks and Tatars, was probably an important station for the spread of buckwheat cultivation, as suggested by words for the plant derived from Greek (Gryka, Grečka, Grechiha) or Tatar. After the 14th century, buckwheat had thus spread over most of Central Europe. As an undemanding and low-input crop, it was used as a contingency food source and a lifesaver, especially for the poor. Much later, buckwheat cultivation was spread by migrants from Europe, or directly from Asia, to North and South America and Australia. Recently, some buckwheat is also grown in Africa.

Through the centuries, different domestic cultivars of buckwheat developed—depending on the diversity of environment, the manner of cultivation, and the utilization value, with a plethora of various genes. Such material is an important natural and cultural heritage of mankind. It is important to preserve these precious materials, not only in long-term seed storage facilities ("gene banks"), but also by cultivation and development of new cultivars, based on the diversity of the preserved genetic material and its utilization value. Part of the important natural heritage are also the wild relatives of cultivated buckwheat. Some new discoveries of the wild relatives, also presented in this book, show that the wild buckwheat species are still not sufficiently investigated and utilized in buckwheat breeding.

The previous book *Molecular Breeding and Nutritional Aspects of Buckwheat* (Editors M. Zhou, I. Kreft, S. H. Woo, N. Chrungoo, and

G. Wieslander), published in 2016 by Elsevier (and its imprint Academic Press), was presented at the 13th International Symposium on Buckwheat and the General Assembly of International Buckwheat Research Association, Cheongju and Bongpyeong, South Korea (September 7−11, 2016) and it received a lot of attention and appreciation.

At the 13th International Symposium on Buckwheat and IBRA Assembly some other decisions were made, which are important for further international cooperation in the field of buckwheat. The president of IBRA for the period 2016−19 became Sun Hee Woo, professor at the Department of Crop Science at the Chungbuk National University, Cheongju, South Korea. The 14th International Symposium on Buckwheat and IBRA Assembly will take place in Shillong and Shrinagar in India, which is represented by Nikhil K. Chrungoo and Jai Chand Rana.

Toshiko Matano, Ohmi Ohnishi, and Kiyokazu Ikeda and their colleagues were greatly appreciated for editing Fagopyrum (scientific journal on buckwheat research), which was published in the period 1995−2016 in Japan (Ina-Minamiminowa, Kyoto, and Kobe respectively); the suggestion of the previous editors of Fagopyrum to move the head office of the journal in 2017 back to Ljubljana, Slovenia, under the editorship of I. Kreft and his colleagues, was confirmed. It was also confirmed that September 3rd is designated as "the World Day of Buckwheat" to commemorate the day in 1980 when in Ljubljana, Slovenia, the International Buckwheat Research Association (IBRA) was established.

We expect that the present book, an important source in buckwheat research, the chapters of which have been well written by scientists who have contributed significantly to the research of buckwheat plant genetic resources, will be of much advantage and pleasure to readers interested in buckwheat genetic resources and related fields.

**Meiliang Zhou, Ivan Kreft, Galina Suvorova,**
**Yu Tang, and Sun Hee Woo**

# Chapter One

# Overview of Buckwheat Resources in the World

**Meiliang Zhou[1], Yu Tang[2], Xianyu Deng[3], Chen Ruan[3], Ivan Kreft[4], Yixiong Tang[5], and Yanmin Wu[5]**

[1]*Institute of Crop Sciences, Chinese Academy of Agricultural Sciences, Beijing, China,* [2]*Sichuan Tourism University, Chengdu, China,* [3]*Kunming University of Science and Technology, Kunming, Yunnan, China,* [4]*Nutrition Institute, Ljubljana, Slovenia,* [5]*Biotechnology Research Institute, Chinese Academy of Agricultural Sciences, Beijing, China*

Buckwheat is a dicotyledonous herb which belongs to the Polygonaceae family and *Fagopyrum* genus. Buckwheat is a joint name of two cultivated species, common buckwheat (*F. esculentum* Moench) and Tartary buckwheat (*F. tataricum* Gaertn). Although it is actually not a kind of triticeae, the usage of buckwheat seeds (achenes) are quiet similar to graminaceous crops, so the agriculturists judge it as a kind of triticeae. Various popular names given to buckwheat have been used to trace its migration through Asia and Europe and are still used to confirm the origin of buckwheat. Today common buckwheat is called *ogal* in India, *mite phapar* in Nepal, *jare* in Bhutan, and *grecicha kul'turnaja* in Russia. *grechka* in Ukrainian and *gryka, tatarka gryka,* or *poganka* in Poland. It is called *pohanka* in the Czech Republic and Slovakia. In Sweden it is *bovete,* in Denmark *boghvede,* and in Finland common buckwheat is *tattari.* In Slovenia it is *ajda, hajdina,* or *idina,* in Bosnia, Serbia, Montenegro, and Croatia it is *heljda.* In French it is called *sarrasin, blé noir, renouée, bouquette*; in Breton (North-Western France) it is *gwinizh-du,* in Italy *fagopiro, grano saraceno, sarasin, faggina,* and in Germany *Buchweizen* or *Heidekorn* (Hammer, 1986). In Korean it is *maemil.* It is referred to as *soba* in Japan where the same word also is used for buckwheat noodles. In Mandarin common buckwheat is called *tian qiao mai* while Tartary buckwheat is referred to as *ku qiao mai.*

*Buckwheat Germplasm in the World.* DOI: https://doi.org/10.1016/B978-0-12-811006-5.00001-X
© 2018 Elsevier Inc. All rights reserved.

Buckwheat is widely cultivated around the world and in some areas it is a major crop. Buckwheat seeds are rich in proteins with well-balanced amino acid composition, fibers, vitamins, and minerals. The content of flavonoids, a kind of bioactive substance, is also substantial. Treated as a functional food, buckwheat has drawn the attention of the wider world.

## THE ORIGIN AND DIFFUSENESS OF CULTIVATED BUCKWHEAT

Over a century ago, the Swiss plant taxonomist De Candolle (1883) raised the theory of a buckwheat origin area, claiming that buckwheat originated from Siberia and the northern part of China. In 1957, Nakao (1957) pointed out that De Candolle was not correct, for there were a lot of wild buckwheat species widely spread in the southern part of China, which indicated that the southern part of China may be the original place of buckwheat rather than Siberia or the northern part of China.

Since the 1980s, Chinese agriculturists and buckwheat researchers discovered a lot of wild buckwheat species in the southwestern part of China, and based on these findings developed some new perspectives. Jiang and Jia (1992) suggested that the Daliangshan region was one of the places of origin of Tartary buckwheat, based on the large amount of wild buckwheat, special ecological conditions, folklore, and customs regarding buckwheat in this region. Li and Yang (1992) suggested that, based on the research of buckwheat history, human history, and the national history of Yunnan, buckwheat should have originated from the southwestern part of China, on the east side of the Himalayas. More specifically, Yunnan and the Daliangshan and Xiaoliangshan regions, which were the borderlands of Yunnan and Sichuan, should be the place of origin of buckwheat. Ye and Guo (1992) suggested that from the comment of botany, the southwestern part of China was not only the differentiation and spread center, but also the original place of the *Fagopyrum* Mill.

Ohnishi (1995,1998a,1998b,2004) studied the *F. esculentum* ssp. *ancestrale* using morphology, reproductive biology, isozyme analysis, RAPD, and AFLP, and he confirmed the wild relative species *F. esculentum* ssp. *ancestrale* as the ancestor of cultivated species *F. esculentum* Moench. Tsuji and Ohnishi (2000,2001a,2001b) studied the relationship between the wild relative species *F. tataricum* ssp. *potanini* and the cultivated species *F. tataricum* Gaertn using isozyme analysis, RAPD, and AFLP. Based on their research they suggested that the eastern part of Tibet and the joint area of Yunnan and Sichuan are the places of origin of cultivated Tartary buckwheat.

The widespread of wild buckwheat species in the southwest of China, the cultivated history of buckwheat, the national history, and the molecular systematical studies, all proved that the southwest of China was not only the distribution center and diversity center of buckwheat, but also the place of origin of the cultivated species (common buckwheat and Tartary buckwheat). These comments are now widely accepted.

Murai and Ohnishi (1995) proposed that after being domesticated in the southwestern part of China, buckwheat was spread through two routes: one from the southwest to the north of China, then further to the Korean peninsula and later to Japan; and the other one from China through Tibet to Bhutan, Nepal, and India, and then spread to Poland through Kashmir. Buckwheat had a long cultivation history in all the East Asian countries and their neighborhoods. In the days of 800 AD, buckwheat was the most important food in Japan. Around 1200 to 1300 AD, buckwheat spread to Europe through Siberia and the south of Russia. Ukraine, Germany, and Slovenia were perhaps among the first countries with cultivated buckwheat in Europe, and then Belgian, French, Italian, and English people began to cultivate buckwheat in the 17th century. After the 17th century, the Dutch brought buckwheat to America. Now, buckwheat is common in many countries which cultivate grain crops.

## GERMPLASM RESOURCES OF BUCKWHEAT IN THE WORLD

After thousands of years of cultivation and spread, cultivated buckwheat can be found in all continents except Antarctica. According to the data provided by FAO in 2014, 25 countries cultivated buckwheat at more than 50 $hm^2$; total buckwheat cultivated acreage reached 2,008,694 $hm^2$; and the total production was 2,056,585 t. Ranked by cultivated acreage, these countries were Russia, China, Ukraine, United States, Kazakhstan, Poland, Japan, Brazil, Lithuania, France, Tanzania, Belgium, Nepal, Latvia, Bhutan, South Korea, Slovenia, Czech Republic, Estonia, Bosnia and Herzegovina, South Africa, Hungary, Croatia, Georgia, Moldova, and Kyrgyzstan. Russia was the country with the largest cultivated acreage, and the acreage in China was about the same and over 700,000 $hm^2$. The second group of countries was Ukraine, United States, Kazakhstan, Poland, Japan, Brazil, Lithuania, and France, acreage reaching about 300,000 $hm^2$. These countries all had large cultivated acreage and a long history of buckwheat cultivation and breeding, so the cultivated buckwheat resources were plentiful and multiple varieties and types were formed. The Chinese Academy of Agricultural Science and Russia's All-Russian Research Institute of Plant Industry had already collected more than 2000 samples of buckwheat germplasm resources each.

The countries mentioned above have mostly cultivated *F. esculentum*, and only a few countries have cultivated *F. esculentum* and *F. tataricum* at the same time, such as China, Nepal, Bhutan, Pakistan, and India. Among them China had the largest *F. tataricum* cultivated acreage and reached 250,000 $hm^2$ every year.

Wild buckwheat species are mainly distributed in China and some South Asian countries such as Nepal, Bhutan, Pakistan, and India. In China, wild buckwheat species were distributed in the southern provinces including Sichuan, Yunnan, Guizhou, Chongqing, Tibet, Shaanxi, Hunan, Hubei,

Zhejiang, Anhui, Fujian, Guangdong, Guangxi, and Hainan, especially in Southwestern China including Sichuan, Yunnan, Guizhou, Chongqing, and Tibet, all the wild species can be found in this region, so this region was considered as not only the distribution center and diversity center of buckwheat, but also the birthplace of cultivated species (common buckwheat and Tartary buckwheat).

## THE CURRENT BUCKWHEAT GERMPLASM RESEARCH

Worldwide organized collection of buckwheat began in 1980s. Subsidized by the International Board for Plant Genetic Resources (IBPGR), botanists searched and collected wild buckwheat resources in the Himalaya Region. Since then, over 10,000 samples of buckwheat resources have been collected and half of them have come from South and East Asia. The sample numbers collected by different countries are presented in Table 1.1. (Campbell, 1997, Joshi,1999, Chauhan et al., 2010). These resources were stored in long-term storage conditions ($-20°C$) or middle-term storage conditions ($5°C$, RH40%).

## CHARACTERIZATION AND EVALUATION

The characterization and evaluation of buckwheat has been carried out in many countries. China, India, Nepal, Japan, and North Korea measured the

**TABLE 1.1** The Sample Numbers Collected by Different Countries

| Country | Number of Samples |
|---|---|
| Nepal | 683 |
| Bhutan | 48 |
| India | 954 |
| Japan | 1448 |
| North Korea | 413 |
| South Korea | 245 |
| Mongolia | 30 |
| China | 2804 |
| Russia | 2200 |
| Slovenia | 361 |
| Germany | 202 |
| Canada | 1100 |
| USA | 190 |

agronomic traits of buckwheat such as period of duration, plant height, number of blades, number of branches, number of flowers, thousand seed weight, plant type, color of stem, color of flowers, color of seeds, and shape of seeds. Furthermore, China, Japan, India, and some other countries did research about the quality characters of buckwheat (Yang, 1992, Chauhan et al., 2010), China investigated 1500 samples of buckwheat for the content of protein, amino acids, VE, VPP, P, Ca, Fe, Zn, Mn, Cu, and Se, and the evaluation of rutin was also continuously carried out. Recently, with the development and extensive use of molecular biology, the characterization and evaluation of buckwheat resources using molecular markers became popular. The characterization and evaluation of buckwheat established many mutant types and a group of excellent or special buckwheat resources were screened out, which makes a great contribution to the breeding of new varieties of buckwheat.

## THE RESEARCH ABOUT THE RELATIONSHIP OF *FAGOPYRUM* MILL

Mill Steward (1930) classified the Polygonaceae plants in Asia, there were 10 species belonging to *Fagopyrum* Mill and they were described. Later Ye and Guo (1992) and Li (1998) proved and described these species and put them into *Fagopyrum* Mill, too. From then on, the argument about the position of *Fagopyrum* Mill became unified, and the *Fagopyrum* Mill was set up. In recent years multiple buckwheat wild species and wild-types of cultivated buckwheat were investigated in the southwest of China, (Ohnishi, 1991,1995; Ohsako and Ohnishi, 1998,2000; Li, 1998; Chen, 1999; Liu et al., 2008; Tang et al., 2010; Shao et al., 2011; Hou et al., 2015; Zhou et al., 2015), and the number of species in *Fagopyrum* Mill increased to more than twenty species. The investigation of wild species enriched the buckwheat germplasm resources and was of great benefit to buckwheat systematic research about the origin of cultivated buckwheat and the relationship between species in *Fagopyrum* Mill.

The buckwheat germplasm resources are not only the important foundation of genetics, breeding, and biotechnology but also the essential material for research into classification, systematic evolution, origin of species, and biological diversity of buckwheat. The various kinds of cultivated and wild buckwheat species contain a lot of valuable genes which are important for breeding improvement, and they also contain abundant nutritional value and can be used to improve the quality of meal and diet structure. Recently, related countries have made great progress in research into buckwheat germplasm resources. In the past 30 years results of the research into the germplasm resources of buckwheat have been remarkable. The local varieties of cultivated buckwheat species have been widely collected and carefully stored, new species of wild buckwheat have been found successively, and their good characters were digging and made good use of. We believe that, with the joint efforts of buckwheat researchers from different countries, the study of buckwheat germplasm resources will be intensified and there will be a large breakthrough in the area of breeding new varieties and exploiting new buckwheat resources.

## ACKNOWLEDGMENTS

This research was supported by the Investigation of Forage Germplasm in Central China (grant number 2017FY100604).

## REFERENCES

Campbell, CG., 1997. Buckwheat. *Fagopyrum esculentum* Moench. Promoting the Conservation and Use of Underutilized and Neglected Crops, Vol 19. Institute of Plant Genetics and Crop Plant Research, Gatersleben/International Plant Genetic Resources Institute, Rome, Italy.

Chauhan, R., Gupta, N., Sharma, S.K., Sharma, T.R., Rana, J.C., Jana, S., 2010. Genetic and genome resources in Buckwheat- present and future perspectives. Eur. J. Plant Sci. Biotechnol. 4, 33–44.

Chen, Q.F., 1999. A study of resources of Fagopyrum (Polygonaceae) native to China. Bot. J. Linnean Soc. 130, 54–65.

De Candolle, A. 1883. L'Origine des plantes cultivees, Paris.

Hammer, K., 1986. Polygonaceae. In: Schultze-Motel, J. (Ed.), Rudolf Mansfelds Verzeichnis landwirtschaftlicher und gärtnerischer Kulturpflanzen (ohne Zierpflanzen). Akademie-Verlag, Berlin, pp. 103–122.

Hou, L.L., Zhou, M.L., Zhang, Q., et al., 2015. *Fagopyrum luojishanense*, a New Species of Polygonaceae from Sichuan, China. Novon. 24 (1), 22–26.

Jiang, J.F., Jia, X., 1992. Daliangshan region in Sichuan province is one of the habitats of tartary buckwheat. In: Proceedings of 5th International Symposium on Buckwheat at Taiyuan, China. Agricultural Publishing House, pp. 17–18.

Joshi, B.D., 1999. Status of Buckwheat in India. Fagopyrum 16, 7–11.

Li, A.R., 1998. Flora of China, Polygonaceae. Science Press, Beijing 25(1), 108–117.

Li, Q.Y., Yang, M.X., 1992. Preliminary investigation on buckwheat origin in Yunnan, China. In: Proceedings of 5th International Symposium Buckwheat at Taiyuan. Agricultural Publishing House, pp.44–46.

Liu, J.L., Tang, Y., Mingzhong, X., et al., 2008. *Fagopyrum crispatifolium* J. L. Liu, a new species of Polygonaceae from Sichuan, China. J. Syst. Evol. 46 (6), 929–932.

Murai M, Ohnishi O., 1995. Diffusion routes of Buckwheat Cultivation in Asia Revealed by RAPD Markers. Current Advances in Buckwheat Research. In: Proceedings of 6th International Symposium on Buckwheat in Shinshu, Japan, pp. 163–173.

Nakao, S., 1957. Transmittance of cultivated plants through Sino-Himalayan rout. In: Kihara, H. (Ed.), Peoples of Nepal Himalaya. Fauna and Flora Research Society, Kyoto, pp. 397–420.

Ohnishi, O., 1991. Discovery of wild ancestor of common buckwheat. Fagopyrum 11, 5–10.

Ohnishi, O., 1995. Discovery of new *Fagopyrum* species and its implication for the studies of evolution of *Fagopyrum* and of the origin of cultivated buckwheat. In: Matano, T., Ujihara, A., Eds. Current Advances in Buckwheat Research. Vol. I–III. In: Proceedings of 6th International Symposium On Buckwheat in Shinshu, 24–29 August 1995. Shinshu University Press, pp. 175–190.

Ohnishi, O., 1998a. Search for the wild ancestor of buckwheat. I. Description of new *Fagopyrum* species and their distribution in China. Fagopyrum 18, 18–28.

Ohnishi, O., 1998b. Search for the wild ancestor of buckwheat.III. The wild ancestor of cultivated common buckwheat, and of Tartary buckwheat. Econ. Bot. 52, 123–133.

Ohnishi, O., 2004. On the origin of cultivated buckwheat. In: Proceedings of 9th International Symposium on Buckwheat in Prague, pp. 16–21.

Ohsako, T., Ohnishi, O., 1998. New *Fagopyrum* species revealed by morphological and molecular analyses. Genes Genet. Syst. 73, 85−94.

Ohsako, T., Ohnishi, O., 2000. Intra and inter-specific phylogeny of the wild Fagopyrum (Polygonaceae) species based on nucleotide sequences of noncoding regions of chloroplast DNA. Am J. Bot. 87, 573−582.

Shao, J.R., Zhou, M.L., Zhu, X.M., et al., 2011. *Fagopyrum wenchuanense* and *Fagopyrum qiangcai*, Two new species of Polygonaceae from Sichuan, China. Novon 21, 256−261.

Steward, A.N., 1930. The Polygoneae of eastern Asia. Cont. Gray Herb. 88, 1−129.

Tang, Y., Zhou, M.L., Bai, D.Q., et al., 2010. *Fagopyrum pugense* (Polygonaceae), a new species from Sichuan, China. Novon. 20, 239−242.

Tsuji, K., Ohnishi, O., 2000. Origin of cultivated Tartary buckwheat *(Fagopyrum tataricum* Gaertn.) revealed by RAPD analyses. Genet. Resour. Crop Evol. 47, 431−438.

Tsuji, K., Ohnishi, O., 2001a. Phylogenetic position of east Tibetan natural populations in Tartary buckwheat (*Fagopyrum tataricum* Gaertn.) revealed by RAPD analyses. Genet. Resour. Crop Evol. 48, 63−67.

Tsuji, K., Ohnishi, O., 2001b. Phylogenetic relationships among wild and cultivated Tartary buckwheat *(Fagopyrum tataricum* Gaertn.) populations revealed by AFLP analyses. Genes Genet. Syst. 76, 47−52.

Yang, K.L., 1992. Research on cultivated buckwheat germplasm resources in China. In: Proceedings of 5th International Symposium on Buckwheat at Taiyuan, China. Agricultural Publishing House, pp. 55−59.

Ye, N.G., Guo, G.Q., 1992. Classification, origin and evolution of genus *Fagopyrum* in China. In: Proceedings of 5th International Symposium on Buckwheat at Taiyuan, China. Agricultural Publishing House, pp. 19−28.

Zhou, M.L., Zhang, Q., Zheng, Y.D., Tang, Y., et al., 2015. *Fagopyrum hailuogouense* (Polygonaceae), One new species from Sichuan, China. Novon 24 (2), 222−224.

## FURTHER READING

Gross, M.H., 1913. Remarques sur les Polygonees de I'Asie Orientale. Bull. Geogr. Bot 23, 7−32.

Nakai, T., 1926. A new classification of Linnean Polygonum. Rigakkai 24, 289−301 (in Japanese).

Ohnishi, O., 1983. Isozyme variation in common buckwheat *Fagopyrum esculentum* and its related species. In: Nagatomo, T, Adachi T. (Eds.), Proceedings of 2nd International Symposium Buckwheat in Miyazaki. Kuroda-toshado Printing Co. Ltd. Miyazaki, pp. 39−50.

Ohnishi, O., Matsuoka, Y., 1996. Search for the wild ancestor of buckwheat. II. Taxonomy of *Fagopyrum* (polygonaceae) species based on morphology, isozymes and cpDNA variability. Genes Genet. Syst. 71, 383 390.

Ohsako, T., Yamane, K., Ohnishi, O., 2002. Two new *Fagopyrum* (Polygonaceae) species *F. gracilipedoides* and *F. jinshaense* from Yunnan, China. Genes Genet. Syst. 77, 399−408.

Yang, K.L., Lu, D.B., 1992. The quality appraisal of buckwheat germplasm resources in China. In: Proceedings of 5th International Symposium on Buckwheat at Taiyuan, China. Agricultural Publishing House, pp. 90−97.

Zhou, M.L., Bai, D.Q., Tang, Y., et al., 2012. Genetic diversity of four new species related to southwestern Sichuan buckwheats as revealed by karyotype, ISSR and allozyme characterization. Plant Syst. Evol. 298, 751−759.

# Chapter Two

# Classification and Nomenclature of Buckwheat Plants

**Meiliang Zhou[1], Yu Tang[2], Xianyu Deng[3], Chen Ruan[3], Yixiong Tang[4], and Yanmin Wu[4]**

[1]*Institute of Crop Sciences, Chinese Academy of Agricultural Sciences, Beijing, China,* [2]*Sichuan Tourism University, Chengdu, China,* [3]*Kunming University of Science and Technology, Kunming, Yunnan, China,* [4]*Biotechnology Research Institute, Chinese Academy of Agricultural Sciences, Beijing, China*

## TAXONOMIC POSITION OF *FAGOPYRUM*

*Fagopyrum* belongs to Polygonaceae, but the determination of *Fagopyrum*'s generic name has undergone many modifications. The earliest name was determined by Tourn (1742), i.e., *Fagopyrum* Tourn ex Hall. Later, when Linnaeus established *Polygonum Linn* in 1753, he included *Fagopyrum* into *Polygonum* Linn so it no longer existed as an independent genus. In 1754, Miller established the genus *Fagopyrum* again, i.e., *Fagopyrum* Miller. Moench also established the genus *Fagopyrum* in 1756, who named it as *Fagopyrum* Moench. But in 1826, Meisner again included *Fagopyrum* into *Polygonum* and treated it as a group of *Polygonum*, i.e., *Fagopyrum* sect. Meisn. But he soon thought *Fagopyrum* should be treated as an independent genus. Afterwards, opinions differed as to the position of *Fagopyrum*, and no unanimous conclusion can be drawn. The opinions mainly included are as follows: (1) Samuelsson (1929), Stewad (1930), and Komarov (1936) proposed that *Fagopyrum* should be included into *Polygonum* Linn. and be treated as a group of it. (2) Graham (1965) and Wu (1983) proposed that the broad *Polygonum* should be kept and buckwheat plants should be included into an independent genus in view of them possessing distinct characteristics. (3) Hedberg (1946), Ye and Guo (1992), Ohnishi and Matsuoka (1996), and Li (1998) proposed that the broad *Polygonum* can be divided into many small genera, and *Fagopyrum* is just one of them.

*Buckwheat Germplasm in the World.* DOI: https://doi.org/10.1016/B978-0-12-811006-5.00002-1
© 2018 Elsevier Inc. All rights reserved.

The key to the above different opinions is the variance degree of *Fagopyrum* and broad *Polygonum* in morphology, palynology, and cytology. Graham thought the difference between *Fagopyrum* and *Polygonum* lies in the fact that *Fagopyrum*'s perianth isn't accrescent, and its embryo is located in the endosperm; the cotyledon of *Fagopyrum* curls around the radicle, and the inflorescence of *Fagopyrum* is more or less corymbose, so it's obvious that *Fagopyrum* is an independent genus. Hedberg divided the broad *Polygonum* into ten types after studying the morphology of pollen grain of the broad *Polygonum*. Among them, the *Fagopyrum*'s pollen grain furrow is perforated, rough in exine and granular, which is different from other types. According to the study of the predecessor and his own findings, Yukio pointed out that the basic chromosome number of *Polygonum* is $n = 10$, 11, or 12, whereas that of *Fagopyrum* is $n = 8$. In recent years, most scholars have confirmed the basic chromosome number of *Fagopyrum* is 8 in cytological terms. Hence, *Fagopyrum* should be viewed as an independent genus according to the morphological, palynological, and cytological studies.

Previously, there were many genus names for *Fagopyrum*: *Fagopyrum* Tourn, *Fagopyrum* Moench, *Fagopyrum* Miller, etc. Now, the unanimous conclusion is *Fagopyrum* Miller. The feature description of *Fagopyrum* Miller is as follows: It is an annual or perennial herb or subshrub, the stem of which has thin furrows; the leaves are alternate, which is triangular, sagittate, cordate, or hastate; the peduncle is inarticulate; the inflorescence is compound, i.e., the floral axis is branched or unbranched, which bears clustered incomplete cymes that are spicate, corymbose, or coniform. Every incomplete cyme has one or more flower(s), and there is a bract(s) outside the cyme, every flower also has a small membranous bract. The flower is bisexual, whose perianth is white, pale red, or yellowish-green, and has five deep lobes, but isn't accrescent. Every flower has eight stamens, among which five are inner ones and three are outer ones. The pistil consists of three carpels, which can be divided up into a prismatoidal ovary and three styles (heterostylous or homostylous), etc. The achene is prismatoidal, which is exsert from the outside of persistent perianth, obviously or not. The embryo is situated in the center of endosperm; the cotyledon is wide and more or less plaited. The furrows of pollen grain are perforated and rough in exine, which forms the granular pattern. The basic chromosome number is $n = 8$. The main features that distinguish *Fagopyrum* from *Polygonum* are that the former has the embryo that curls up in the center of endosperm, broad cotyledon(s) that is (are) more or less flat, the persistent perianth that is not accrescent, the coarse pollen exine that is granular and the basic chromosome number is 8.

## THE SPECIES OF *FAGOPYRUM*

Because of the changeable (taxonomic) position of *Fagopyrum*, there have been various classifications of species within *Fagopyrum* for a long time. When Gross classified the Asian Polygonaceae in 1913, he included some proven buckwheat species into 2 groups of the *Polygonum*. Specifically, common buckwheat, Tartary buckwheat and *Fagopyrum cymosum* were included into the Eufagopyrum group together, while *Fagopyrum urophyllum* and some buckwheat-like species were included into the Tiniara group. Although Steward (1930) didn't treat buckwheat as an independent genus, he had conducted pretty accurate divisions towards the Asian Polygonaceae plants and included ten buckwheat species into the buckwheat group. Among these ten species, nine of them were distributed in the southwest of China, and only one species—*Fagopyrum suffruticosum* F. Schmidt—grew in the far east of Russia.

In 1992, Ye and Guo classified buckwheat plants on the basis of confirming (the taxonomic position of) *Fagopyrum* Miller. They divided the buckwheat plants at that time into ten species, i.e., *Fagopyrum esculentum* Moench, *Fagopyrum tataricum* Gaertn., *Fagopyrum gracilipes* (Hemsl.) Dammer ex diels, *F. cymosum* (Trev.) Meisn, *Fagopyrum lineare* (Sam.) Haraldson, *F. urophyllum* Bur et. Franch., *Fagopyrum gilesii* (Hemsl.) Hedberg, *Fagopyrum statice* Levl., *Fagopyrum leptopodum* (Diels) Hedberg, and *Fagopyrum caudatum* (Sam.) A. J. Li. Among them, except for *F. caudatum*, nine other species were the same as the version of Steward. In "Flora of China", Li (1998) further confirmed the (taxonomic) positions of Polygonaceae and *Fagopyrum* Miller, and demonstrated and described the ten species that had been divided by Ye and Guo.

Ohnishi (1991, 1998a, b), Ohsako and Ohnishi (1998), Ohsako and Matsuoka (1996), Ohsako et al. (2002) have also done lots of research into buckwheat species. They have reported one wild closely-related species of the cultivated common buckwheat (sub-species), and eight wild buckwheat species, on the basis of confirming the existing buckwheat species, i.e., *F. esculentum* ssp. *ancestrale* Ohnishi, *Fagopyrum homotropicum* Ohnish, *Fagopyrum capillatum* Ohnishi, *Fagopyrum pleioramosum* Ohnishi, *Fagopyrum callianthum* Ohnishi, *Fagopyrum macrocarpum* Ohsako et Ohnishi, *Fagopyrum rubifolium* Ohsako et Ohnishi, *Fagopyrum jinshaense* Ohsako et Ohnishi, and *Fagopyrum gracilipedoides* Ohsako et Ohnishi. These species were distributed in Sichuan (China), Yunnan (China), and the surrounding areas. Ohnishi and Matsuoka (1996) proposed that two groups (the Cymosum group and the Urophyllum group) could be established under *Fagopyrum* according to the differences in morphology, isozymes, and DNA polymorphism, etc. The main characteristic of Cymosum group is the bigger, lackluster achene whose base is covered by the tepal(s). This group includes two cultivars (*F. tataricum* (L.) Gaertn., *F. esculentum* Moench) and two wild types (*F. cymosum* Meisn., *F. homotropicum* Ohnishi). The main characteristic of the Urophyllum group is the smaller, glossy achene whose surface is covered tightly by the tepal(s). This group includes *F. statice*, *F. leptopodum*, *F. urophyllum*, *F. lineare*, *F. gracilipes*, *F. gracilipes* var. *odontopterum*, *F. capillatum* Ohnishi, *F. pleioramosum* Ohnishi, and *F. callianthm* Ohnishi, etc. In fact, there are so many achene types of *Fagopyrum*, which vary considerably in size, that there are some undistinguishable transition types between the Cymosum group and the Urophyllum group. So it is necessary to collect more buckwheat types and combine the evidences of reproductive biology and molecular biology etc. to confirm these transition types. Later, multiple new wild species of buckwheat that had been discovered in the southwest of China were reported successively. In 1999, Chen discovered three wild buckwheat species in the Sichuan-Tibet region: *Fagopyrum zuogongense* Chen, *Fagopyrum pilus* Chen and *Fagopyrum megaspartanum* Chen. Moreover, Liu et al. (2008), Tang et al. (2010), Shao et al. (2011), Hou et al. (2015), Zhou et al. (2015) discovered six new wild species of buckwheat in China's Sichuan province respectively: *Fagopyrum crispatofolium* Liu, *F. pugense* Yu, *Fagopyrum wenchuanense* Shao, *Fagopyrum qiangcai* Bai, *Fagopyrum luojishanense* Shao and *Fagopyrum hailuogouense* Shao, Zhou, and Zhang.

According to the statistical records of Zipcode Zoo (http://zipcodezoo.com/) and The Plant List (TPL) (http://www.theplantlist.org/tpl/), the existing Fagopyrum plants have at least more than 70 species names, but only 20 of them have been accepted, and the others are mainly synonyms or duplicate names, such as *F. cymosum*, where some people adopt the species name *Fagopyrum dibotrys* (D. Don) Hara (Li, 1998) (to refer to *F. cymosum*). There are also a number of other (species) names (*Fagopyrum acutatum* etc.), but the one that has been widely accepted is *F. cymosum* Meisner. Some buckwheat plants are very similar with each other in terms of morphological characteristics, for example, there is no difference between *Fagopyrum sagittatum* Gilib. and *F. esculentum* Moench in morphology, so they are viewed as the same species. Similarly, *Fagopyrum kashmirianum* and *F. tataricum* are viewed as the same. Although there is not much difference both between *F. cymosum* and *F. megaspartanum*, and between *F. leptopodum* and *F. jinshaense* in morphology, it requires further identification in reproductive biology, cytology, and molecular biology to determine whether they are the same biological species or not. Moreover, Krotov and Dranenko (1973) named the hybrid nurtured by interspecific hybridization *Fagopyrum giganteum* Krotov. This is a hybrid that is still in the process of formation, whose stability is hard to predict.

## MORPHOLOGY KEY TO SPECIES OF *FAGOPYRUM*

The early classification of *Fagopyrum* was mainly dependent on morphological characteristics. With the continuous development of science and technology, especially the molecular biological techniques that have been widely used, nowadays the classification of *Fagopyrum* is conducted with the combination of palynology, cytology, reproductive biology, and molecular biology besides the morphological characteristics of plants (i.e., the traditional classification way) to get more accurate classification results. But whatever research method is used, the morphological characteristics of plants are always the key evidences for species identification. The morphology keys to species of *Fagopyrum* of different times and different scholars are listed as below for reference: (Table 2.1−2.4).

For the morphological descriptions of *Fagopyrum* species, see the following chapters.

**TABLE 2.1** Classification of *Fagopyrum* Species by Stewad (1930)

| | |
|---|---|
| a. Surfaces of the achenes grooved, the angles rounded at the bases and keeled towards the tips | b |
| b. Plant annual and glabrous or nearly so. stem usually simple and erect | *Polygonum tataricum* |
| b. Plant perennial: stem shrubby at the base and decumbent | *Polygonum suffruticosum* |
| a. Surfaces of the achenes flat or concave, the angles acute | c |
| c. Branches leafless above, flower-bearing branches never (or rarely) from leaf axils | d |
| d. Inflorescence congested and head-like | *Polygonum gilesii* |
| d. Inflorescence of spike-like branches | e |
| e. A coarse herb from a perennial woody caudex | *Polygonum statice* |
| e. Low herbs without woody caudex | f |
| f. Leaf blades triangular, bases sagittate or truncate | *Polygonum leptopodum* |
| f. Leaf blades linear, bases hastate | *Polygonum lineare* |
| c. Branches more or less leafy above, flower-bearing branches often from leaf axils | g |
| g. Plant woody | *Polygonum urophyllum* |
| g. Plant herbaceous | h |
| h. Flowers laxly arranged and often on drooping branches | *Polygonum gracilipes* |
| h. Flowers densely arranged and usually on erect branches | i |
| i. Cultivated annual, inflorescence in sub-capitate cymes | *Polygonum fagopyrum* |
| i. Perennial from a woody caudex, inflorescence of rather distinct spikes | *Polygonum cymosum* |

**TABLE 2.2** Classification of *Fagopyrum* Species by Ye and Guo (1992)

1. It is perennial herbaceous plants having shrubby stem at the base or underground tuber, heterostyly.

  2. It is a tall plant, more than 1 m, with cauline leaves, broader blades, mostly longer than 5 cm, terminal and axillary inflorescences, articulate pedicle.

    3. Leaves are triangular, broader with hastate and acute base; inflorescences branch in corymbose cyme, achene is longer than 5 mm, exerting more than twice the length of the persistent perianthes     *Fagopyrum cymosum*

    3. Leaf forms are diversified, longer with auriculate, rounded base, inflorescences branch in lax, paniculate spike, achene is less than 5 mm, exerting less than 2 times the length of the persistent perianthes     *Fagopyrum urophyllum*

  2. It is a short plant, about 50 cm high, leaves are mostly basal with blades less than 5 mm, inflorescences are mostly terminal, filiform, pedicel is nonarticulate     *Fagopyrum statice*

1. It is an annual plant with herbaceous stem, without underground tuber, heterostyly or homostyly

  4. Heterostyly, smooth or concave achene, acute angle

    5. Cultivated plants with larger achene, longer than 5 mm, exserting more than twice the length of the persistent perianthes     *Fagopyrum esculentum*

    5. Wild plants with smaller achene, less than 5 mm, more or less exserting or including in the persistent perianthes

      6. Triangular leaves, truncate or sagittate base     *Fagopyrum leptopodum*

      6. Linear leaves, hastate base     *Fagopyrum lineare*

  4. Homostyly, smooth, concave, or furrow achene, acute or obtuse angle.

    7. Cultivated plants with larger blades, wider than 5 cm, yellow—green flower, larger achene, longer than 5 mm, with furrow and obtuse angle, exserting more than twice the length of the persistent perianthes     *Fagopyrum tartaricum*

    7. Wild plants with smaller blades, less than 5 cm in width, white or pink flower, smaller achene, less than 5 mm, being smooth or concave, acute or winged angle, more or less exerting or including in the persistent perianthes

      8. Terminal or axillary, filiform inflorescences, arrange in laxspicate cyme

        9. Leaves are mostly cordate—ovate, winged (variety) or not winged achene

        9.Leaf forms are diversified, triangular—sagittate to lanceolate (or linear)—hastate, apex is more or less caudate, achene is not winged     *Fagopyrum gracilipes*

          *Fagopyrum caudatum*

      8.Terminal inflorescences, are tightly capitated     *Fagopyrum gilesii*

**TABLE 2.3** Classification of *Fagopyrum* Species by Ohnishi (1995–1998)

| | | |
|---|---|---|
| (1) | Thick plaited cotyledons lie in the center of the achene (*Fagopyrum*) | (2) |
| (2) | a. Cotyledons horizontally long, large lusterless achene is partially covered with persistent perianths | (3) |
| | b. Cotyledons laterally long or round, small lustrous grains are completely covered with persistent perianths | (4) |
| (3) | a. Cotyledons in endosperm are colorless, blade veins are not transparent | (5) |
| | b. Cotyledons in endosperm are yellowish, blade veins are transparent | (6) |
| (5) | a. Heterostylous, cross-pollinating species | *Fagopyrum esculentum* |
| | b. Homostylous, self-fertilizing species | *Fagopyrum homotropicum* |
| (6) | a. Surface of achene is smooth | *Fagopyrum cymosum* |
| | b. Surface of achene is rough, with a canal in the center | *Fagopyrum tartaricum* |
| (4) | a. Five perianths are equal in size, the lower two lack a green stripe | (7) |
| | b. Perianths consist of two smaller and three larger, the lower small perianths have greenish stripes | (8) |
| (7) | a. Perennial with well developed roots | *Fagopyrum statice* |
| | b. Annual with a poor root system | (9) |
| (8) | a. Ochrea is green and not transparent | *Fagopyrum urophyllum* |
| | b. Ochrea is transparent with greenish stripes | (10) |
| (9) | a. Achens are relatively large and plants are vigorous | *Fagopyrum jinshaense* |
| | b. Achens are very small and plants are small and slim | (11) |

| | | |
|---|---|---|
| (11) | a. Blades are ovate or cordate | *Fagopyrum leptopodum* |
| | b. Blades are linear | *Fagopyrum lineare* |
| (10) | a. Ochrea is not pubescent, main blade vein number is 5 | (12) |
| | b. Ochrea is pubescent, main blade vein number is 7 | (13) |
| (12) | a. Plants are erect | *Fagopyrum callianthum* |
| | b. Many branches are creeping on the ground | *Fagopyrum macrocarpum* |
| (13) | a. Ochrea is not heavily pubescent, stem is not pubescent, it has many creeping branches | *Fagopyrum pleioramosum* |
| | b. Ochrea are heavily pubescent, stems are also pubescent | (14) |
| (14) | a. Ochrea and stems are heavily pubescent, blades are cordate or sagittate, inflorescence are drooping | (15) |
| | b. Pubescence in ochrea and stems is not so heavy as (a), blade cordate or ovate, branches are erect | *Fagopyrum capillatum* |
| (15) | a. Only the main veins of blades visible | *Fagopyrum rubifolium* |
| | b. Veinlets connecting the lateral veins are clearly visible | *Fagopyrum gracilipes* |

**TABLE 2.4** Classification of *Fagopyrum* Species by Meiliang Zhou and Yu Tang

1. Stem is within 1 m, the upper internode is very long and almost without any leaves, cymes cluster at the top of common peduncle and form the head-like shape — *Fagopyrum gilesii*

1. The plants that are unlike the above

2. Perennial with underground stems, heterostylous or homostylous

3. The stem base is fleshy (i.e., thick and soft) or a little woody, homostylous, blades are cordate — *Fagopyrum hailuogouense*

3. The stem base is woody, heterostylous, blades are not cordate

4. It is high plant, with cauline leaves, broader blades, mostly longer than 5 cm, sometimes can reach 10 cm, terminal and axillary inflorescences, the pedicle is obviously articulate, the flowers can fall out from the articulation

5. Leaves are smaller, nearly the equilateral triangular, often with hastate base, inflorescences branch into corymbose cyme, achene is longer than 5 mm, exserting more than twice the length of the persistent perianthes — *Fagopyrum cymosum*

5. Leaf forms are diversified, longer, with auriculate, rounded base, inflorescences branch into lax, paniculate spike, achene is smaller, whose length is about 3.5 mm, exserting less than 2 times the length of the persistent perianthes — *Fagopyrum urophyllum*

4. It is short plant, leaves are mostly basal with blades less than 5 cm, inflorescences are terminal and filiform, the flowers can fall out from the base of receptacle — *Fagopyrum statice*

2. It is an annual plant without underground tuber, heterostyly or homostyly

6. Heterostyly, smooth or concave achene, acute angle.

7. Cultivated plants with larger achenes, longer than 5 mm, exserting more than twice the length of the persistent perianthes — *Fagopyrum esculentum*

7. Wild plants with smaller achenes, less than 5 mm in length, more or less exserting or including in the persistent perianthes.

8. Stem is erect, leaves are nearly triangular or linear

9. Leaves are nearly triangular, whose base is truncate or sagittate — *Fagopyrum leptopodum*

9. Leaves are nearly linear, whose base is hastate ......... *Fagopyrum lineare*

8. Stems are sub-erect or creeping, leaf forms are diversified, which is fleshy (i.e., thick and soft) or papery (i.e., thin and dry), with or without gray-white patches on the leaf surface

10. The leaves are fleshy (i.e. thick and soft), slightly fleshy, or thickly papery with gray-white patches on the leaf surface ......... *Fagopyrum qiangcai*

10. The leaves are papery, without gray-white patches on the surface ......... *Fagopyrum caudatum*

6. Homostyly, smooth, concave, or furrow achene, acute or obtuse angle.

11. Cultivated plants with larger achene, longer than 5 mm, exserting more than twice the length of the persistent perianthes ......... *Fagopyrum tataricum*

11. Wild plants with smaller achene, less than 3mm, more or less exserting or including in the persistent perianthes

12. There are vesicular projections on the on the leaf surface, the corrugated leaf margin with irregular corrugated crenation, crenation or small ......... *Fagopyrum crispatofolium* crenation, cymes cluster on the rachis

12. The leaf surface is relatively flat, with fine lines or vesicular projections, the leaf margin is entire or sinuolate, cymes are laxly or relatively laxly arranged on the rachis

13. The whole plant is densely covered with short hairs or long hairs, the stem is relatively vigorous, the articulation is relatively dense, the leaf ......... *Fagopyrum pugense* surface with fine lines or vesicular projections

13. The whole plant is densely covered with micro-rough hairs or nearly glabrous, the stem is relatively slim, the articulation is relatively sparse, the leaf surface is relatively flat

14. Achene with acute angle is winged or not, the wings of young achene are green-white ......... *Fagopyrum gracilipes*

14. Achene is winged and with acute angle, the wings of young achene are red ......... *Fagopyrum luojishanense*

## ACKNOWLEDGMENTS

This research was supported by the National Natural Science Foundation of China (grant number 31572457) and the Investigation of Forage Germplasms in Central China (grant no. 2017FY100604).

## REFERENCES

Graham, S.A, 1965. The Genera of Polygonaceae in the Southeastern United States. J.Arn. Vol.46 (2), 91−121.

Gross, M.H, 1913. Remarques sur les polygonees del′Asie orientale. Bull. Torrey Bot. Club 23, 7−32.

Hedberg, O, 1946. Pollen Morphology in the Genus Polygonum L.(S.lat.) and Its Taxonomical Significance. Svensk Bot.Tidskr. 40, 371−404.

Hou, L.L., Zhou, M.L., Zhang, Q., Qi, L.P., Yang, X.B., Tang, Y., et al., 2015. *Fagopyrum luojishanense*, a new species of Polygonaceae from Sichuan. China. Novon 24 (1), 22−26.

Komarov, V.L.,1936. Flora USSR Mosqva & Leningrad.

Krotov, A.S., Dranenko, E.T.,1973. Amphidiploid grechikha F.giganteum Krotov sp.nova. Byulleten Vsesoyuznogo Ordena Lenina Instituta Rastenievodstva Imeni N.I.Vavilova. 30:41-45.

Li, A.R., 1998. Flora of China, Polygonaceae. Science Press, Beijing 25(1):108-117.

Linnaeus, C., 1753. Species plantarum I:359, Holmiae.

Liu, J.L., Tang, Y., Xia, M.Z., Shao, J.R., Cai, G.Z., Luo, Q., et al., 2008. *Fagopyrum crispatofolium* J. L. Liu, a new species of Polygonaceae from Sichuan, China. J. Syst. Evol. 46 (6), 929−932.

Meisner, C.F., 1826. Monographiae Generis Polygoni Prodromus, Genevae.

Miller, P.H., 1754. The Gardeners Dictionary. Abridged(edition4), London.

Ohnishi, O., 1991. Discovery of wild ancestor of common buckwheat. Fagopyrum 11, 5−10.

Ohnishi, O. 1995. Discovery of new Fagopyrum species and its implication for the studies of evolution of Fagopyrum and of the origin of cultivated buckwheat. Pp. 175-190 in Current Advances in Buckwheat Research. Vol. I-III. Proc. 6th Int. Symp. On Buckwheat in Shinshu, 24−29 August 1995 (T. Matano and A. Ujihara, eds.). Shinshu University Press.

Ohnishi, O., 1998a. Search for the wild ancestor of buckwheat. I. Description of new Fagopyrum species and their distribution in China. Fagopyrum 18, 18−28.

Ohnishi, O., 1998b. Further study on wild buckwheat species, their distribution and classification. In: Advances in Buckwheat Research. Vol.VI Proceedings of 6th International Symposium on Buckwheat in Winnipeg, Canada, pp. 175−190.

Ohnishi, O., Matsuoka, Y, 1996. Search for the wild ancestor of buckwheat. II. Taxonomy of Fagopyrum (Polygonaceae) species based on morphology, isozymes and cpDNA variability. Genes & Genet. Syst. 71, 383−390.

Ohsako, T., Ohnishi, O., 1998. New Fagopyrum species revealed by morphological and molecular analyses. Genes Genet. Syst. 73, 85−94.

Ohsako, T., Yamane, K., Ohnishi, O., 2002. Two new Fagopyrum (polygonaceae) species *F. gracilipedoides* and *F. jinshaense* from Yunnan. China. Genes Genet. Syst. 77, 399−408.

Samuelsson,G., 1929. Polygonaceae. H.Handel-Mazzetti:Symbolae Sinicae.Wien.

Shao, J.R., Zhou, M.L., Zhu, X.M., Wang, D.Z., Bai, D.Q., 2011. *Fagopyrum wenchuanense* and *Fagopyrum qiangcai*, two new species of Polygonaceae from Sichuan. China. Novon 21, 256−261.

Stewad, A.N., 1930. The Polygoneae of eastern Asia. Cont. Gray Herb. Harvard Univ. 88, 1−129.

Tang, Y., Zhou, M.L., Bai, D.Q., Shao, J.R., Zhu, X.M., Wang, D.Z., et al., 2010. *Fagopyrum pugense* (Polygonaceae), a new species from Sichuan. China. Novon 20, 239–242.

Wu, Z.Y., 1983. Flora of Tibet. Science Press, Beijing.

Ye, N.G. and Guo, G.Q., 1992. Classification, origin and evolution of genus Fagopyrum in China. Proc. 5th Int. Symp on Buckwheat at Taiyuan, China. Agricultural Publishing House. Pp. 19–28.

Zhou, M.L., Zhang, Q., Zheng, Y.D., Tang, Y., Li, F.L., Zhu, X.M., et al., 2015. *Fagopyrum hailuogouense* (Polygonaceae), one new species of Polygonaceae from Sichuan. China. Novon 24 (2), 222–224.

## FURTHER READING

Yukio, D., 1960. Cytological studies in polygonum and related Genera I, Bot. Mag. Tokyo, 37. pp. 337–340.

# Chapter Three

# Distribution of Cultivated Buckwheat Resources in the World

**Galina Suvorova[1] and Meiliang Zhou[2]**
[1]*All-Russia Research Institute of Legumes and Groat Crops, Orel, Russia*
[2]*Institute of Crop Sciences, Chinese Academy of Agricultural Sciences, Beijing, China*

## INTRODUCTION

The cultivated buckwheat species and their wild relatives have become attractive to researchers of many countries in the last few decades. Discovery of the new *Fagopyrum* species has changed the idea about the buckwheat origin. According to present opinion the genus *Fagopyrum* Mill. consists of more than 20 species, with two species cultivated among them. The common buckwheat *Fagopyrum esculentum* is wildly grown on all continents of the world, while the Tartary buckwheat *Fagopyrum tataricum* Gaertn. is traditionally cultivated in the mountainous areas of China and the Himalayas. Discovery of the wild ancestor of common buckwheat *F. esculentum* ssp. *ancestrale* led to the conclusion that cultivated common buckwheat originated in the Yunnan-Tibet border area of southwestern China (Ohnishi, 2009). The same area is also considered to be the original birthplace of cultivated Tartary buckwheat whose wild ancestor was shown to be the wild Tartary buckwheat (*F. tataricum* ssp. *potanini*) (Ohnishi, 2013). The original birthplace of common buckwheat and the center for the wild ancestor is located almost in the middle of the center of wild buckwheat species in the northwestern part of Yunnan province in China (Ohnishi, 2012). After the birth of cultivated buckwheat, it diffused to southern and northern China, arrived at the Korean peninsula, spread to the Japanese archipelago, went west entering the Central Asia via the Silk Road, entering Tibet and Bhutan, and on to Nepal and India (Ohnishi, 2016). From China it was brought during the 9—12th centuries to Russia and in the 13—15th centuries to Central and Northern Europe (Kreft et al., 2016). Not so long ago the common buckwheat was introduced to

*Buckwheat Germplasm in the World.* DOI: https://doi.org/10.1016/B978-0-12-811006-5.00003-3
© 2018 Elsevier Inc. All rights reserved.

American and Australian continents. The cultivated Tartary buckwheat has diffused many times through different routes (Tsuji and Ohnishi, 2001), but the species has a narrow range of cultivation comparative to the common buckwheat. The distribution of buckwheat genetic resources in the world mainly coincides with the center of its origin and continued on the routes of their migration. The present paper describes the buckwheat genetic resources based on the analyses of the recent publications concerning buckwheat germplasm evaluation, exploitation, and conservation.

## *Fagopyrum esculentum* Genetic Resources

The world buckwheat production in 2014 exceeded 2 million tons (FAOSTAT, 2016). The largest buckwheat producers with an annual production of more than 700 thousand tons were the Russian Federation and the People's Republic of China. They were followed by Ukraine, France, Kazakhstan, and Poland. Traditionally buckwheat is grown in Japan, Bhutan, Nepal, Korea, Belorus, Czech Republic, Slovenia, and some other countries.

China, as the buckwheat birthplace, was the earliest country to cultivate buckwheat. Archeological studies show that the Chinese people cultivated buckwheat for 2000 years. In the main growing area of buckwheat and the minority living area, buckwheat was the main food (Wei, 1995). The utilization of buckwheat is changing from just traditional use into more diversified and value-added products due to buckwheat's health enhancing properties (Zhang et al., 2004). Both common and Tartary buckwheat are cultivated in China, the share of Tartary buckwheat is about 30% of the total buckwheat production. Common buckwheat is mainly distributed in Inner Mongolia, Shaanxi, Gansu, Ningxia, and Shanxi, and Tartary buckwheat is grown in Yunnan, Sichuan, and Guizhou provinces (Lin and Chai, 2007). Hence buckwheat has been cultured for thousands of years in China, and the germplasm resources are extremely rich, and the varieties numerous (Tang et al., 2016). There are more than 150 all-year-round cultivated local varieties of common buckwheat (Zhao et al., 2004). The local varieties have high adaptability and resistance to the barren conditions of cold weather and poor soil, which are typical of most buckwheat planting areas.

In the Tibet Autonomous Region buckwheat is a valuable crop in remote and food-deficient mountain areas. It is mostly cultivated in areas with sloping infertile land on which other crops cannot be grown (Scheuche, 2004). The unique local varieties—with rosy, red, or dark pink flower color and never white, and adaptable to the marginal areas in high altitudes—are grown by farmers in the region. But the cultivated area of buckwheat in Tibet is gradually declining and buckwheat production in Tibet should be encouraged (Scheuche, 2004). The Tibetans use buckwheat mainly to prepare different kinds of bread—the original traditional fermented pancake is made at the New Year and at Festivals. It is interesting that a similar dish was popular among the Russians in the past centuries.

The tendency to reduce buckwheat planting areas and the conversion of the buckwheat land to growing corn was observed in the Sichuan province. Even people who acknowledged buckwheat's nutritional superiority often painted it as a food from a more difficult past, associated with manual labor, subsistence farming, and mountain poverty, rather than modern economic dynamism and prosperity (Saunders, 2010).

Between 2000 and 2010, buckwheat research has been listed in one research subject of the State Scientific Support Plan of China, which is implemented to study the integration of buckwheat germplasm, new buckwheat varieties, farming practices, and production of healthcare buckwheat foods (Lin and Chai, 2007). According to the statistics, between 1998−2015, 45 buckwheat varieties, among them 12 of common buckwheat, have been examined and approved by national or provincial committees (Li et al., 2016). Replacement of traditional varieties by the advanced breeding ones may increase buckwheat yield and production but on the other hand may cause the loss of the unique local germplasm.

In the 1980s, the Crops Genetic Resources Institute of the Chinese Academy of Agricultural Sciences (CAAS), together with other institutes in 24 provinces, began to collect buckwheat resources, which are stored in the germplasm bank of CAAS. The local varieties of cultivated buckwheat from the provinces have been collected and carefully stored (Tang et al., 2016). Thus China has remained one of the main buckwheat producers in the world as well as a major source of buckwheat genetic resources.

Buckwheat is a traditional food in Japan. Noodles (soba-kiri) made from buckwheat flour are a traditional, popular food there. In addition to noodles, there are various buckwheat products such as confectioneries, cookies, and spirits (Ikeda et al., 2001). Buckwheat has been considered one of Japans favorite foods and one of the indigenous crops in some districts of Japan (Onjo and Park, 2001). The traditional buckwheat noodle preparation methods have been reported in Japanese history for about 400 years or more and developed in Edo (Tokyo) (Ikeda and Ikeda, 2016).

Buckwheat production in Japan in 2014 was around 33 thousand tons (FAOSTAT, 2016). But the amount is not sufficient for domestic consumption. As a result, Japan is now importing approximately 120,000 tons of buckwheat grain per year (Suzuki, 2003; Katsube-Tanaka, 2016). Buckwheat planting areas in Japan have drastically decreased since the beginning of the last century. One of the reasons why buckwheat acreage continued to decrease during this period was the expansion of rice; another reason was the unwillingness of people to grow lower yielding buckwheat (Suzuki, 2003). Many producers associated buckwheat cultivation with exploitation of barren lands.

Hokkaido is the most important area for buckwheat production in Japan (Morishita et al., 2016). Buckwheat is also grown in Yamagata, Fukushima, Ibaragi, and Nagano prefectures and in some prefectures of the southwestern district (Katsube-Tanaka, 2016).

Breeding work with buckwheat in Japan has been active for a long time. The variety Hashigamiwase was bred in 1919, Botansoba was released in 1930, and Shinano No.1 was created in 1944 (Campbell, 2003). 42 common buckwheat varieties, including the tetraploid and determinate types, were released in Japan as of July 2016 (Hayashi, 2016).

Nevertheless the local buckwheat varieties and populations can still be found in Japan. In Kagoshima area two kinds of buckwheat were distributed: Kanoya (diploid), considered as a native variety, and Miyazakiotsubu (tetraploid) developed in Miyazaki University (Onjo and Park, 2001). In some areas of Kagoshima a prosperity festival is celebrated in connection with buckwheat. There is evidence of common buckwheat traditional culture in Minamiminova village in Japan not later than 70 years. Approximately 30% of farmers purchased the commercial seeds, while the others continued the home seed-raising for 5−20 years (Inoue et al., 1995). Currently, many indigenous varieties have been classified into three agroecotypes. Late-summer, intermediate, and summer types, based on the photosensitivity of flowering, are distributed throughout Japan (Iwata et al., 2005).

Due to the governmental policy to subsidize the farmers growing other crops instead of rice, the cropping acreage of buckwheat has firmly increased in Japan for the last two decades. Moreover, the Genebank of the National Institute of Agrobiological Sciences conserved landrace accessions of 218 common buckwheat and eight Tartary buckwheat varieties originating from Hokkaido in the north to Kagoshima in the south, and disclosed detailed information on various characteristics of the accessions (Katsube-Tanaka, 2016).

Buckwheat is an important crop in the hilly areas of Nepal. Around 10,000 tons of buckwheat were produced in Nepal in 2014 (FAOSTAT, 2016). Buckwheat in Nepal is grown in places with less transport accessibility, less availability of other foods, unavailability of alternate crop, and poorer economic settings (Dongol et al., 2004). Two buckwheat species are cultivated in Nepal, but farmers prefer common-type mainly because Tartary buckwheat is bitter and difficult to dehull (Joshi et al., 2010). People in Nepal use buckwheat grain for bread-making purpose, and the leaves for vegetables (Chaudhary et al., 1995).

The local people try to survive with locally available resources, and have developed extremely sensitive landraces suitable for the stress conditions under which they are grown (Baniya et al., 2000). Different types of landraces are maintained by different ethnic communities for their different purposes, in various socio-cultural conditions, from their birth to death (Dongol et al., 2004). The Government of Nepal supported the activity on conservation and utilization of the genetic resources. The Hill Crops Research Programme, established in 1986, is aimed at the research on buckwheat and some other crops (Dongol et al., 2004). Nepal established the National Agriculture Genetic Resource Centre (NAGRC), the alias of the Genebank, in Khumaltar in 2010, which collects, identifies, and preserves the plant genetic resources. Among them 230 buckwheat accessions are available (Paudel et al., 2016).

Buckwheat in India is cultivated in the Himalayan region extending from Jammu and Kashmir in the west to Arunachal Pradesh in the east (Joshi, 1999). The cultivated landraces of common and Tartary buckwheat are mainly distributed in the western Himalayan region than north-eastern region that can be due to their migration from China and Tibet along the several trade routes (Rana et al., 2016a). In the most northern and western states of India, buckwheat flour is consumed on fasting days. Buckwheat is also pounded and boiled like rice and consumed as a substitute for rice (Rana et al., 2016a). Buckwheat has been traditionally grown in the Himalayas for the centuries by the resources-poor mountain farmers and is considered to be a staple food for some tribal people (Hore and Rathi, 2002; Dutta, 2004). The buckwheat crop in Kashmir is thus regarded to be the crop of the poor (Tahir and Farooq, 1998).

In earlier days, buckwheat was an integral part of the cropping systems of the hills but the developmental activity led to the drastic decline of buckwheat cultivation (Joshi, 1999; Rana et al., 2016b). This can cause a problem of loss of buckwheat diversity and related traditions in the Himalayan region. In order to reintroduce the buckwheat cultivation in the western Himalayas many farmers and local political bodies adopted a resolution that at least one buckwheat recipe will be served by every household during any religious or family event and at least some buckwheat crops grown in its field (Rana et al., 2016b).

The National Bureau of Plant Genetic Resources (NBPGR) is working on the management of germplasm in India. In the past four decades, the NBPGR has built up 857 germplasm accessions from more than 70 buckwheat growing sites including farmers' fields, seed stores, and local markets. Under the All India Coordinated Network Project on Under-Utilized Crops, five high-yielding buckwheat varieties have been released (Rana et al., 2016b).

Buckwheat is an indispensable food in Bhutan in the non rice-growing regions, especially at high altitudes (Norbu, 1995). In 2014 Bhutan produced more than 4000 tons of buckwheat (FAOSTAT, 2016). Common and Tartary buckwheat are both grown in Bhutan and frequently mixed together in fields and not distinguished when they are consumed (Ohnishi, 1992). Buckwheat flour is used for the making of noodles, cooked dough, pancakes, and bread. People in the southern belt consume buckwheat during fasting days (Norbu, 1995). Buckwheat was considered simply the poor man's food, but within a few decades buckwheat in the form of pancakes and noodles has acquired the status of restaurant food (Norbu and Roder, 2001). The germplasm diversity was observed in Bhutan and some accessions were collected. So far, the germplasm screening work has not yet been initiated and therefore all the varieties grown by Bhutanese farmers are local indigenous varieties (Norbu, 1995).

Buckwheat has been cultivated in Korea for a long time, probably from the beginning of the 8−9th century (Hun and Hun, 2000). Besides the traditional noodles, buckwheat is used as a vegetable and medicinal plant (Choi et al., 1995). Buckwheat production in 2014 in Korea was about 2000 tons, and approximately the same quantity of buckwheat grain is imported annually (FAOSTAT, 2016). The buckwheat cultivation area has gradually decreased recently because of the low profit to farmers (Yoon et al., 2004). Many farmers in Korea still grow self-produced landraces, most varieties recommended for production in Korea have been developed from the foreign varieties, as the domestic genetic resources are limited (Park et al., 2004). Allozyme and SSR analyses did not reveal the high level of genetic variation among the common buckwheat local populations (Hun and Hun, 2000; Song et al., 2011). In spite of the reduction of buckwheat production in Korea the cultivation area, as a landscape of buckwheat blossoms is attractive for various festival events, has significantly increased (Yoon et al., 2004).

The largest buckwheat producer in the world is the Russian Federation. The first historical charters mentioning buckwheat in Russia were dated from the 15−16th centuries (Krotov, 1975). Since that time buckwheat has become a favorite food for the Russians and other nationalities inhabiting Russia. The main buckwheat dish in Russia is a "kasha" prepared from the boiled buckwheat grain, which is usually eaten with milk, or with vegetables or meat or as a separate dish. Buckwheat pancakes, fermented or not, were popular in Russia in the last centuries.

For the long period of adaptation to severe climates the buckwheat plants have changed their physiology and habitat. Early maturing and neutral to the day length populations were formed in the northern latitudes (Fesenko, 1990). The tendency towards a reduction of growth in the south-to-north direction was linked with the limited secondary branching (Fesenko et al., 2016).

The breeding work with buckwheat in Russia started at the beginning of the last century. The first breeding variety—Bogatyr—created by mass selection from the local population, was released in 1938 and has been grown until now. Breeding work in Russia has been developed in different ways: early maturing, large grain, determinate growth, limited branching, and green or red flowers. The State Register of varieties approved for use in the Russian Federation includes 48 buckwheat varieties adapted to various regions of the country. Most buckwheat is planted in the Altai, the Orenburg region, the Republic of Bashkortostan, and the Central region (Zotikov, 2013). Buckwheat in Russia is grown on an industrial scale and the registered commercial varieties are cultivated only. But the local landraces grown early were not lost and have been carefully collected since the 1940s and stored at the N. I. Vavilov Research Institute of Plant Genetic Recourses (VIR), the former N. I. Vavilov Research Institute of Plant Industry, in St. Petersburg (Romanova et al., 2007). The VIR collection maintains more than 2000 buckwheat accessions—among them 1700 are local varieties from former USSR (Fesenko, 1990).

In Ukraine buckwheat is grown in the Forest-Steppe and Polissya regions for its demand for moisture (FAO, 2008). About 200,000 tons were produced in Ukraine in 2014 (FAOSTAT, 2016). Twenty varieties are registered to be approved in Ukraine. Different types of genetic recombinogenesis and phenotypic selection were the basic methods of their creation with the use of the induced mutagenesis method (Taranenko et al., 2004). Buckwheat germplasm collections are preserved in the V. Y. Yuriev Institute of Plant Production (1600 accessions) and the Podillya State Agricultural University (900 accessions) (FAO, 2008).

Belarus produced 30,000 tons of buckwheat in 2014 (FAOSTAT, 2016). The most resultative breeding direction in the Belarus is the polypoid buckwheat. The area occupied with tetraploid buckwheat in some years reached 50% of the total areas under buckwheat in Belarus (Dubovik, 2004). The State Register of varieties recommended for growing in Belarus includes the 12 buckwheat varieties.

Buckwheat reached Europe in the Middle Ages and became popular due to its low requirement for soil fertility (Jacquemart et al., 2012). Being popular in western Europe during 16−18th centuries, buckwheat vanished from the fields in competition with wheat (Weislander, 2016). Presently buckwheat is grown mostly in eastern European countries.

Poland is one of the main buckwheat producers of the European Union with annual production of around 100,000 tons (FAOSTAT, 2016). Buckwheat breeding in Poland has a long history from the selection of local populations to the creation of high-yielding Polish varieties, but in 2004 buckwheat was removed from the official crop list, so foreign seeds flooded the Polish market (Kwiatkowski, 2013). Buckwheat has been cultivated primarily by small farms. Sometimes the big companies sow buckwheat as an insurance crop in case of cereal damage. However, in spite of numerous problems with buckwheat cultivation in Poland, its acreage is relatively stable as it serves as an important component of the diet of Poles (Grabinski, 2016).

Buckwheat was a favorite food on the territory of the Czech Republic in the 16−17th centuries, and then growing decreased due to expansion of bakery products and potato popularity (Michalova, 2001). About 2500 tons per year are produced by the Czech Republic (FAOSTAT, 2016). The domestic and foreign varieties are cultivated there. Buckwheat is mostly consumed as a kasha, and also buckwheat noodles are becoming popular with Czech consumers (Janovska and Chepkova, 2016). Under the auspices of the National Programme of Plant Genetic Resources, the buckwheat germplasm collection was established in the genebank of the Crop Research Institute in Prague Ruzyne (Stehno et al., 1998.). The collection includes 170 accessions, most of which were obtained from foreign gene banks (Cepková et al., 2009). The interest in buckwheat growing has increased since the 1990s and buckwheat has taken up position as the main crop of the Czech organic farming systems (Janovska et al., 2007).

Buckwheat in Slovenia was first mentioned in the middle of the 15th century (Kreft, 2001). Now Slovenia produced around 1000 tons of buckwheat annually (FAOSTAT, 2016). The commercial buckwheat varieties, including the determinate ones, are cultivated in Slovenia. In Slovenia buckwheat is primary grown as a stubble-crop following vegetables, barley, or wheat, and seldom as a full-season crop (Bavec et al., 2002). Slovenia is very rich in buckwheat germplasm, and possibly some forms which disappeared in their original locations are grown in Slovenia (Kreft, 1980). The first buckwheat collection was made primarily for breeding or teaching activities at the Biotechnical Faculty, University of Ljubljana (FAO, 1996). In recent years buckwheat has come back to Slovenian eating culture. Some traditional dishes from buckwheat like kasha, bread, and cakes have been preserved in Slovenia and modern recipes have been developed (Vombergar et al., 2016).

Buckwheat was historically cultivated in Italy in the mountain areas of the Alps, but as of now the crop has almost completely disappeared (Borghi, 1995), although it is still grown in a small area where the flour is the basic ingredient in the preparation of several local foods. However, buckwheat has now become more attractive with respect to its high nutrient value as a functional food and as an ideal crop for bioagriculture (Brunori et al., 2016).

Buckwheat is cultivated to a greater or lesser extent in other European countries such as France, Finland, Sweden, Slovakia, and Bosnia and Herzegovina. A certain amount of buckwheat is produced in the USA, Canada, Brazil, and Australia. But the short period of buckwheat growing does not imply any local genetic diversity in the countries where the crop was introduced or reintroduced recently.

### *Fagopyrum tataricum* Genetic Resourcers

Tartary buckwheat has been traditionally cultivated by the local people inhabiting the mountainous areas of southern China and the Himalayas. In China, Tartary buckwheat is a national crop of Yi people and it has been cultivated by them since the 2nd century B.C. (Zhang et al., 2004). The Yi people belonging to the Tibet−Burmy linguistic group domesticated Tartary buckwheat in the process of their migration from Eastern Tibet to the southwestern part of China (Ohnishi, 2004). Tartary buckwheat cultivation and consumption has been integrated into the Yi's living habits, so it has developed a unique Buckwheat food culture. Yi people eat the Tartary buckwheat on holidays and at weddings and funerals (Lin et al., 2010). As Tartary buckwheat is the main component of Yi diet they are free from lifestyle related diseases (Kano et al., 2004).

Tartary buckwheat is a cold and drought-tolerant crop and due to its short growing period can be grown at elevations up to 4500 m in the Tibetan area. Both sweet and bitter buckwheat flour are used for making bread in Tibet. (Scheuche, 2004).

Tartary Buckwheat has been traditionally cultivated by the Sherpas people living in the high hills of Nepal at a height of over 3500 m (Kano et al., 2004). The people have been adapting themselves to the severe environmental conditions and developed a self-sufficient style of agriculture. Tartary buckwheat in the highlands produces a considerable volume of anti-oxidative agents and protects as people and animals from strong ultraviolet rays. The Sherpa tribe migrated from east Tibet to the Nepal region between the 12th and 14th centuries and their living customs and culture have been inherited until today throughout these hundreds years (Kano et al., 2004).

People in Nepal use buckwheat grain for bread-making purposes and the leaves for vegetables. As reported by the majority of farmers, the leafy vegetables of Tartary buckwheat leaves are very tasty, but continuous consumption for long periods creates body pain (Chaudhary et. al., 1995).

A drawback of Tartary buckwheat is its bitter taste because of the high content of rutin. Traditionally farmers follow two techniques to reduce its bitterness: they put groat in boiling water to cook, as with rice, or they soak groat in water overnight and then remove the yellowish water (Joshi et al., 2010). In the traditional preparation of bread or pancakes in Nepal the Tartary buckwheat flour is kneaded in the boiled water (Kano et al., 2004).

Tartary buckwheat genetic resources are widely distributed in the areas where it is cultivated. Distributed mainly in provinces south of the Yantze River in China, Tartary buckwheat has over 300 cultivated local varieties (Zhao et al., 2004). In Nepal, farmers maintain different landraces of Tartary buckwheat and some of them are very unique (Joshi et al., 2011). Nepalese buckwheat varies in seed color and shape, flower and leaf color, height, seed productivity, and duration of growth period. A peculiar rice-type called "bhate", with non-adhering hull, was found by the National Hill Crops Research Program Nepal team in 1991 (Hirose et al., 1995). Two types of rice buckwheat, namely Kalo Kishe (black seeded) and Seto Kishe (white seeded), are cultivated in Nepal in the Dolpa district only (Joshi at al., 2011).

The Liangshan Autonomous Prefecture of Yi Nationality in Sichuan, in China, and the western mountain region of Nepal were selected for the realization of the project on in situ conservation of buckwheat genetic resources (Zang et al., 2004). But today many people for example the Yi have moved from the summit of the mountain to the plains, and the basis of their lifestyle has changed (Lin et al., 2010). If buckwheat populations are geographically unique, significant genetic losses may occur when villages discontinue cultivation (Saunders, 2010).

The *F. tataricum* species consists of cultivated landraces, wild subspecies, and a weedy type. The weedy type is a weed having accompanied the cultivated landraces and possibly having been originated from a hybrid between the wild subspecies and the cultivated landraces (Tsui and Ohnishi, 2009). The unique collection of weedy Tartary buckwheat containing forty accessions found in the USSR as an admixture of common buckwheat is preserved at the VIR Genebank (Romanova, 2004). Tartary buckwheat accessions are also available in all known Genebank collections of the buckwheat germplasm.

Tartary buckwheat has been found in the nature of the Hokkaido and Sakhalin islands (Honda et al., 2004, 2006). Whereas the wild Tartary buckwheat, Ishisoba, had been cultivated in some northern districts of Hokkaido, probably it was domesticated early on.

Tartary buckwheat has a high nutritional value due to the high content of protein, vitamins, lipids, flavonoids, and other functional elements (Lin et al., 2010). Food quality and other technological and medical properties of buckwheat are the reasons for its increase in popularity (Vombergar et al., 2016). The crop has become more popular in China, Japan, Korea, and Europe in recent decades.

Many new buckwheat products are introduced every year in China, such as Tartary buckwheat tea, noodles, cakes, and wine. Among them, Tartary buckwheat tea is one of the most popular buckwheat products (Zhao et al., 2016). The tea is made from roasting the grain, which leads to lower levels of bitter taste (Ikeda et al., 2012). The progress of Tartary buckwheat products in China had passed several stages: from local foods made by conventional family workshops, to public foods made by factories with machines (Lin, 2004). Additionally, the Education Centre Piramida Slovenia has developed more than 30 bakery and confectionery products from Tartary buckwheat, including pasta products and also ice cream with Tartary buckwheat groats (Vombergar et al., 2016).

The main producer of Tartary buckwheat in the world is China, where about a third of the buckwheat planting area is occupied by Tartary buckwheat. Thirty-one varieties of Tartary buckwheat have been developed in China in recent times (Li et al., 2016). Thirteen Tartary buckwheat varieties were released in Japan by 2016 (Hayashi, 2016). The Himpriya variety was released by the National Bureau of Plant Genetic Resources in India (Dutta, 2004). Tartary buckwheat came back to Slovenia, was introduced to Sweden, and continued to grow in Bosnia and Herzegovina in Europe (Ikeda et al., 2012). It is believed that Tartary buckwheat is a crop of the future (Lin, 2004).

## CONCLUSION

Summarizing this chapter we may conclude that genetic resources of the genus *Fagopyrum* are distributed around the world. Buckwheat resources include the wild species and subspecies, the landraces, and the advanced breeding varieties. The area of wild species of *Fagopyrum* is mostly limited to some districts of Yunnan and the Sichuan provinces of China. The local landraces have been still distributed in several highland areas of China, Nepal, India, and Bhutan which are economically less developed. The processes of agricultural industrialization and replacement of the local landraces by modern commercial varieties cannot be turned back. So research activity in the collection and evaluation of available buckwheat germplasm is very important regarding the preservation of the unique germplasm with optimal gene combination. Local varieties have been developed by humans over the centuries through selections that made them adaptable to any unfavorable environment. In situ conservation of buckwheat genetic resources may be useful for wild species. In regard to the local varieties it may work until the farmers continue buckwheat cultivation at the same region. The commercial varieties developed at different times by various methods may illustrate the progress of intensive breeding for the last century and should also be preserved. The safer method to preserve the existent genetic diversity is considered to be the collection and long-time conservation of buckwheat seeds in the national or international genebanks. The largest national genebanks are involved in the preservation of buckwheat genetic resources. They are the Crops Genetic Resources Institute of the Chinese Academy of Agricultural Sciences (CAAS), the N. I. Vavilov Research Institute of Plant Genetic Recourses (VIR) in Russia, the National Bureau of Plant Genetic Resources (NBPGR) in India, the National Agriculture Genetic Resource Centre (NAGRC) in Nepal, and the Genebank of the Crop Research Institute in Prague Ruzyne, Czech Republic. Buckwheat food culture has been carefully saved in every country and has continued to be developed. So we share the view of Professor Adachi (2016) that era of a Buckwheat Renaissance is coming.

## ACKNOWLEDGMENTS

This research was supported by the Investigation of Forage Germplasm in Central China (grant number 2017FY100604) and National Program on Key Basic Research Project (973 Program) (grant number 2014CB138701).

## REFERENCES

Baniya, B.K., Dongol, D.M.S., Vaidya, M.L., Upadhyay, S.R., Nemoto, K., 2000. Buckwheat-based cropping patterns in Upper Mustang of Nepal. Fagopyrum 17, 1−7.

Bavec, F., Pusnik, S., Rajcan, I., 2002. Yield performance of two buckwheat genotypes grown as full-season and stubble-crop. Rostlina Vyrova 48 (8), 351−355.

Borghi, B., 1995. Buckwheat in Italy, state of the art and preliminary results. Proceedings of 6th International Symposium. Buckwheat, Japan, pp. 65−70.

Brunori, A., Nobili, C., Procacci, S., 2016. Toward the use of buckwheat as an ingredient for the preparation of functional food, chapter 17. In: Zhou, M., Kreft, I., Woo, S.-H., Chrungoo, N., Wieslander, G. (Eds.), Molecular Breeding and Nutritional Aspects of Buckwheat. Elsevier, pp. 219−227.

Campbell, C., 2003. Buckwheat crop improvement. Fagopyrum 20, 1−6.

Cepková, P.H., Janovská, D., Stehno, Z., 2009. Assessment of genetic diversity of selected tartary and common buckwheat accessions. Spanish J. Agric. Res. 7 (4), 844−854.

Chaudhary, N.K., Namai, H., Goto, T., 1995. Ecological genetic study on collection, evaluation and conservation of buckwheat (*Fagopyrum tataricum* Gaertn.) genetic resources in Nepal. Proceedings of 6th International Symposium. Buckwheat, Japan, pp. 79−85.

Choi, B.H., Park, K.Y., Park, R.K., Current status of buckwheat culture technology in Korea, In: Proceedings of 6th International Symposium, 1975, Buckwheat, Japan, pp. 27-38.

Dongol, Baniya, B.K., Joshi, B.K., 2004. Psychosocial basis of cultivation ad food preference of Nepalese buckwheat growers. Proceedings of 9th International Symposium. Buckwheat, Prague, pp. 285−290.

Dubovik, E.I., 2004. Breeding of polyploid buckwheat in belarus: results, problems, directions. Proceedings of 9th International Symposium. Buckwheat, Prague, pp. 202–206.

Dutta, M., 2004. Buckwheat improvement in India: current status and future prospects. Proceedings of 9th International Symposium. Buckwheat, Prague, pp. 302–312.

FAO, 1996, Slovenia: country report to the fao international technical conference on plant genetic resources.

FAO, 2008. Second country report on the state of plant genetic resources for food and agriculture in Ukraine.

FAOSTAT data, 2016. Online database. Available online: <http://faostat.fao.org>.

Fesenko, A.N., Fesenko, N.N., Romanova, O.I., Fesenko, I.N., 2016. Crop evolution of buckwheat in Eastern Europe: microevolutionary trends in the secondary center of buckwheat genetic diversity, chapter 8. In: Zhou, M., Kreft, I., Woo, S.-H., Chrungoo, N., Wieslander, G. (Eds.), Molecular Breeding and Nutritional Aspects of Buckwheat. Elsevier, Burlington, MA, pp. 99–107.

Fesenko, N.V., 1990. Research and breeding work with buckwheat in USSR (Historic survey). Fagopyrum 10, 47–50.

Grabiński, J., 2016. The problems in cultivation of buckwheat in Poland. Proceedings of 13th International Symposium. Buckwheat, Korea, pp. 151–157.

Hayashi, H., 2016. Buckwheat productivity in Japan. Proceedings of 13th International Symposium. Buckwheat, Korea, pp. 193–199.

Hirose, T., Yoshida, M., Nemoto, K., Kitabayashi, H., Minami, M., Matano, T., et al., 1995. Diversity of grain character of tartary buckwheat in Nepal. Proceedings of 6th International Symposium. Buckwheat, Japan, pp. 385–388.

Honda, Y., Mukasa, Y., Suzuki, T., Abe, N., 2004. Stone buckwheat, genetic resource of tartary buckwheat in Japan. Proceedings of 9th International Symposium. Buckwheat, Prague, pp. 184–189.

Honda, Y., Suzuki, T., Sabitov, A., Romanova, O.I., 2006. Collaborative exploration and collection of resources crops including Tartary buckwheat, *Fagopyrum tataricum* L, in Sakhalin, Russia. Plant Inheritance Resour. Search Investig. Rep. 22, 1–99.

Hore, D., Rathi, R.S., 2002. Collection, cultivation and characterization of buckwheat in Northeastern region of India. Fagopyrum 19, 11–15.

Hun, H.W., Hun, M.K., 2000. Allozyme variation and population structure of common buckwheat F.E. in Korea. Fagopyrum 17, 21–27.

Ikeda, K., Ikeda, S., 2016. Factors Important for Structural Properties and Quality of Buckwheat Products Chapter 15. In: Zhou, M., Kreft, I., Woo, S.-H., Chrungoo, N., Wieslander, G. (Eds.), Molecular Breeding and Nutritional Aspects of Buckwheat. Elsevier, Burlington, MA, pp. 193–202.

Ikeda, K., Arai, R., Fujiwara, J., Asami, Y., Kreft, I., 2001. Food-scientific characteristics of buckwheat products. Proceedings of 8th International Symposium. Buckwheat, Korea, pp. 489–493.

Ikeda, K., Ikeda, S., Kreft, I., Lin, R., 2012. Utilization of Tartary buckwheat. Fagopyrum 29, 27–30.

Inoue, N., Matano, T., Ujihara, A., 1995. Traditional culture of common buckwheat around of Minamiminova village in Japan. Proceedings of 6th International Symposium. Buckwheat, Japan, pp. 699–703.

Iwata, H., Imon, K., Tsumura, Y., Ohsawa, R., 2005. Genetic diversity among Japanese indigenous common buckwheat (*Fagopyrum esculentum*) cultivars as determined from amplified fragment length polymorphism and simple sequence repeat markers and quantitative agronomic traits. Genome 48, 367–377.

Jacquemart, A.-L., Cawoy, V., Kinet, J.-M., Ledent, J.-F., Quinet, M., 2012. Is buckwheat (*Fagopyrum esculentum* Moench.) still a valuable crop today?. Eur. J. Plant Sci. Biotechnol. 6 (Special Issue 2), 1–10.

Janovská, D., Cepková, H.P., 2016. Nutritional aspects of buckwheat in the Czech Republic, Chapter 14. In: Zhou, M., Kreft, I., Woo, S.-H., Chrungoo, N., Wieslander,

G. (Eds.), Molecular Breeding and Nutritional Aspects of Buckwheat. Elsevier, Burlington, MA, pp. 177—192.

Janovska, D., Stehno, Z., Cepkova, P., 2007. Evaluation of common buckwheat genetic resources in Czech Gene Bank. Proceedings of 10th International Symposium. Buckwheat, China, pp. 31—40.

Joshi, B.D., 1999. Status of buckwheat in India. Fagopyrum 16, 7—11.

Joshi, B.K., Okuno, K., Bimb, H.P., Sharma, D.R., 2010. Farmers' knowledge on and on-station characterization of Bhate Phapar (Rice Tartary Buckwheat). Proceedings of 11th International Symposium. Buckwheat, Orel, pp. 86—93.

Joshi, B.K., Okuno, K., Ohsawa, R., Kawase, M., Otobe, C., Hayashi, H., et al., 2011. Biplot analysis of multiple traits of Bhate (rice) and non-Bhate (normal) types of Tartary buckwheat. Fagopyrum 28, 9—15.

Kano, M., Kizaki, T., Inasawa, T., 2004. The state of cultivating Tartary buckwheat (*Fagopyrum tataricum*) and its traditional dishes along everest trekking route, Nepal. Proceedings of 9th International Symposium. Buckwheat, Prague, pp. 547—552.

Katsube-Tanaka, T., 2016. Buckwheat Production, consumption, and genetic resources in Japan, chapter 5. In: Zhou, M., Kreft, I., Woo, S.-H., Chrungoo, N., Wieslander, G. (Eds.), Molecular Breeding and Nutritional Aspects of Buckwheat. Elsevier, Burlington, MA, pp. 61—80.

Kreft, I., 1980. Starting points for buckwheat breeding in Yugoslavia. Proceedings of 1st International Symposium. Buckwheat, Ljubljana, pp. 69—73.

Kreft, I., 2001. Buckwheat research, past, present and future perspectives - 20 years of internationaly coordinated research. Proceedings of 8th International Symposium. Buckwheat, Korea, pp. 361—366.

Kreft, I., Wieslander, G., Vombergar, B., 2016. Bioactive flavonoids in buckwheat grain and green parts, chapter 12. In: Zhou, M., Kreft, I., Woo, S.-H., Chrungoo, N., Wieslander, G. (Eds.), Molecular Breeding and Nutritional Aspects of Buckwheat. Elsevier, Burlington, MA, pp. 161—167.

Krotov, A.C., 1975. Buckwheat. In: Krotov, A.C. (Ed.), Cultured flora of USSR. Leningrad, Kolos, pp. 7—118.

Kwiatkowski, J., 2013. Buckwheat breeding and seed production in Poland. Proceedings of 12th International Symposium. Buckwheat, Slovenia, pp. 188-161.

Li, F.-L., Ding, M.-Q., Tang, Y., Tang, Y.-X., Wu, Y.-M., Shao, J.-R., et al., 2016. Cultivation of Buckwheat in China, Chapter 24. In: Zhou, M., Kreft, I., Woo, S.-H., Chrungoo, N., Wieslander, G. (Eds.), Molecular Breeding and Nutritional Aspects of Buckwheat. Elsevier, Burlington, MA, pp. 321—324.

Lin, R., 2004. The development and utilization of tartary buckwheat resources. Proceedings of 9th International Symposium. Buckwheat, Prague, pp. 252—258.

Lin, R., Chai, Y., 2007. Production, research and academic exchanges of China on buckwheat seeds. Proceedings of 10th International Symposium. Buckwheat, China, pp. 7—12.

Lin, R., Inasawa, T., Sun, Y., 2010. Tartary buckwheat food culture of Yi nationality. Proceedings of 11th International Symposium. Buckwheat, Orel, pp. 540—544.

Michalova, A., 2001. Buckwheat in the Czech Republic and in Europe. Proceedings of 8th International Symposium. Buckwheat, Korea, pp. 702—709.

Morishita, K.T., Suzuki, T., Mukasa, Y., Honda, Y., 2016. The breeding and characteristics of a new common buckwheat variety. Proceedings of 13th International Symposium. Buckwheat, Korea, pp. 327—330.

Norbu, S., 1995. Buckwheat in Bhutan. Proceedings of 6th International Symposium. Buckwheat, Japan, pp. 55—60.

Norbu, S., Roder, W., 2001. Traditional uses of buckwheat in Bhutan. Proceedings of 8th International Symposium. Buckwheat, Korea, pp. 670—672.

Ohnishi, O., 2009. On the origin of cultivated common based on allozyme analyses of cultivated and wild populations of common buckwheat. Fagopyrum 26, 3—9.

Ohnishi, O., 2012. On the distribution of natural populations of wild buckwheat species. Fagopyrum 29, 1—6.

Ohnishi, O., 2004. On the origin of cultivated buckwheat. Proceedings of 9th International Symposium. Buckwheat, Prague, pp. 16−21.

Ohnishi, O., 2013. Distribution of wild buckwheat species and perspective for their utilization. Fagopyrum 30, 9−14.

Ohnishi, O., 2016. On the diffusion of buckwheat cultivation and the diffusion of consumption of buckwheat noodles. Proceedings of 8th International Symposium. Buckwheat, Korea, pp. 77−82.

Ohnishi, O., 1992. Buckwheat in Bhutan. Fagopyrum 12, 5−13.

Onjo, M., Park, B.J., 2001. Production and utilization of buckwheat in Kagoshima Area in Japan. Proceedings of 8th International Symposium. Buckwheat, Korea, pp. 161−167.

Park, B.J., Park, J.I., Chang, K.J., Park, C.H., 2004. Characteristics of genetic resources in Tartary Buckwheat (Fagopyrum tataricum. Procccdings of 9th International Symposium. Buckwheat, Prague, pp. 342−345.

Paudel, M.N., Joshi, B.K., Ghimire, K.H., 2016. Management status of agricultural plant genetic resources in Nepal. Agronomy J. Nepal 4, 75−91.

Rana, J.C., Singh, M., Yadav, R., 2016a. Germplasm Resources of Buckwheat in India. Proc. 13th Int. Symp. Buckwheat, Korea, pp. 49−63.

Rana, J.C., Singh, M., Chauhan, R.S., Chahota, R.K., Sharma, T.R., Yadav, R., et al., 2016b. Genetic resources of buckwheat in India, Chapter 9. In: Zhou, M., Kreft, I., Woo, S.-H., Chrungoo, N., Wieslander, G. (Eds.), Molecular Breeding and Nutritional Aspects of Buckwheat. Elsevier, Burlington, MA, pp. 109−135.

Romanova, O., 2004. Northern populations of Tartary buckwheat with respect to day length. Proceedings of 9th International Symposium. Buckwheat, Prague, pp. 173−178.

Romanova, O.I., Kurtseva, A.F., Matveeva, G.V., Malinovski, B.N., 2007. Genetic diversity of millet, buckwheat, sorghum and maize, biological science development and breeding for groat properties. Bull. Appl. Bot. Genet. Plant Breed. 164, 142−152.

Saunders, M., 2010. Losing ground: an uncertain future for buckwheat farming in its center of origin. Proceedings of 11th International Symposium. Buckwheat, Orel, pp. 60−68.

Scheuche, S., 2004. Buckwheat in Tibet (TAR). Proceedings of 9th International Symposium. Buckwheat, Prague, pp. 295−298.

Song, J.Y., Lee, G.-A., Yoon, M.-S., Ma, K.-H., Choi, Y.-M., Lee, J.-R., et al., 2011. Analysis of genetic diversity and population structure of buckwheat (Fagopyrum esculentum Moench) landraces of Korea using SSR markers. Korean J. Plant Res. 24 (6), 702−711.

Stehno, Z., Vlasák, M., Holubec, V., Michalová, A., 1998. evaluation and utilization of genetic resources collections of wheat, triticale, winterbarley, wild triticeae, buckwheat, millet and amaranthus. National Programme on PlantGenetic Resources Conservation and Utilization in the Czech Republic, Prague 22−28.

Suzuki, I., 2003. Production and usage of buckwheat grain and flour in Japan. Fagopyrum 20, 13−16.

Tahir, I., Farooq, S., 1998. Buckwheat research in Kashmir. Proc. 6th Int. Symp. Buckwheat, Japan, pp. 1-83−1-87.

Tang, Y., Ding, M.-Q., Tang, Y.-X., Wu, Y.-M., Shao, J.-R., Zhou, M.-L., 2016. Germplasm Resources of Buckwheat in China, Chapter 2. In: Zhou, M., Kreft, I., Woo, S.-H., Chrungoo, N., Wieslander, G. (Eds.), Molecular Breeding and Nutritional Aspects of Buckwheat. Elsevier, Burlington, MA, pp. 13−20.

Taranenko, L.K., Yatsishen, O.L., Karazhbej, P.P., Taranenko, P.P., 2004. State and Prospects of Buckwheat Selection in. Proceedings of 9th International Symposium. Buckwheat, Prague, pp. 391−396.

Tsui, K., Ohnishi, O., 2009. Morphological characters in cultivated, wild and weedy Tartary buckwheat. Fagopyrum 26, 11−19.

Tsuji, K., Ohnishi, O., 2001. Phylogenetic relationships among wild and cultivated Tartary buckwheat (Fagopyrum tataricum Gaert.) populations revealed by AFLP analyses. Genes Genet. Syst. 76, 47−52.

Vombergar, B., Kreft, I., Horvat, M., Vorih, S., Germ, M., Tašner, L., et al., 2016. Buckwheat in Food Culture of Slovenia. Proceedings of 13th International Symposium. Buckwheat, Korea, pp. 519−526.

Wei, V., 1995. Buckwheat Production in China. Proceedings of 6th International Symposium. Buckwheat, Japan, pp. 7–10.

Wieslander, G., 2016. Some case reports in chronic disease on buckwheat health effects after prolonged daily intake. Proceedings of 13th International Symposium. Buckwheat, Korea, pp. 665–670.

Yoon, Y.H., Jang, D.C., Jeong, J.C., 2004. Effect of Soil Moisture Condition on Some Growth Characteristics Related With Landscape and Yield of Buckwheat. Proceedings of 9th International Symposium. Prague, Buckwheat, pp. 465–469.

Zhang, Z., Upadhayay, M.P., Zhao, Z., Lin, R., Zhou, M.V, Rao, R., et al., 2004. Conservation and Use of Buckwheat Biodiversity for the Livelihood Development. Proceedings of 9th International Symposium. Buckwheat, Prague, pp. 422–431.

Zhao, G., Tang, Y., Wang, A., Hu, Z., 2004. China's Buckwheat Resources and Their Medicinal Values. Proceedings of 9th International Symposium. Buckwheat, Prague, pp. 630–632.

Zhao, G., Peng, L.-X., Zhao, J.-L., Xiang, D.-B., Zou, L., 2016. Investigation and development of buckwheat in China. Proceedings of 13th International Symposium. Buckwheat, Korea, pp. 813–820.

Zotikov, V., 2013. State and perspectives of buckwheat production in Russia. Proceedings of 12th International Symposium. Buckwheat, Slovenia, pp. 19–21.

## FURTHER READING

Taiji, A., 2016. Buckwheat Renaissance: Tradition and prospect for Breakthrough of future cultivation. Proceedings of 13th International Symposium. Buckwheat, Korea, pp. 201–204.

# Chapter Four

# Chromosomes and Chromosome Studies in Buckwheat

**Alexander Betekhtin, Agnieszka Rybicka, Joanna Wolna, and Robert Hasterok**

*University of Silesia in Katowice, Katowice, Poland*

## INTRODUCTION

Cytogenetic analyses of plants involve various aspects of their genome organization at the chromosome level, such as karyotype structure, which includes revealing the number, size, shape, and more recently even various aspects of the molecular structure of chromosome. The latter is ensured by using a powerful cytomolecular methodology such as DNA:DNA fluorescence *in situ* hybridization (FISH), which enables various DNA sequences along the chromosomes to be detected and visualized under an epifluorescence microscope.

The *Fagopyrum* genus includes two species that are important crops—common buckwheat (*Fagopyrum esculentum* Moench) and Tartary buckwheat (*Fagopyrum tataricum* Gaertn.). While the first produces higher yields, the second is more tolerant to adverse environmental conditions, mainly low temperatures. Due to its agronomical importance, numerous genetic studies and breeding treatments have been undertaken to improve the existing buckwheat varieties and to create new ones. One of the goals is to obtain plants that combine the useful traits of common and Tartary buckwheat. One very important aspect of such studies is connected with cytogenetic analyses, which not only allow the chromosome numbers in various buckwheat species to be determined but also permit them to be compared with the chromosomes in related species or interspecific hybrids, preferably with the possibility of identifying individual chromosomes. However, to date, representatives of *Fagopyrum* genus have not been well analyzed cytogenetically as most publications have focused on simple analyses of the chromosome number and size (Chen, 2001; Wang et al., 2005). In this chapter, we extend this information

**37**

*Buckwheat Germplasm in the World.* DOI: https://doi.org/10.1016/B978-0-12-811006-5.00004-5
© 2018 Elsevier Inc. All rights reserved.

with the results of FISH with ribosomal DNA (rDNA) probes and demonstrate an improved methodology for karyotyping. We also review chromosome behavior in different buckwheat species during meiosis.

## SOMATIC CHROMOSOMES IN BUCKWHEAT

Currently, there are 21 known representatives of the *Fagopyrum* genus whose chromosome numbers have been determined. However, for a variety of reasons, more detailed characteristics of their karyotypes are only available for seven of them (Table 4.1). One of the most prominent crops in the *Fagopyrum* genus is common buckwheat (*F. esculentum*), which in some parts of the world (Asia, Eastern Europe, USA, Brazil, France) is of significant agronomical importance. Its karyotype is characterized by the presence of 16 chromosomes that are very similar in their size and shape. The diploid status of common buckwheat was identified by Taylor (1925) and then confirmed by Quisenberry (1927), who analyzed different varieties of this species. It was also Quisenberry who drew attention to the need for more detailed cytogenetic analyses within the genus *Fagopyrum* as a prerequisite to obtaining useful information for breeding studies and strategies.

As mentioned earlier, one striking feature of the common buckwheat chromosomes is their great morphometric uniformity, since all of them are approximately 2 μm long during the mitotic metaphase (Zu et al., 1984). Further comparative analyses revealed that other buckwheat species also have relatively small chromosomes, which are less than 3.5 μm in length. According to the chromosome classification proposed by Levan et al. (1964), the majority of *Fagopyrum* chromosomes are metacentric and some are

**TABLE 4.1** Cytogenetic Characteristics of the Selected *Fagopyrum* Species

| Species | 2 n | Karyotype formula | References |
|---|---|---|---|
| *Fagopyrum esculentum* Moench | 16 | $12m + 4m^{SAT}$ | Taylor (1925); Zu et al. (1984) |
| *Fagopyrum tataricum* Gaertn. | 16 | $12m + 4sm$; $12m + 2sm + 2sm^{SAT}$ | Chen et al. (2004); Quisenberry (1927) |
| *Fagopyrum cymosum* (Trev.) Meisn. | 32 | $20m + 10sm + 2m^{SAT}$ | Chen et al. (2004) |
| *Fagopyrum pilus* Q.-F. Chen | 16 | $12m + 2m^{SAT} + 2sm^{SAT}$ | Chen (2001) |
| *Fagopyrum megaspartanium* Q.-F. Chen | 16 | $8m + 4sm + 4m^{SAT}$ | Chen (2001) |
| *Fagopyrum giganteum* Krotov. | 32 | $22m + 6sm + 4m^{SAT}$ | Chen et al. (2004) |
| *Fagopyrum zuogongense* Q.-F. Chen | 32 | $24m + 4sm + 4m^{SAT}$ | Chen (2001) |

submetacentric. No acrocentric or telocentric chromosomes have been identified in any buckwheat species.

Some buckwheat species (*Fagopyrum cymosum*, *Fagopyrum giganteum*, and *Fagopyrum zuogongense*, Table 4.1) have somatic chromosome numbers (2*n*) of 32. This may be associated with polyploidy, which is a very common phenomenon in the plant kingdom. Polyploidy involves the coexistence of more than two genomes in the somatic cell nucleus. It is estimated that about 70% of angiosperms and 95% of ferns underwent at least one round of polyploidization in their evolutionary history (Soltis and Soltis, 1995, 1999). Polyploids can be divided into distinct categories depending on the mechanisms of their formation. One of these are autopolyploids (e.g., *Coffea arabica* and *Solanum tuberosum*), which result from the multiplication of the same genome. In contrast, allopolyploids (e.g., *Triticum aestivum*, *Avena sativa*, *Nicotiana tabacum*, *Gossypium hirsutum*) arise through interspecies or intergeneric hybridization and therefore contain more than two genomes that are of different origin in their somatic cells. The formation of polyploids may occur due to an abnormal course of mitosis or meiosis (Soltis and Soltis, 1995; Wendel, 2000).

Due to the similarity of chromosomes in various buckwheat species, it is difficult to determine the origin of individual polyploids. However, this is not the case for *F. giganteum*, which is an artificial amphidiploid that was created by crossing a tetraploid specimen of the annual *F. tataricum* with the perennial buckwheat *F. cymosum* Meissn. Another interesting species is *Fagopyrum homotropicum*, which was found for the first time in the Yunnan province (southwest China) in 1992. It did not take long for this species to draw attention for its strong resemblance to the cultivated species of buckwheat. Further cytogenetic observations of other populations of *F. homotropicum* have revealed the presence of diploid and tetraploid forms within this species (Ohnishi and Asano, 1999). Due to the high degree of morphological and genetic similarity of the diploid forms of *F. homotropicum* to *F. esculentum*, Chen (2001) proposed that they should be defined as *F. esculentum var. homotropicum* with only the tetraploid forms being referred to as *F. homotropicum* (Chen et al., 2004; Matsui et al., 2004). The results of phylogenetic analyses of the nuclear gene AGAMOUS suggest that the wild species *F. esculentum ssp. ancestralis* is the ancestor of the tetraploid *F. homotropicum* (Tomiyoshi et al., 2012).

Another interesting species in cytogenetic terms is *F. cymosum*, of which both diploid and tetraploid forms were found by Ohnishi and Matsuoka (1996). An analysis of its chloroplast DNA revealed the existence of two types of allotetraploids, thus indicating that the polyploidization process in this species occurred at least twice. Moreover, observations of the meiotic chromosomes in different populations of the tetraploid forms of *F. cymosum* revealed the presence of not only allopolyploids among them but also autopolyploids. The formation of polyploid complexes or series is a common phenomenon in angiosperms, and some good examples are the *Triticum* species, which have three different ($2x$, $4x$, $6x$) ploidy levels or members of the European complex of *Cardamine pratensis* in which diverse ploidy levels can also be observed (Feldman and Levy, 2005; Marhold et al., 2002; Soltis and Soltis, 1999).

The difficulties with analyzing the *Fagopyrum* karyotypes based only on the morphometric parameters of chromosomes have led to the necessity of using more advanced methods, which can provide extra markers, thereby enabling a more effective identification of individual chromosomes. One such karyotype analysis methods is based on the use of atomic force microscopy (AFM). To date, there have not been many studies that demonstrate the successful application of this methodology to identify plant chromosomes, though some of them have reported detailed karyotype analyses for barley (Schaper et al., 2000; Sugiyama et al., 2003; Yoshino et al., 2002), wheat (McMaster et al., 1996), and common soapwort (Di Bucchianico et al., 2008). The first report of the cytogenetic karyotype analysis of *F. esculentum* and *F. tataricum* using AFM was carried out by Neethirajan et al. (2011), who demonstrated the morphological characteristics of the chromosomes in these two species in terms of their size, volume, arm lengths, and arm ratios. AFM analysis revealed that the height of the *F. esculentum* and *F. tataricum* chromosomes was approximately 350 and 150 nm, while their calculated volumes ranged from 1.08 to 2.09 $\mu m^3$ and from 0.49 to 0.78 $\mu m^3$, respectively. Those authors stressed that *F. esculentum* is an ancient species compared to *F. tataricum* and that the genome sizes of both species are proportional to their chromosomal volumes. Another advanced technique that has been used to karyotype *Fagopyrum* chromosomes is fluorescent chromosome *in situ* PCR using nuclear genes of a chloroplast origin and then utilizing these data to reveal the *Fagopyrum* species phylogeny (Li et al., 2013). These studies permitted the introgression of chloroplast genes such as 16S rDNA, 4.5S rDNA and *psbA* into the nuclear genome to be tracked and used their chromosomal location to construct a physical map. Based on these results and on the findings of their previous study (Chen, 2012), these authors determined the phylogenetic relationships within the big-achene group of buckwheat species. According to this determination, diploid species such as *Fagopyrum megaspartanium* (wild perennial, genome MM) and *Fagopyrum pilus* (wild perennial, genome PP) are the most primitive ones. *F. esculentum* (wild, genome E′E′) and *F. tataricum* (wild, genome T′T′) evolved from *F. megaspartanium* and *F. pilus* and the diploid *F. tataricum* (small rice buckwheat, genome T″T″) seem to be a progeny of wild *F. esculentum* and cultivated *F. tataricum* (genome TT). *F. zuogongense* (genome EE E′E′) is an amphidiploid hybrid of the cultivated and the wild *F. esculentum* (genomes EE and E′E′, respectively), whereas the tetraploid perennial *F. cymosum* (wild, genome EE MM) is a hybrid between the perennial *F. megaspartanium* and the cultivated *F. esculentum* (genome EE). Finally, the human-made tetraploid *F. giganteum* (genomes MM TT) is the product of the interspecific hybridization between *F. megaspartanium* and the cultivated *F. tataricum*.

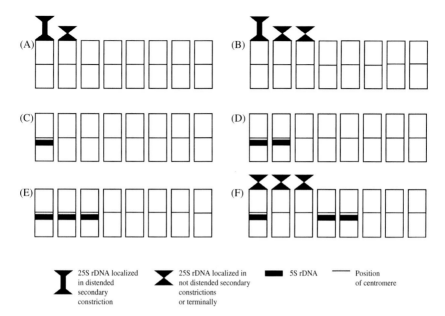

FIGURE **4.1** Ideograms demonstrating the results of FISH with 25S rDNA (A, B, F) and 5S rDNA (C, D, E, F) probes hybridizing with *F. esculentum* Moench chromosomes ($n = 8$; for simplification the ideograms show the haploid number of the chromosomes). Relative lengths of the chromosomes as well as the localizations of the centromeres and rDNA sequences along the chromosomes are only approximate.

In our research, we analyzed the root-tip cell chromosomes of the *F. esculentum* karyotype by FISH using two types of ribosomal DNA sequences—25S rDNA and 5S rDNA—as the probes. We observed the presence of interindividual polymorphisms in the number of chromosomes that carried these sequences (Fig. 4.1). The 25S rDNA probe hybridized with two pairs of chromosomes that contained secondary constrictions and satellites that were localized in the terminal part of their top arms and one pair of these secondary constrictions appeared to be clearly more distended than the other (Fig. 4.1A). Furthermore, in the case of some individuals, the presence of an additional pair of 25S rDNA signals was observed. These signals had a terminal localization on the chromosome (Fig. 4.1B, F).

In most cases, the 5S rDNA probe hybridized with two pairs of chromosomes (Fig. 4.1D), but occasionally individuals with only one pair (Fig. 4.1C) or as many as three pairs of chromosomes (Fig. 4.1E) bearing 5S rDNA loci were also observed. Regardless of their number, the sites of 5S rDNA were always localized in the pericentromeric regions of the chromosomes, most often in the bottom (long) chromosome arm. For the individuals with the maximal number of both 25S rDNA and 5S rDNA loci in the case of one chromosome pair, these sequences were closely linked on the same chromosome (Fig. 4.1F).

## MEIOTIC CHROMOSOMES IN BUCKWHEAT

Cytogenetic observations not only enable the number and morphology of chromosomes to be determined but also permit their behavior during meiosis to be analyzed. The structural similarity of homologous chromosomes ensures their proper pairing, formation of bivalents, and the correct distribution of homologues to the daughter cells. Observations of meiosis in polyploids are particularly instructive since they may provide important information on the origin of these species (Grandont et al., 2013). In autopolyploids, the presence of more than one pair of homologous chromosomes often leads to the situation where multivalents and univalents are observed during prophase I, which may disturb chromosome segregation and result in decreased fertility. Observations of the course of meiosis in artificially created autopolyploids of *F. esculentum* and *F. tataricum* demonstrated a positive correlation between the number of univalents that were present during metaphase I and the number of micronuclei at telophase II. Furthermore, a negative correlation between these two factors and the fertility of pollen was also observed. It was also revealed that the autotetraploids that had been obtained from the pure homozygous lines had a more irregular meiosis than the tetraploids with a higher degree of heterozygosity. This is particularly important if an induced autopolyploid is to be used in breeding studies and crossed with other species (Chen et al., 2007).

Chromosomes in allopolyploids behave similarly to those in diploids during meiosis. However, in allopolyploids that have resulted from crosses between closely related species, the chromosomes that belong to different ancestral genomes often share a high degree of homology during meiosis and in some cases may also be subjected to an unwanted intergenomic pairing. Such chromosomes are known as homeologous chromosomes. Based on the frequency of bivalent formation and the appearance of univalents as well as the different kinds of multivalents, it is possible to conclude whether the analyzed species are auto- or allopolyploids (Ohsako et al., 2002).

Analysis of chromosome pairing can be an indicator of phylogenetic relationships between the genomes in many cases. Observations of the course of meiosis in an amphihaploid hybrid that was created by crossing diploid forms of *F. homotropicum* with *F. esculentum* ssp. *ancestralis* proved that the chromosomes that had originated from these two species successfully pair and form bivalents, which is a good evidence for their very close relationship (Ohsako et al., 2002). A partial genomic homology was also detected in similar studies using an interspecific hybrid of *F. esculentum* $\times$ *F. zuogongense*, which led to the conclusion that the genomes of these two species are very closely related (Chen et al., 2004).

## REFERENCES

Chen, Q.-F., 2001. Karyotype analysis of five *Fagopyrum* species native to China. Guihaia 21 (2), 107–110.
Chen, Q.F., 2012. Plant Sciences on Genus *Fagopyrum*. Science Press, Beijing.

Chen, Q.-F., Hsam, S.L.K., Zeller, F.J., 2004. A study of cytology, isozyme and interspecific hybridisation on the big-achene group of buckwheat species (*Fagopyrum*, Polygonaceae). Crop Sci. 44, 1511–1518.

Chen, Q.-F., Hsam, S.L.K., Zeller, F.J., 2007. Cytogenetic studies on diploid and autotetraploid common buckwheat and their autotriploid and trisomics. Crop Sci. 47, 2340–2345.

Di Bucchianico, S., Venora, G., Lucretti, S., Limongi, T., Palladino, L., Poma, A., 2008. *Saponaria officinalis* karyology and karyotype by means of image analyzer and atomic force microscopy. Microsc Res. Technol. 71, 730–736.

Feldman, M., Levy, A.A., 2005. Allopolyploidy - a shaping force in the evolution of wheat genomes. Cytogenet Genome Res. 109, 250–258.

Grandont, L., Jenczewski, E., Lloyd, A., 2013. Meiosis and its deviations in polyploid plants. Cytogenet. Genome Res. 140, 171–184.

Levan, A., Fredga, K., Sandberg, A.A., 1964. Nomenclature for centromeric position on chromosomes. Hereditas 52, 201–220.

Li, F.L., Zeller, F.J., Huang, K.F., Shi, T.X., Chen, Q.F., 2013. Improvement of fluorescent chromosome *in situ* PCR and its application in the phylogeny of the genus *Fagopyrum* Mill. using nuclear genes of chloroplast origin (cpDNA). Plant Syst. Evol. 299, 1679–1691.

Marhold, K., Huthmann, M., Hurka, H., 2002. Evolutionary histroy of polyploid complex of *Cardamine amara* (Brassicaceae): isozyme evidence. Plant Syst. Evol. 233, 15–28.

Matsui, K., Nishio, T., Tetsuka, T., 2004. Genes outside the S supergene suppress S functions in buckwheat (*Fagopyrum esculentum*). Ann Bot. 94, 805–809.

McMaster, T.J., Miles, M.J., Winfield, M.O., Karp, A., 1996. Analysis off cereal chromosomes by atomic force microscopy. Genome 39, 439–444.

Neethirajan, S., Hirose, T., Wakayama, J., Tsukamoto, K., Kanahara, H., Sugiyama, S., 2011. Karyotype analysis of buckwheat using atomic force microscopy. Microsc. Microanal. 17, 572–577.

Ohnishi, O., Asano, N., 1999. Genetic diversity of *Fagopyrum homotropicum*, a wild species related to common buckwheat. Genet. Res. Crop Evol. 46, 389–398.

Ohnishi, O., Matsuoka, M., 1996. Search for the wild ancestor of buckwheat II. Taxonomy of *Fagopyrum* (Polygonaceae) species based on morphology, isozymes and cpDNA variability. Genes Genet. Syst. 71, 383–390.

Ohsako, T., Yamane, K., Ohnishi, O., 2002. Two new *Fagopyrum* (Polygonacae) species, *F. gracilipedoides* and *F. jinshaense* from Yunnan, China. Genes Genet. Syst. 77, 399–408.

Quisenberry, K.S., 1927. Chromosome numbers in buckwheat species. Bot. Gazette 83, 1.

Schaper, A., Rossle, M., Formanek, H., Jovin, T.M., Wanner, G., 2000. Complementary visualization of mitotic barley chromatin by field-emission scanning electron microscopy and scanning force microscopy. J. Struct. Biol. 129, 17–29.

Soltis, D.E., Soltis, P.S., 1995. The dynamic nature of polyploid genomes. Proc. Natl. Acad. Sci. USA 92, 8089–8091.

Soltis, D.E., Soltis, P.S., 1999. Polyploidy: recurrent formation and genome evolution. Trends Ecol. Evol. 14, 348–352.

Sugiyama, S., Yoshino, T., Kanahara, H., Kobori, T., Ohtani, T., 2003. Atomic force microscopic imaging of 30 nm chromatin fiber from partially relaxed plant chromosomes. Scanning 25, 132–136.

Taylor, R., 1925. Chromosome constrictions as distinguishing characteristics in plants. Am. J. Bot. 12 (4), 238–244.

Tomiyoshi, M., Yasui, Y., Ohsako, T., Li, C.Y., Ohnishi, O., 2012. Phylogenetic analysis of AGAMOUS sequences reveals the origin of the diploid and tetraploid forms of self-pollinating wild buckwheat, *Fagopyrum homotropicum* Ohnishi. Breed Sci. 62, 241–247.

Wang, J.-S., Chai, Y., Zhao, X.-T., Ji, W.-Q., 2005. Karyotype analysis of Chinese buckwheat cultivars. Acta Bot. Boreali-Occidentalia Sin. 25 (6), 1114–1117.

Wendel, J.F., 2000. Genome evolution in polyploids. Plant Mol. Biol. 42, 225−249.

Yoshino, T., Sugiyama, S., Hagiwara, S., Ushiki, T., Ohtani, T., 2002. Simultaneous collection of topographic and fluorescent images of barley chromosomes by scanning near-field optical/atomic force microscopy. J. Electron Microsc. (Tokyo) 51, 199−203.

Zu, F.X., Lin, R.F., Li, Q.Y., Liu, D.K., 1984. Preliminary study on chromosome of various types of buckwheat. Chin. J. Cytobiol. 3, 130−131.

# Chapter Five

# Description of Cultivated Tartary Buckwheat

**Meiliang Zhou[1], Yu Tang[2], Xianyu Deng[3], Chen Ruan[3], Mengqi Ding[4], Jirong Shao[4], Yixiong Tang[5], and Yanmin Wu[5]**

[1]Institute of Crop Sciences, Chinese Academy of Agricultural Sciences, Beijing, China, [2]Sichuan Tourism University, Chengdu, China, [3]Kunming University of Science and Technology, Kunming, Yunnan, China, [4]Sichuan Agricultural University, Yaan, China, [5]Biotechnology Research Institute, Chinese Academy of Agricultural Sciences, Beijing, China

Tartary buckwheat is one of the cultivars of *Fagopyrum*, whose species name is *Fagopyrum tataricum* (L) Gaertn and whose English name is Tartary buckwheat. Tartary buckwheat is called *phapar* in India, *tite phapar* in Nepal, and *bjo* in Bhutan. In China and Nepal, it is also called bitter buckwheat. Tartary buckwheat is mainly grown in the south of China, India, the southern Himalayas, Nepal, Bhutan, and Pakistan, etc. The grains of Tartary buckwheat are rich in proteins, fats, vitamins, and minerals, as well as rutin, quercitin, and other flavonoids that other Gramineae crops don't contain. Hence, Tartary buckwheat has considerable nutritional and medicinal values, which is considered as an ideal functional food source for humans.

## THE MORPHOLOGICAL DESCRIPTIONS OF *FAGOPYRUM TATARICUM* (L) GAERTN

Herbs annual. Stems erect, green. 30−70 (100) cm tall, branched, striate, papillate. Petioles ca. as long as blade; leaf blade broadly triangular, 2.7 cm × 2.8 cm, both surfaces papillate along veins, base cordate or truncate, margin entire; apex acute; ocrea brown, ca. 5mm, membranous, oblique. Inflorescence terminal or axillary, racemes, several racemes together paniculate, lax; bracts ovate, 2−3 mm, apex acute, each 2−4-flowered. Pedicel 3−4 mm, articulate at middle. Perianth white, yellow-green or greenish; tepals elliptic, ca.

**45**

© 2018 Elsevier Inc. All rights reserved.

2 mm. Stamens included. Stigmas capitate. Achenes much exceeding persistent perianth, black, black-brown, brown, narrowly ovoid, trigonous, 5−6 mm, surfaces grooved; angles rounded below middle sharply acute above, sometimes sinuate-dentate along angles.

This species is cultivated in mountainous regions at altitudes of 400−4400 m. This species includes many cultivars and landraces, whose achenes vary considerably in form and size. Some achenes are winged or spinous on the angles, others can split between the angles of the hull when mature (Fig. 5.1).

**FIGURE 5.1** Phenotype of *Fagopyrum tataricum*. (A) The field plants of *F. tataricum*; (B) The ripe fruits of *F. tataricum*; (C) The single plant of *F. tataricum*; (D) The inflorescences of *F. tataricum*; (E) The flowers of *F. tataricum*.

The plants of *F. tataricum* are usually less vigorous than those of *Fagopyrum esculentum*. Generally, *F. tataricum* is more branched in the stems and more sagittate in the leaves. The flower of *F. tataricum* is smaller with inconspicuous greenish-white sepals and it does not appear to be attractive to insects. It is homomorphic, self-fertile and cleistogamous with the pollination occurring before flowering. The chromosome number of *F. tataricum* is $2n = 16$.

## THE PROFILE OF *FAGOPYRUM TATARICUM* CULTIVATION

*Fagopyrum tataricum* originated in the southwest of China, Asia, where it has a long history of cultivation. Nowadays, China is a country with large-scale cultivation of *F. tataricum*, which ranks number one in the world for cultivated area and production, with 200,000–300,000 $hm^2$ of annual cultivated area and 300,000–400,000 t of total production. The major production provinces of *F. tataricum* in China include Yunnan, Sichuan, Guizhou, and Chongqing, which are located in the southwest of China. But *F. tataricum* cultivars and/or wild *F. tataricum* were also distributed in other provinces, such as Gansu, Guangxi, Hebei, Hubei, Hunan, Qinghai, Shaanxi, Shanxi, Tibet, Sinkiang, Jilin, Liaoning, and Heilongjiang. Among the countries in Asia, Nepal, Bhutan, Pakistan, India, Afghanistan, Burma, and other South Asian countries there is also a long history of cultivating *F. tataricum*, which includes both cultivars and wild species. Moreover, Kazakhstan, Kyrgyzstan, Mongolia, Tajikistan, and Russia also have small-scale cultivation of *F. tataricum*. In recent years, *F. tataricum* has also been introduced to Japan, South Korea, and some countries in Europe and North America for cultivation.

Although *F. tataricum* can grow at altitudes of 400–4400 m, it is usually planted in the frigid alpine areas and the highland regions with altitudes of 1500–3000 m, due to its remarkable adaptability. *F. tataricum* belongs to the short-day crop, whose reproductive growth can be accelerated by short-light induction to complete the lifecycle in shorter time. Its growing days are about 60–110 days with a normal per-hectare yield of 900–2250 kg and a maximum per-hectare yield of 4500 kg.

## THE COLLECTION AND PRESERVATION OF *FAGOPYRUM TATARICUM* GERMPLASM RESOURCES

*Fagopyrum tataricum* has been cultivated for more than 2000 years, which has informed the abundant germplasm resources. *F. tataricum* is not widespread around the world, so only a few countries such as China, Nepal, India, and Bhutan have conducted the collection of *F. tataricum*. Since the 1980s, China has conducted the collection of *F. tataricum* germplasm resources nationally and introduced *F. tataricum* species from Nepal, Japan, and other countries. At the beginning of the 21st century, 1086 materials had been collected from 18 provinces, among which Sichuan, Yunnan, Gansu, and Shanxi predominated, reaching 278 (materials), 176, 133, and 112, respectively. Moreover, 37 materials were collected from Nepal and 13 from Japan. All these seeds of *F. tataricum* resources have been well preserved in the National Long-term Storage Gene Bank for Crop Germplasm ($-20°C$), CAAS (Yang, 1995). During the 1980s−1990s, IPGRI (International Plant Genetic Resources Institute) Office for South Asia conducted the collection of buckwheat resources in the Himalayan region of South Asia, where a large number of buckwheat resources has been collected for cultivation in Nepal, India, and Bhutan. The National Bureau of Plant Genetic Resources (NBPGR) of India has collected and preserved hundreds of buckwheat resources including *F. tataricum*, *F. esculentum*, and *Fagopyrum cymosum*. In Shimla and other places alone, 583 materials were preserved, including nearly 200 *F. tataricum* materials (Baniya et al., 1995). The National Hill Crop Research Programme of Nepal has also collected 534 buckwheat resources in Kabre, which were preserved in the mid-term gene bank (5°C, R H 40%), 196 materials of which are *F. tataricum* (Arora et al., 1995). Moreover, Canada, Japan, and Russia have also collected quite a few *F. tataricum* germplasm resources that were preserved in the mid-term gene bank (Campbell, 1997).

## THE CHARACTERIZATION AND EVALUATION OF *FAGOPYRUM TATARICUM* RESOURCES

*Qualitative characters.* Buckwheat varieties usually differ from each other in plant type, stem color, leaf color, flower color, grain color, and grain shape, etc. After the analysis of *F. tataricum*'s identifying data, two plant types of *F. tataricum* were discovered, i.e., the compact and the loose. Stem colors vary considerably, including reddish, pink, red, reddish green, yellowish green, green, greenish red, greenish, dark green, mauve, purple, purplish red, and brown. Chinese scholars found that among the stems with various colors green stems are more common than others, accounting for more than 50%, while the reddish ones come second. Baniya et al. (1995) reported that 67% of the *F. tataricum*'s stems they observed are red during flowering. While the colors of *F. tataricum*'s leaves have a distinction of greenish, green, and dark green, the main colors are green. Among them, the green predominates, which accounts for more than 60%. The flower colors are mainly greenish white, greenish, yellowish green, yellow, and green, etc., while few are white. Grain colors also vary considerably, which include grayish, gray, dark gray, brown, dark brown, grayish black, and black, etc. Baniya et al. (1995) pointed out that 42% of the seeds are gray, 30% are brown, 18% are grayish black, and 10% are black. The grain shapes are mainly in the form of a long cone, a short cone or a rectangle, which were determined with the aspect-ratio measuring method adopted by Yang (1992), showing that 99% of species are of long cone.

*Agronomic characters.* The agronomic characters of *F. tataricum* are the main indices of identification and evaluation, which include days of growth period, plant height, node on main stem, branches on main stem (i.e., number of primary branches), grain weight/plant weight, 1000-grain weight, number of leaves on main stem, number of flower clusters, and number of seeds/cyme etc. Yang (1995) reported the identification results of 550 *F. tataricum* materials' (Note:550 accessions) agronomic characters (Table 5.1). Joshi et al. (2011) also reported the identification results of 192 *F. tataricum* materials' agronomic characters (Table 5.2). So it can be seen that the variation range of *F. tataricum*'s agronomic characters is very wide, representing its abundant diversity.

*Nutritional composition.* Tartary buckwheat germplasms are also rich in various nutrients including proteins, fats, different amino acids, vitamins, minerals, trace elements, and bioactive substances. During 1986–1990, Chinese researchers measured the contents of amino acids, Ve, Vpp, and trace elements (Se, P, Mn, Zn, Ca, etc.) in more than 500 *F. tataricum* germplasm materials. There are 18 kinds of amino acids in the grains, whose average content is 10.79,

**TABLE 5.1** The Agronomic Characters of Chinese *Fagopyrum tataricum* Germplasm Resources (550 Materials)

| Character | Mean | The Range of Character | Number | The Percentage Share in the Total Number (%) |
|---|---|---|---|---|
| Days of growth period | 90 | <70 | 36 | 6.7 |
| | | 70–90 | 318 | 59.6 |
| | | >90 | 180 | 33.7 |
| Plant height (cm) | 107.9 | <80 | 97 | 18.2 |
| | | 80–100 | 115 | 21.5 |
| | | 100.1–120 | 129 | 24.2 |
| | | 120.1–140 | 86 | 16.1 |
| | | >140 | 107 | 20.0 |
| Nodes of main stem | 18.2 | <12 | 19 | 3.6 |
| | | 12.1–15 | 113 | 21.2 |
| | | 15.1–18 | 220 | 41.2 |
| | | >18 | 182 | 34.1 |
| Branches on main stem | 5.84 | <3 | 8 | 1.5 |
| | | 3–5 | 227 | 42.5 |
| | | 5.1–7 | 219 | 41.0 |
| | | >7 | 80 | 15.0 |
| Grain weight/plant (g) | 4.56 | <3 | | 38.7 |
| | | 3–5 | 205 | 27.0 |
| | | 5.1–7 | 143 | 18.7 |
| | | >7 | 99 | 15.7 |
| | | | 83 | |
| 1000-seed weight (g) | 19.7 | <15 | 66 | 12.4 |
| | | 15.1–20 | 319 | 59.7 |
| | | 20.1–25 | 147 | 27.5 |
| | | >25 | 2 | 0.4 |

**TABLE 5.2** The Minimum, Maximum and Mean Values for Agronomic Traits of 192 Tartary Buckwheat Landraces Grown at Khumaltar, Karthmandu Kabre, Nepal, 2008

| Trait | Minimum | Maximum | Mean | CV (Coefficient of Variation) (%) |
|---|---|---|---|---|
| Number of cymes per plant | 5 | 27 | 13 | 35.5 |
| Number of seeds per cyme | 5 | 37 | 18 | 32.9 |
| Number of seeds per cluster | 5 | 37 | 18 | 32.5 |
| Days to flowering | 22 | 37 | 30 | 11.1 |
| Number of primary branches | 2 | 8 | 5 | 23.9 |
| Number of leaves on main stem | 5 | 29 | 8 | 25.8 |
| Plant height (cm) | 21 | 76.5 | 41.8 | 23.3 |
| Days to maturity | 60 | 118 | 88 | 11.4 |
| Seed length (mm) | 3.1 | 6.5 | 4.7 | 12.4 |
| Seed width (mm) | 2.1 | 3.5 | 2.7 | 11.4 |
| No. of plants/plot | 15 | 272 | 136.7 | 36.6 |
| Grain yield (kg/ha) | 201 | 1492 | 756.2 | 26.9 |
| Grain yield (g/plant) | 0.31 | 5.36 | 1.39 | 57.84 |
| 1000-grain weight (g) | 3.2 | 24.4 | 15.24 | 24.15 |

ranging from 7.04 to 15.8 (Table 5.3). The average content of Ve is 0.99 mg/ 100 g with the range of 0.21−4.58 mg/100 g, and the average content of Vpp is 3.42 mg/100 g with the range of 0.46−9.69 mg/100 g. The contents of minerals and/or trace elements—Se, Zn, Fe, Mn, Ca, Cu, and P—were 0.05, 29.0, 115.7, 11.8, 308.4, 9.2, and 3762 ppm respectively (Yang and Lu, 1992).

*Rutin.* Rutin is one of the important bioactive substances in *F. tartaricum.* Rutin has many physiological and biological properties, including antioxidation, antiinflammation, antihypertension, vasoconstrictive effect, and positive inotropic effect (Griffith et al.,1944; Campbell, 1997; Przybylski and Gruczynska, 2009; Vojtiskova et al., 2012). A large number of studies have shown that the rutin content of *F. tartaricum* is more than that of *F. esculentum* (Campbell, 1997. Yan et al., 2004; Gupta et al., 2012). There are several measuring methods to determine the rutin content of *F. tartaricum*: spectrophotometry, high performance thin-layer chromatography (HPTLC), and HPLC, etc. To a certain extent, the measuring results vary according to the measuring method one adopts. Generally, the variation range is considerable among *F. tartaricum* germplasms. Tang and Zhao (1990) measured the rutin contents of the grains of 50 buckwheat landraces from China's Sichuan

**TABLE 5.3** Average Contents of Various Amino Acids in Tartary Buckwheat (%)

| Amino Acids | Mean | Min | Max | CV |
|---|---|---|---|---|
| Threonine | 0.43 | 0.24 | 0.62 | 13.7 |
| Valine | 0.58 | 0.26 | 1.18 | 14.0 |
| Methionine | 0.17 | 0.03 | 0.34 | 27.5 |
| Isoleucine | 0.46 | 0.05 | 0.71 | 16.4 |
| Leucine | 0.75 | 0.08 | 1.07 | 13.1 |
| Phenylalanine | 0.55 | 0.05 | 0.99 | 15.3 |
| Lysine | 0.64 | 0.12 | 1.86 | 17.9 |
| Tryptophan | 0.13 | 0.06 | 0.18 | 15.9 |
| Aspartic acid | 1.06 | 0.60 | 1.84 | 18.2 |
| Histidine | 0.27 | 0.03 | 0.50 | 16.3 |
| Arginine | 1.09 | 0.11 | 1.61 | 15.2 |
| Serine | 0.58 | 0.05 | 0.90 | 16.1 |
| Glutamic acid | 2.05 | 0.60 | 3.12 | 16.2 |
| Glycine | 0.64 | 0.07 | 2.54 | 18.6 |
| Alanine | 0.50 | 0.26 | 0.79 | 12.7 |
| Cystine | 0.16 | 0.03 | 0.44 | 40.6 |
| Tyrosine | 0.33 | 0.07 | 0.55 | 19.8 |
| Proline | 0.41 | 0.11 | 1.22 | 24.6 |
| Total | 10.79 | 7.04 | 15.83 | 14.2 |

province by means of spectrophotometry, discovering the average value of 27 . *F. tartaricum* species is 2.02% (d.w.) with the range of 1.22%−2.91% (d.w.). Accordingly, the average value of rutin content of the *F.e* grain is only 0.23% (d.w.) with the range of 0.01%−0.71% (d.w.). Chauhan et al. (2010) measured the rutin contents of seeds of 200 *F. tartaricum* germplasms with spectrophotometry, which ranges from 0.6% to 2.0% (d.w.), while that of *F. esculentum* is only 0.07% (d.w.). Gupta (2012) measured the rutin contents of seeds of 195 *F. tartaricum* accessions with HPLC, which has a wide variation range of 6−30 μg/mg (d.w.). In most of the accessions (81%), rutin contents range from 10 to 16 μg/mg, while 14% accessions show considerably higher rutin content (17 μg/mg to 30 μg/mg) and few (5%) accessions contain less rutin (≤10 μg/mg). Through two years of observation and investigation, Gupta et al. (2011) found that the germplasms with a high rutin content are stable, which implies the genetic differences that actually exist in *F. tartaricum* accessions and which can be used and preserved for relevant cultivations.

## ACKNOWLEDGMENTS

This research was supported by the Investigation of Forage Germplasm in Central China (grant number 2017FY100604).

## REFERENCES

Arora, R.K., Baniya, B.K., Joshi, B.D., 1995. Buckwheat genetic resources in the Himalayas: their diversity, conservation and use. In: Current Advances in Buckwheat Research, Vol. I–III. Proceedings of 6th International Symposium on Buckwheat in Shinshu, Japan. Shinshu University Press, pp. 39–45.

Baniya, B.K., Dongol D.M.S., Dhungel N.R., 1995. Further characterization and evaluation of Nepalese buckwheat (*Fagopyrum* spp.) landraces. In: Current Advances in Buckwheat Research. Vol. I–III. Proceedings of 6th International Symposium on Buckwheat in Shinshu, Japan. Shinshu University Press, pp. 295–304.

Campbell, C.G., 1997. Buckwheat. Fagopyrum esculentum Moench. Promoting the conservation and use of underutilized and neglected crops, Vol. 19. IPK, Bioversity International, Germany and IPGRI, Rome, Italy.

Chauhan, R.S., Gupta, N., Sharma, S.K., Rana, J.C., Sharma, T.R., Jana, S., 2010. Genetic and genome resources in Buckwheat–present and future perspectives. Eur. J. Plant Sci. Biotechnol. 4, 33–34.

Griffith, J.Q., Couch, J.F., Lindauer, M.A., 1944. Effect of rutin on increased capillary fragility in man. Proc. Soc. Exptl. Biol. Med. 55, 228–229.

Gupta, N., Sharma, S.K., Rana, J.C., Chauhan, R.S., 2011. Expression of flavonoid biosynthesis genes vis-à-vis rutin content variation in different growth stages of Fagopyrum species. J. Plant Physiol. 168, 2117–2123.

Gupta, N., Sharma, S.K., Rana, J.C., Chauhan, R.S., 2012. AFLP fingerprinting of tartary buckwheat accessions (*Fagopyrum tataricum*) displaying rutin content variation. Fitoterapia 83, 1131–1137.

Joshi, B.K., Okuno, K., Ohsawa, R., Hayashi, H., Otobe, C., Kawase, M., 2011. Characterization and evaluation of Nepalese Tartary buckwheat accessions simultaneously in augmented design. Fagopyrum 28, 23–41.

Przybylski, R., Gruczynska, E., 2009. A review of nutritional and nutraceutical components of buckwheat. Eur. J. Plant Sci. Biotechnol. 3, 10–22.

Tang, Y., Zhao, G., 1990. A preliminary study on the nutritious quality of buckwheat varieties in Sichuan province. Buckwheat Trend 13 (2), 20–24.

Vojtiskova, P., Kristyna, K., Kuban, V., Kracman, S., 2012. Chemical composition of Buckwheat plant (*F. esculentum*) and selected Buckwheat products. J. Microbiol. Biotechnol. Food Sci. 1, 1011–1019.

Yan, C., Baili, F., Yin gang H., Jinfeng, G., Xiaoli G., 2004. Analysis on the variation of rutin content in different buckwheat genotypes. In: Proceedings of the 9th International Symposium on Buckwheat, August 18–22. Prague, Czech Republic, pp. 688–691.

Yang, K.L., 1992. Research on cultivated buckwheat germplasm resources in China. In: Proceedings 5th International Symposium on Buckwheat, 20–26 August 1992, Taiyuan, China. Agricultural Publishing House, pp. 55–59.

Yang, K.L., 1995.Current status sna prospects of buckwheat genetic resources in China. In: Current Advances in Buckwheat Research. Vol. I–III. Proceedings of 6th International Symposium on Buckwheat in Shinshu, Japan. Shinshu University Press, pp. 91–95.

Yang, K.L., Lu, D.B., 1992. The quality appraisal of buckwheat germplasm resources in China. In: Proceedings of 5th International Symposium on Buckwheat, 20–26 August 1992, Taiyuan, China. Agricultural Publishing House, pp. 90–97.

# Chapter Six

# Description of Cultivated Common Buckwheat

**Meiliang Zhou[1], Yu Tang[2], Xianyu Deng[3], Chen Ruan[3], Mengqi Ding[4], Jirong Shao[4], Yixiong Tang[5], and Yanmin Wu[5]**

[1]*Institute of Crop Sciences, Chinese Academy of Agricultural Sciences, Beijing, China,* [2]*Sichuan Tourism University, Chengdu, China,* [3]*Kunming University of Science and Technology, Kunming, Yunnan, China,* [4]*Sichuan Agricultural University, Yaan, China,* [5]*Biotechnology Research Institute, Chinese Academy of Agricultural Sciences, Beijing, China*

Common buckwheat is the buckwheat cultivar that has the largest cultivated area and is most widely distributed in the world. The species name of common buckwheat is *Fagopyrum esculentum* Moench, while common buckwheat is its English name. Moreover, common buckwheat is also called *tian'qiao* in China and *mite phapar* in Nepal, which means sweet buckwheat. Common buckwheat is an important crop in many countries that has become one of the human staples. People usually grind the fruits of common buckwheat (i.e., the achenes) into flour, which is then processed into buckwheat noodles, buckwheat pancakes and other pasta. Buckwheat groats that are processed from grains are the most widely used form, employed to make porridge, soup, kasha, and other staples both in Europe and America. Moreover, common buckwheat is also one of the important nectar plants and relief crops.

## ACCEPTED BOTANICAL NAMES AND SYNONYMS OF THE CULTIVATED SPECIES

*Fagopyrum esculentum* Moench, Methodus (1794) 290; *Polygonum fagopyrum* L., Sp.Pl. (1753) 522; *vulgare* Hill, Brit. Herb. (1756) 486, nom. illeg.; *Polygonum tataricum* Lour., Fl. cochinch. (1790) 242, non L.; *Fagopyrum sagittatum* Gilib., Exerc. phyt. 2 (1792) 435, nom. illeg.; *Polygonum cereale* Salisb., Prodr. (1796) 259; *Fagopyrum sarracenicum* Dumort., Fl. Belg.

**53**

*Buckwheat Germplasm in the World.* DOI: https://doi.org/10.1016/B978-0-12-811006-5.00006-9
© 2018 Elsevier Inc. All rights reserved.

Prodr. (1827) 18; *Fagopyrum cereale* (Salisb.) Raf., Fl. Tellur. 3(1836) 10; *Kunokale carneum* Raf., l.c., 12; *Phegopyrum esculentum* (Moench) Peterm., Fl. Bienitz (1841) 92; *Fagopyrum fagopyrum* Karst., Deutschl. Fl. (1883) 522; *Helxine fagopyrum* Kuntze, Revis. 2 (1891) 553.

## THE BOTANICAL DESCRIPTION

The phenotype of *F. esculentum* is described in Fig. 6.1. Herbs annual. Stems green or red when mature, erect, 30—90 cm tall, branched above, glabrous or

**FIGURE 6.1** Phenotype of *Fagopyrum esculentum*. (A) The field plants of *F. esculentum*; (B) The single plant of *F. esculentum*; (C) The inflorescences of *F. esculentum* (long-style flowers); (D) Long-style flowers; (E) The inflorescences of *F. esculentum* (short-style flowers); (F) Short-style flowers.

papillate on one side. Petiole 1.5−5 cm; leaf blade triangular, 2.5−7 × 2−5 cm, both surfaces papillate along veins, base cordate or nearly truncate, apex acuminate; ocrea caducous, ca. 5 mm, membranous, oblique, not ciliate. Inflorescence axillary or terminal, racemose or corymbose; peduncles 2−4 cm, papillate along one side; bracts green, ovate, 2.5−3 mm, margin membranous, each 3- or 5-flowered. Pedicels longer than bracts, not articulate. Perianth pink or white; tepals elliptic, 3−4 mm. Anthers pinkish. Styles heterostylous. Achenes exceeding persistent perianth, dark-brown, opaque, ovoid, sharply trigonous, 5−6 mm, surfaces flat. Fl. May-Sep, fr. June−October $2n = 16$.

The flowers of common buckwheat are dimorphic. While one of which has long styles and short stamens (i.e., short filaments), the other has short styles and long stamens (i.e., long filaments). But one plant can only bear one flower type. Usually the same-type flowers can't pollinate each other, i.e., the pollination and fruiting can only happen between the heterostyly flowers. Hence, common buckwheat is a cross-pollinated crop that is self-incompatible. In general, the proportion of the plants with long and short styles is 1:1 (one to one) within one community. But people also found the existence of short-style and short-stamen types in common buckwheat, which can realize self-pollination (Esser 1953; Marshall 1969; Fesenko and Antonov 1973).

## THE CULTIVATION PROFILE OF COMMON BUCKWHEAT

Common buckwheat is the more widespread around the world of the two buckwheat cultivars, and is planted in Asia, Europe, Africa, North America, South America, and Oceania. In general, the vertical distribution of common buckwheat ranges in altitude from 600 m to 1000 m, and can reach a maximum altitude of 4100 m and a minimum altitude of 100 m. At least 50 countries have cultivated common buckwheat as a crop. A century ago, the Russian Empire was the world leader in buckwheat production. Growing areas in the Russian Empire were estimated at 6.5 million acres (2,600,000 ha), followed by those of France at 0.9 million acres (360,000 ha) (Taylor and Belton, 2002). In the 1970s, an estimated 4.5 million acres (1,800,000 ha) of buckwheat were grown in the Soviet Union. In recent years, the cultivated area of buckwheat around the world has decreased. But so far, Russia has been the country with the largest cultivated area. China comes second, while Ukraine, United States, Kazakhstan, Poland, Japan, Brazil, Lithuania, and France, etc. also have lots of cultivated area. In general, the yield per hectare of common buckwheat ranges from 200 kg to 700 kg, which can reach a maximum yield of 2000 kg. Common buckwheat is a short-duration crop with a growth period of 60−120 days, which is so remarkably adaptable that it can be planted in almost any kind of soil, including infertile land that is not suitable for other crops. The sowing time of common buckwheat is also very flexible, which can be spring, summer, or autumn in different countries and different regions.

## THE COLLECTION AND PRESERVATION OF COMMON BUCKWHEAT GERMPLASM RESOURCES

There are abundant germplasm resources of common buckwheat around the world. Since the institute of Crop Germplasm Resources (ICGR) conducted the work of common buckwheat germplasm resources around the world extensively in 1980, many countries have collected lots of buckwheat resources, among which many are common buckwheat. Until 1996, the countries that have collected more than 100 common buckwheat germplasms were: Russia (more than 2000 germplasms), China (1821 germplasms), Canada (572 germplasms), India (376 germplasms), Japan (nearly 1000 germplasms), Democratic People's Republic of Korea (160 germplasms), Republic of Korea (245 germplasms), Nepal (388 germplasms), Slovenia (361 germplasms), and the USA (161 germplasms). Generally, the buckwheat seeds will lose viability at room temperature in several years and cannot germinate. So the collected buckwheat germplasms should be preserved under certain conditions to prevent the germplasms from eroding. Nowadays, many counties choose to preserve germplasms in mid-term gene banks (5°C,R H 40%) or long-term gene banks ($-20$°C). For example, China, has had 1821 collected common buckwheat germplasms preserved in the nationally-built gene bank; Japan has had more than 300 germplasms preserved in the long-term gene bank while the rest are put in the mid-term gene bank; USA has had 161 germplasms all preserved in the long-term gene bank, while Canada and Slovenia have had all germplasms preserved in the mid-term gene bank (Yang, 1995; Arora et al., 1995; Campbell,1997; Chauhan et al., 2010).

## THE CHARACTERIZATION AND EVALUATION OF COMMON BUCKWHEAT RESOURCES

The main qualitative characters of common buckwheat—plant type, stem color, leaf color, flower color, grain color, and grain shape—are basically the qualitative characters, which vary considerably. The plant types of common buckwheat are mainly erect and sub-erect (i.e., loose), the stem colors include reddish, red, green and greenish, etc. Baniya et al. (1995) found that 68% (of the stem) are erect and 80% (of the stem colors) are red. The flower colors of common buckwheat are mainly red, pink, and white. Although Baniya reported that about 42% (of the flowers) are pink, 18% are red, and 36% are white, Chinese scholars reported that 60% (of the flowers) of the common buckwheat germplasms they collected are red or pink, while 40% of them are white. The grain colors are mainly black, brown, and gray. Baniya reported that among the germplasms they collected, 50% of them are grayish black, 36% brown, and the rest gray or black. The grain shapes include triangular pyramid, long cone, and short cone. It was identified that common buckwheat is mainly the long cone.

The agronomic characters of common buckwheat mainly include days of growth period, plant height, node on main stem, branches on main stem

**TABLE 6.1** The Agronomic Characters of Chinese Common Buckwheat Germplasm Resources (964 Materials)

| Character | Mean | The Range of Character | Number | The Percentage Share in the Total number (%) |
|---|---|---|---|---|
| Days of growth period | 76.9 | <70 | 204 | 21.2 |
| | | 70–90 | 623 | 64.4 |
| | | >90 | 137 | 14.2 |
| Plant height (cm) | 107.4 | <80 | 69 | 7.2 |
| | | 80–100 | 285 | 29.6 |
| | | 100.1–120 | 354 | 36.7 |
| | | 120.1–140 | 212 | 22.0 |
| | | >140 | 44 | 4.6 |
| Nods of main stem | 14.0 | <12 | 221 | 22.9 |
| | | 12.1–15 | 464 | 48.1 |
| | | 15.1–18 | 217 | 22.5 |
| | | >18 | 62 | 6.4 |
| Branches on main stem | 4.7 | <3 | 191 | 19.8 |
| | | 3–5 | 463 | 48.1 |
| | | 5.1–7 | 224 | 23.3 |
| | | >7 | 85 | 8.8 |
| Grain weight/plant (g) | 3.3 | <3 | 490 | 52.1 |
| | | 3–5 | 293 | 31.1 |
| | | 5.1–7 | 66 | 7.0 |
| | | >7 | 92 | 9.8 |
| 1000-seed weight (g) | 26.3 | <25 | 352 | 36.8 |
| | | 25.1–30 | 384 | 40.1 |
| | | 30.1–35 | 208 | 21.7 |
| | | >35 | 13 | 1.4 |

(i.e., number of primary branches), grain weight/plant weight, 1000-grain weight, number of leaves on main stem, number of flower clusters, number of seeds/cyme, etc. CAAS has identified the agronomic characters of 964 common buckwheat accessions (Yang and Lu, 1992), and the results are shown in Table 6.1. Baniya et al. (1995) also reported the results of agronomic character identification for 309 common buckwheat germplasms (Table 6.2).

*Nutritional composition of common buckwheat*

The grains of common buckwheat contain 18 kinds of amino acids that differ from each other in content significantly. Yang (1992) reported the results of amino acids in grains of 906 common buckwheat germplasms measured by Chinese researchers (Table 6.1). Among these amino acids, glutamic acid predominates in content, which is usually more than 2%, while that of tryptophan is the lowest with a mean value of 0.12%−0.13%. The mean value of total amino acids is 11.1% ± 1.72% with a maximum of 16.5% and a minimum of 7.18% (Table 6.3).

**TABLE 6.2** Minimum, Maximum, and Mean Values for Agronomic Traits of Common Buckwheat Landraces Grown at Kabre, Nepal, Autumn Season, 1992

| Trait | Minimum | Maximum | Mean | S.E. (standard error) of mean |
|---|---|---|---|---|
| Days to 50% flowering | 26.0 | 45.0 | 28.3 | 0.16 |
| Days to 95% maturity | 67.0 | 98.0 | 80.0 | 0.37 |
| Grain-filling days | 33.0 | 69.0 | 51.6 | 0.37 |
| Plant height (cm) | 25.0 | 116.4 | 7.0 | 0.96 |
| No. of primary branches | 1.4 | 14.0 | 4.5 | 0.13 |
| No. of leaves on main stem | 2.0 | 18.4 | 7.9 | 0.11 |
| No. of flower clusters | 1.0 | 6.0 | 4.2 | 0.05 |
| No. of seeds/cymes | 7.0 | 50.2 | 19.7 | 0.40 |
| 1000-grain weight (g) | 10.2 | 31.8 | 20.7 | 0.22 |
| Grain yield (g/plot) | 1.0 | 423.0 | 83.9 | 2.93 |
| No. of plants/plot | 2.0 | 315.0 | 83.4 | 3.82 |
| Grain yield (g/plant) | 0.0 | 0.0 | 1.6 | 0.08 |

Source: Baniya, B.K., Dongol, D.M.S., Dhungel, N.R., 1995. Further characterization and evaluation of Nepalese buckwheat (Fagopyrum spp.) landraces. In: Current Advances in Buckwheat Research. Vol. I–III. Proceedings of 6th International Symposium on Buckwheat in Shinshu, Japan. Shinshu University Press, pp. 295–304.

Moreover, Yang (1992) also reported the measured results of vitamins (Ve and Vpp) and minerals (Se, Zn, Fe, Mn, Cu, Ca, and P, etc.) in 906 samples of common buckwheat grains: the average content of Ve is 1.42 mg/100 g, ranging from 0.09 mg/100 g to 8.51 mg/100 g; the average content of Vpp is 3.11 mg/100 g with the range of 0.84−9.84 mg/100 g; for minerals Se, Zn, Fe, Mn, Cu, Ca, and P, the average contents are 0.054 ppm, 27.2 ppm, 156.4 ppm, 17.1 ppm, 437.1 ppm, 12.1 ppm, and 3682 ppm, respectively.

*Rutin.* Common buckwheat also contains rutin, but not much research has been done on the rutin content of common buckwheat germplasm resources as its rutin content is much lower than that of the other buckwheat cultivar, Tartary buckwheat. Kitabayashi et al. (1995) evaluated the varietal differences and heritability of rutin content in the seeds and leaves of common buckwheat. The evaluated objects were 27 common buckwheat cultivars from Japan, the average rutin contents of whose seeds are 14.5−18.9 mg/100 g DW (dry weight), which doesn't change greatly. Raina, A.P. et al. (2015) measured the rutin contents of 61 common buckwheat germplasms collected from the Himalayan region with spectrophotometry and high performance thin-layer chromatography (HPTLC). The range of rutin content is 0.023%−0.097% with HPTLC and 0.018%−0.079% with spectrophotometry.

**TABLE 6.3** Average Contents of Various Amino Acids in Common Buckwheat (%)

| Amino acids | Mean | Min. | Max. | CV |
|---|---|---|---|---|
| Threonine | 0.44 | 0.16 | 0.67 | 14.7 |
| Valine | 0.58 | 0.38 | 1.24 | 17.0 |
| Methionine | 0.17 | 0.02 | 0.33 | 26.0 |
| Isoleucine | 0.46 | 0.04 | 1.65 | 21.6 |
| Leucine | 0.77 | 0.49 | 1.83 | 15.5 |
| Phenylalanine | 0.57 | 0.31 | 1.79 | 18.5 |
| Lysine | 0.66 | 0.06 | 1.96 | 16.3 |
| Tryptophan | 0.12 | 0.06 | 0.18 | 20.0 |
| Aspartic acid | 1.09 | 0.10 | 2.34 | 18.9 |
| Histidine | 0.26 | 0.17 | 0.40 | 15.9 |
| Arginine | 1.11 | 0.69 | 1.72 | 15.7 |
| Serine | 0.57 | 0.35 | 0.91 | 16.8 |
| Glutamic acid | 2.21 | 0.67 | 3.49 | 17.4 |
| Glycine | 0.65 | 0.44 | 0.95 | 14.7 |
| Alanine | 0.51 | 0.28 | 0.75 | 15.9 |
| Cystine | 0.17 | 0.03 | 0.33 | 32.2 |
| Tyrosine | 0.32 | 0.09 | 0.55 | 18.0 |
| Proline | 0.44 | 0.20 | 1.06 | 24.2 |
| Total | 11.10 | 7.18 | 16.51 | 15.5 |

There are no significant differences between the two measuring methods in the rutin content, but much more significance in the variation range. Kitabayashi et al. (1995) thought the heritability of seeds' rutin content was relatively higher among the main characters of common buckwheat.

## ACKNOWLEDGMENTS

This research was supported by the Investigation of Forage Germplasm in Central China (grant number 2017FY100604).

## REFERENCES

Arora, R.K., Baniya, B.K., Joshi, B.D., 1995. Buckwheat genetic resources in the Himalayas: their diversity, conservation and use. Current Advances in Buckwheat Research. Vol. I–III. Proceedings of 6th International Symposium on Buckwheat in Shinshu, Japan. Shinshu University Press, Nagano, pp. 39–45.
Baniya, B.K., Dongol, D.M.S., Dhungel, N.R., 1995. Further characterization and evaluation of Nepalese buckwheat (Fagopyrum spp.) landraces. Current Advances in

Buckwheat Research. Vol. I-III. Proceedings of 6th International Symposium on Buckwheat in Shinshu, Japan. Shinshu University Press, Nagano, pp. 295−304.

Campbell, C.G., 1997. Buckwheat. Fagopyrum esculentum Moench. Promoting the conservation and use of underutilized and neglected crops, Vol 19; IPK, Bioversity International, Germany and IPGRI, Rome, Italy.

Chauhan, R.S., Gupta, N., Sharma, S.K., Rana, J.C., Sharma, T.R., Jana, S., 2010. Genetic and genome resources in Buckwheat−present and future perspectives. Eur. J. Plant Sci. Biotechnol. 4, 33−34.

Esser, K., 1953. Genome doubling and pollen tube growth in heterostylous plants [in German]. Z. für ind. Abstammungs und Vererbungslehre 85:25−50.

Fesenko, N.V., Antonov, V., 1973. New homostylous form of buckwheat. Plant Breed. Abstr. 46, 10172.

Kitabayashi, H., Ujihara, A., Hirose, T., Minami, M., 1995. Varietal differences and heritability for rutin content in common buckwheat, *Fagopyrum esculentum* Moench. Jpn J. Breed. 45, 75−79.

Marshall, H., 1969. Isolation of self-fertile, homomorphic forms in buckwheat, *Fagopyrum sagittatum* Gilib. Crop Sci. 9, 651−653.

Raina, A.P., Gupta, V., 2015. Evaluation of buckwheat (*Fagopyrum* species) germplasm for rutin content in seeds. Indian J. Plant Physiol. 20 (2), 167−171.

Taylor, J.R.N., Belton, P.S., 2002. Pseudocereals and Less Common Cereals. Springer, p. 125.

Yang, K.L., 1992. Research on cultivated buckwheat germplasm resources in China. Proceedings of 5th International Symposium on Buckwheat, 20−26 August 1992, Taiyuan, China. Agricultural Publishing House, Beijing, pp. 55−59.

Yang, K.L., 1995. Current status sna prospects of buckwheat genetic resources in China. Current Advances in Buckwheat Research. Vol. I−III Proceedings of 6th International Symposium on Buckwheat in Shinshu, Japan. Shinshu University Press, Nagano, pp. 91−95.

Yang, K.L., Lu, D.B., 1992. The quality appraisal of buckwheat germplasm resources in China. Proceedings of 5th International Symposium on Buckwheat, 20−26 August 1992, Taiyuan, China. Agricultural Publishing House, Beijing, pp. 90−97.

# Chapter Seven

# Perennial Self-Incompatible Wild *Fagopyrum* Species

**Meiliang Zhou[1], Yu Tang[2], Xianyu Deng[3], Chen Ruan[3], Mengqi Ding[4], Jirong Shao[4], Yixiong Tang[5], and Yanmin Wu[5]**

[1]*Institute of Crop Sciences, Chinese Academy of Agricultural Sciences, Beijing, China,* [2]*Sichuan Tourism University, Chengdu, China,* [3]*Kunming University of Science and Technology, Kunming, Yunnan, China,* [4]*Sichuan Agricultural University, Yaan, China,* [5]*Biotechnology Research Institute, Chinese Academy of Agricultural Sciences, Beijing, China*

Typical wild buckwheats of perennial self-incompatible species consist of *Fagopyrum cymosum* (Trev.) Meisn, *Fagopyrum urophyllum* (Bur. et Fr.) H. gross, and *Fagopyrum statice* (Lévl.) H. Gross (Fig. 7.1).

## FAGOPYRUM CYMOSUM

*Fagopyrum cymosum* has been known about and discussed for a long time. The table below shows the different names of *F. cymosum* throughout different ages, although now, most researchers regard *F. cymosum* as its scientific name (Table 7.1).

### The Botanical Description

Herbs perennial. Rhizomes black-brown, stout, ligneous. Stems erect, green or brownish, 40−100 cm tall, much branched, striate, glabrous. Petiole 2−10 cm; leaf blade triangular, 4−12 × 3−11 cm, both surfaces papillate, base nearly hastate, margin entire, apex acuminate; ocrea brown, 5−10 mm, membranous, oblique, apex truncate, not ciliate. Inflorescence terminal or axillary, corymbose; bracts ovate-lanceolate, ca. 3 mm, margin membranous, apex acute, each 4-flowered, rarely 6-flowered. Pedicels equaling bracts,

**61**

© 2018 Elsevier Inc. All rights reserved.

**FIGURE 7.1** Phenotype of *Fagopyrum cymosum*. (A) The wild plants of *F. cymosum*; (B) The inflorescences of *F. cymosum*; (C) The field plants of *F. cymosum*; (D) The immature fruits of *F. cymosum*; (E) The matured fruits of *F. cymosum*; (F) The root of *F. cymosum*.

articulate at middle. Perianth white; tepals narrowly elliptic, ca. 2.5 mm. Stamens included. Styles free; stigmas much exceeding persistent perianth, capitate, opaque. Achenes blackish brown, dull, broadly ovoid, 6−8 mm, trigonous, sometimes narrowly winged, angles smooth to repandous, apex acute. Fl. April−October, fr. May−November $2n = 24$ or 32

## Distribution

*Fagopyrum cymosum* is mainly distributed in Bhutan, India, Kashmir, Myanmar, Nepal, Sikkim, Vietnam, Thailand, and China, etc., among which *F. cymosum* can be found in many southern provinces of China, such as Anhui, Fujian, Gansu, Guangdong, Guangxi, Guizhou, Henan, Hubei, Hunan,

**TABLE 7.1** Different Names of *Fagopyrum cymosum* in Different Ages

| Names | Ages |
|---|---|
| *Polygonum triangulare* | 1689 |
| *Polygonum acutatum* Lehm. | 1821 |
| *Polygonum dibotrys* D. Don. | 1825 |
| *Polygonum cymosum* Trevir. | 1826 |
| *Fagopyrum cymosum* (Trrev) Meisn | 1832 |
| *Polygonum volubile* | 1840 |
| *Polygonum Labordei* H. Lév. et Vaniot | 1902 |
| *Polygonum tristachyum* H. Lév. | 1912 |
| *Fagopyrum dibotrys* (D. Don.) | 1966 |
| *Fagopyrum pilus* Q. F. Chen, Bot. J. Linn. Soc | 1999 |

Jiangsu, Jiangxi, Shaanxi, Sichuan, Tibet, Yunnan, and Zhejiang, etc. *F. cymosum* can grow at altitudes of 300−3200 m.

## The Habitat and Growth Habit

It usually grows in the valleys, on grassy slopes, sills, or grassy roadsides that are pretty damp. This species varies considerably in morphology, which is large and luxuriant with three types (erect, sub-erect, or decumbent) and longer flowering and fruiting (as long as 70−140 days). The flowers of *F. cymosum* are dimorphic (with long and short styles, i.e., heterostylous), cross-pollinated, and self-incompatible. The achenes are black or brown with two types—long triangular pyramid (with acute edge), and short triangular pyramid (with obtuse edge).

This species has been treated as an independent species for a long time. Chen, Q. F. (1999) proposed that, according to the differences in morphology, isozymes, and chromosomes, the current *F. cymosum* (Trrev) Meisn should be treated as a complex species, i.e., *F. cymosum* complex, which is composed of tetraploid *F. cymosum,* diploid *Fagopyrum megaspartanium,* and *Fagopyrum Pilus* that should be classified into different independent species. These three are very similar in botanical morphology. The differences are while diploid *F. megaspartanium* has thicker branches, bigger leaves, flowers, and fruits, less pubescence on the leaves and petioles; tetraploid *F. cymosum* has softer stems, less flowers, and fruits that are not full; and the plants are straight, whose leaves, flowers and fruits are smaller, while the lower lamina surface and petiole are densely covered with pubescence. But so far, Chen's views haven't been commonly agreed upon, which should be confirmed by more researches (Fig. 7.2).

**FIGURE 7.2** Phenotype of *Fagopyrum urophyllum*. (A) The wild plants of *F. urophyllum*; (B) The inflorescences of *F. urophyllum*; (C) The flower of *F. urophyllum*.

## *FAGOPYRUM UROPHYLLUM*

*Fagopyrum urophyllum* has also had different names throughout different ages. In 1891, it was called *Polygonum urophyllum* Bureau & Franchet, in 1909, it was called *Fagopyrum mairei* (H. Léveillé) H. Gross, and in 1913, it was called *Fagopyrum urophyllum* (Bureau & Franchet) H. Gross.

### The Botanical Description

Perennial subshrubs. Stems suberect, 60−90 cm tall, much branched; branches ligneous, bark red-brown, exfoliating, simple, herbaceous, green striate. Petioles 2−5 cm, shortly pubescent; leaf blade greenish abaxially, green adaxially, sagittate or ovate triangular, 2−8 × 1.5−4 cm, both surfaces shortly pubescent along veins, base broadly sagittate, margin entire, apex long acuminate or caudate; ocrea brown, 4−6 mm, membranous, oblique. Inflorescence terminal, paniculate, 15−20 cm; branches spreading, lax; bracts greenish, narrowly funnel-shaped, 2−2.5 mm, apex acute, each 3- or 4-flow-ered. Pedicels longer than bracts, 3−3.5 mm, slender, apex articulate. Perianth white; tepals elliptic, 2−3 mm. Stamens included. Styles free; stig-mas capitate. Achenes exceeding persistent perianth, black-brown, shiny, broadly ovoid, sharply trigonous, 3.4 mm. Fl. April−September, fr. May−Nov.

**FIGURE 7.3** Phenotype of *Polygonum statice*. (A) The single plant of *F. statice*; (B) The seedling of *F. statice*; (C) The root of *F. Statice*.

## Distribution

*Fagopyrum urophyllum* is mainly distributed in Yunnan, Sichuan, and Gansu, etc., which can grow to altitudes of 900−2800 m.

## The Habitat and Growth Habit

*Fagopyrum urophyllum*, a perennial subshrub, usually grows on grassy slopes, shrubberies, gravelly slopes, or rocky areas, and the height of its plant can reach 150−200 cm sometimes, which is of erect type. The stem has many branches with the old branches lignified. The flowering and fruiting lasts longer (as long as 70−150 days).The flowers of *F. urophyllum* are dimorphic (with long and short styles, i.e., heterostylous), cross-pollinated, and self-incompatible (Fig. 7.3).

## *FAGOPYRUM STATICE*

The current scientific name *F. statice* (H. Léveillé) H. Gross was named in 1913, and in 1909 it was named *Polygonum statice* H. Léveillé.

## The Botanical Description

Herbs perennial. Rhizomes black, ligneous. Stems 40−50(−65)cm tall, branched at base, glabrous, slenderly striate, leafless above. Petiole 2−4 cm;

leaf blade broadly ovate or triangular, 2−3 × 1.5−2.5 cm, both surfaces glabrous, veins rather prominent abaxially, base broadly cordate or nearly truncate apex acute; ocrea membranous, oblique, margin not ciliate, apex acute. Inflorescence of several spikes aggregated and panicle-like, lax; bracts funnel-shaped, each 2- or 3-flowered. Pedicels longer than bracts, 2−2.5 mm, slender, apex articulate. Tepals elliptic, 1−1.5 mm. Stamens equaling perianth. Styles free; stigmas capitate. Achenes shiny, ovoid, 2−2.5 mm. Fl. July−October, fr. August−November.

## Distribution

*F. statice* is mainly distributed in Guizhou, Yunnan, and Sichuan, etc., which can grow at altitudes of 1300−2200 m.

## The Habitat and Growth Habit

*F. statice*, a perennial plant, usually grows on grassy slopes or gravelly slopes, and is smaller than perennial *F. cymosum* or *F. urophyllum.* The stem is of the erect type. While most of the branches are centered in the base, the above is almost leafless. The underground stem is a little accrescent, which is lignified and blocky. The flowering and fruiting is briefer, and the flowers and fruits are smaller. The flowers of *F. urophyllum* are dimorphic (with long and short styles, i.e., heterostylous), cross-pollinated, and self-incompatible.

## ACKNOWLEDGMENTS

This research was supported by the Investigation of Forage Germplasm in Central China (grant number 2017FY100604).

## REFERENCE

Chen, Q.F., 1999. A study of resources of Fagopyrum (Polygonaceae) native to China. Bot. J. Linnean Soc. 130 (1), 53−64.

## FURTHER READING

Lin, R.F., 1994. Buckwheat in China. Agricultural Publisher, Peking, China, pp. 97−105.
Ohnishi, O, Matsuoka, Y., 1996. Search for the wild progenitor of buckwheat II. Taxonomy of *Fagopyrum* (Polygonaceae) species based on morphology, isozymes and cpDNA variability. Genes Genet. Syst. 71, 383−390.
Yamane, K, Ohnishi, O., 2001. Phylogenetic relationships among natural populations of perennial buckwheat, *Fagopyrum cymosum* Meisn., revealed by allozyme variation. Genet. Resour. Crop Evol. 48 (1), 69−77.
Yamane, K, Yasui, Y, Ohnishi, O., 2003. Intraspecific cpDNA variations of diploid and tetraploid perennial buckwheat, *Fagopyrum cymosum* (Polygonaceae)[J]. Am. J. Bot. 90 (3), 339−346.
Yasui, Yasuo, Ohnishi, Ohmi, 1998. Phylogenetic relationships among *Fagopyrum* species revealed by the nucleotide sequences of the ITS region of the nuclear rRNA gene. Genes Genetic Syst. 73 (4), 201−210.

# Chapter Eight

# Perennial Self-Compatible Wild *Fagopyrum* Species: *F. hailuogouense*

**Meiliang Zhou[1], Yu Tang[2], Xianyu Deng[3], Chen Ruan[3], Mengqi Ding[4], Jirong Shao[4], Yixiong Tang[5], and Yanmin Wu[5]**

[1]*Institute of Crop Sciences, Chinese Academy of Agricultural Sciences, Beijing, China,* [2]*Sichuan Tourism University, Chengdu, China,* [3]*Kunming University of Science and Technology, Kunming, Yunnan, China,* [4]*Sichuan Agricultural University, Yaan, China,* [5]*Biotechnology Research Institute, Chinese Academy of Agricultural Sciences, Beijing, China*

## THE BOTANICAL DESCRIPTION

Herbs perennial, 30−70 cm tall; succulent, perennial, horizontal, brown or dark brown in color, slightly ligneous, with dense, short internodes, (0.3−)0.5−1.2(−1.5) cm, with adventitious roots from the nodes; all nodes leafy, terrestrial stems erect, rarely obliquely positioned, greenish or reddish brown in color, glabrous, slender, cylindrical in transverse section, with longitudinal stripes and sparse, long internodes, (3.4−)4.9−17.2(−21.5) cm, most branching at or near the base. Leaves simple, alternate, papery (i.e., thin and dry), and cordate, broadly cordate, or ovate, (1.8−)2.2−5.8(−6.2) × (1.5−)1.8−5.3(−6.1) cm, apex acuminate or rarely acute, base cordate, with two broad and rounded lateral blade segments; adaxial blade surface green or deep green, abaxial blade surface green or grayish green, glabrous on both surfaces, midrib veins highlighted with 4−8 pairs of lateral veins, light green leaf margin entire; petioles on basal and cauline leaves, (0.5−)1.6−9.5

**67**

Buckwheat Germplasm in the World. DOI: https://doi.org/10.1016/B978-0-12-811006-5.00008-2
© 2018 Elsevier Inc. All rights reserved.

(−20.5) cm, glabrous, greenish or whitish green, slightly grooved longitudinally; upper leaves sessile, long cordate, apex acuminate or acute; ocrea membranous, dark brown, obliquely tubular, 0.2−3 cm, cleaved on one side and longitudinally striped, apex acuminate. Inflorescences racemose, with 1−3 flowers, erect, axillary, interrupted, flowers lax; peduncle (0.3−) 0.8−4.5(−9) cm; rachis slender, green or yellow-green, glabrous; bracts membranous, funnel-shaped, oblique, apex acuminate. Flowers bisexual, with 5 perianth segments, tepals pink at base and red at apex, elliptic to ovate-elliptic, 1.8−2 × 0.8−1 mm; stamens 8, arranged in 2 whorls (5 outer and 3 inner), anthers purple, oval, 0.1−0.2 mm, filaments linear, slender, white, glabrous, 1−1.5 mm long; ovary superior, ovoid, trigonous, green or yellow-green, 0.3−0.6 × 0.2−0.5 mm, styles filiform, ca. 1 mm; stigmas 3, capitate, exceeding persistent perianth. Achenes brown or dark brown, ovoid or broadly ovoid, trigonous, 2.5−3 × 1.5−1.8 mm, surface smooth; styles persisting and reclinate, persistent on achenes.

## DISTRIBUTION

*Fagopyrum hailuogouense* is only distributed in Luding County, Sichuan, which can grow in the altitude of 3100−3300 m.

**FIGURE 8.1** Phenotype of *Fagopyrum hailuogouense*. (A), (B) The habitat of *F. hailuogouense*; (C) The whole plant (underground stem, leaves, stems, and fruits); (D) Fruits; (E) The spherical underground stems sprout at the top; (F) The seedlings grow from the underground stem; (G) The plant grows from the underground stem.

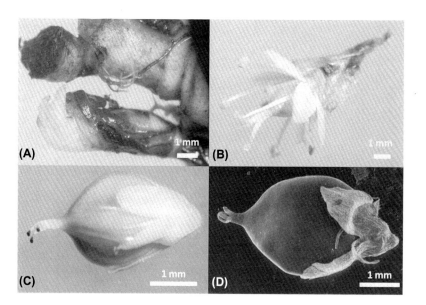

**FIGURE 8.2** scanning electron microscope photos of *Fagopyrum hailuogouense* (A) Rhizome; (B) Flower; (C) Fruit, under LM; (D) Fruit, under SEM (Zhou et al., 2015).

## THE HABITAT AND GROWTH HABIT

*Fagopyrum hailuogouense* usually grows on shady, damp, or grassy streams. Its plant is erect and its stem has many branches. The flowering and fruiting period is July–September. The flowers of *F. hailuogouense* are self-pollinated and self-compatible. The underground stem is a little accrescent, which is long or spherical. The new plants can emerge from the internodes of underground stem every summer (April–May) (Figs. 8.1 and 8.2).

## ACKNOWLEDGMENTS

This research was supported by the Investigation of Forage Germplasm in Central China (grant number 2017FY100604).

## REFERENCE

Zhou, M.L., Zhang, Q., Zheng, Y.D., Tang, Y., Li, F.L., Zhu, X.M., et al., 2015. *Fagopyrum hailuogouense* (polygonaceae), one new species from Sichuan, china. Novon 24 (2), 222–224.

# Chapter Nine

# Annual Self-Incompatible Species

**Ohmi Ohnishi[1] and Meiliang Zhou[2]**

[1]*Kyoto University, Kyoto-fu, Japan,* [2]*Institute of Crop Sciences, Chinese Academy of Agricultural Sciences, Beijing China*

As of December 2016, the *Fagopyrum* genus consists of about 25 species. The annual self-incompatible wild species consist of *Fagopyrum esculentum* ssp. *ancestrale, Fagopyrum wenchuanense, Fagopyrum lineare, Fagopyrum gracilipedoides, Fagopyrum jinshaense, Fagopyrum capillatum, Fagopyrum leptopodum,* and *Fagopyrum gilesii.*

## THE SPECIES OF THE CYMOSUM GROUP WHICH BELONG TO THE CATEGORY OF ANNUAL AND SELF-INCOMPATIBLE SPECIES

### *Fagopyrum esculentum* ssp. *ancestrale*

*Fagopyrum esculentum* ssp. *ancestrale* is the only wild species of the cymosum group in this category (Ohnishi, 2010). It is distributed in southwestern part of China, Yunnan, Sichuan provinces, and eastern Tibct (see Ohnishi, 2012, 2013 for the distribution areas). It was first found on October 20th 1990 in the Wulang river valley of Yonsheng district (Ohnishi, 1991). After its discovery, it was also found in the 1990s in Yanyuan and Muli districts in Sichuan province, in Deqin district and in Lijiang city of Yunnan province, and in Mankang district of eastern Tibet.

This species is heterostyly, and is completely self-incompatible. The plant height is 1−1.5 m, almost the same as cultivated common buckwheat. Although the plants of this subspecies have many branches, each branch has a fewer flowers than cultivated common buckwheat, hence this subspecies has less grains,

*Buckwheat Germplasm in the World.* DOI: https://doi.org/10.1016/B978-0-12-811006-5.00009-4
© 2018 Elsevier Inc. All rights reserved.

**FIGURE 9.1** The plant of *Fagopyrum esculentum* ssp. *ancestrale* (left) and its seeds (the upper part of the right-hand side). Seed grains of cultivated common buckwheat are shown in the lower part of the right-hand side as for comparison.

and the grain weight is 6−8 mg/seed, about 1/5 of cultivated common buckwheat (see Fig. 9.1).

By showing that the natural populations of this subspecies from the Sanjiang area (Ohnishi, 2002) have the closest genetic relationships with cultivated populations of common buckwheat, by analyzing allozyme, AFLP, and SSR variability (Ohnishi, 2009; Konishi et al., 2005; Konishi and Ohnishi, 2007), it was concluded that *F. esculentum* ssp. *ancestrale* Ohnishi is the wild ancestor of the cultivated common buckwheat and the Sanjiang area is the original birthplace of cultivated common buckwheat (Ohnishi, 2004, 2007).

This subspecies is easily crossable with cultivated common buckwheat, hence it has potentiality as a genetic resources for future breeding of cultivated common buckwheat. At present, more than 30 populations have been collected from Yunnan and Sichuan provinces and eastern Tibet (see Fig. 1 of Ohnishi, 2013). Some of them were analyzed for allozyme, AFLP, and SSR variability (Ohnishi, 2009; Konishi et al., 2005; Konishi and Ohnishi, 2007). However, those studies are not for genetic variations for future breeding. In the search for useful genetic germplasm of this subspecies for future breeding, a systematic survey of natural populations should be conducted.

It was clarified that the natural populations of the wild ancestor from the Dongyi river valley and from the Nyiru river valley are huge in population size, and have much more genetic variations in allozymes, AFLP, and SSR than the natural populations in other places (see references mentioned above). A systematic survey should be started by analyzing the natural populations from the Dongyi river valley and the Nyiru river valley for disease resistance and cold tolerance.

However, to the best of the author's knowledge, the wild ancestor has never been used for improving cultivars of common buckwheat. Evaluation of the populations of the subspecies have never practiced systematically.

### *Fagopyrum wenchuanense*

This species was first discovered in Wenchuan county, Aba state, Sichuan province, China, by Dr. Shao, Prof. Tang, and Dr. Zhou (Shao et al., 2011). It is a diploid taxon, with a chromosome number of $2n = 2x = 16$, and all 16 chromosomes are metacentric. As shown in Fig. 9.2, it is closely related to *Fagopyrum gracilipes* (Hemsl.) Dammer ex Diels but differs in its caespitose habit, the plants often with prostrate stems. Its leaves vary from broadly cordate to ovate to hastate or long hastate, the pedicels are basally puberulent, the stamens and pistils are variable in length, and the capsules are

**FIGURE 9.2** The plant of *Fagopyrum wenchuanense*. (A) and (B) Whole plant; (C) Flowers and inflorescences; (D) Root.

ellipsoid-triangular or broadly ovoid-triangular. Plants were observed in flower from August to October and in fruit from September to November. Phylogenetic analysis by internal transcribed spacers (ITSs) and the maturase K(matK) sequences showed that this species belongs to the cymosum group (Zhou et al., 2012, 2014).

## THE WILD SPECIES OF THE UROPHYLLUM GROUP WHICH BELONG TO THE CATEGORY OF ANNUAL AND SELF-INCOMPATIBLE SPECIES

There are many wild species of the urophyllum group in this category. Since the genus *Fagopyrum* is considered as the genus of the species which are heteromorphic and self-incompatible, most of the wild species of the urophyllum group fall into this category, annual and self-incompatible. Only several exceptional species fall into the group of species of self-compatible or partially self-compatible species, as the terminal end species of evolutionally processes.

Among the wild species of which phylogenetic position are known, two annual herbaceous species associated with *Fagopyrum urophyllum*-Dali and *F. urophyllum*-Kunming, that is, *F. lineare* and *F. jinshaense*, fall into this category (see Ohnishi, 2011, 2016). However, the latter species is still suspected being partially self-fertile. Among two pairs of $2x-4x$ complex in the urophyllum group (Ohnishi, 2011), two $2x$ species are self-incompatible, but the $4x$ species are self-fertile. Diploid species, *F. capillatum* and *F. gracilipedoides*, belong to the annual and self-incompatible category of the species.

Among the species in the *Fagopyrum statice*−*F. leptopodum* complex (Ohsako and Ohnishi, 2001; Ohnishi, 2011, 2016), *F. leptopodum* and *F. gilesii* are annual and self-incompatible, hence these two species fall into the present category.

Among the three new species of the urophyllum group found in the upper Min river valley, all three appear to be partially self-fertile.

Below follows a description of each species of the category of annual and self-incompatible in more detail.

### *Fagopyrum lineare*

*Fagopyrum lineare* is a small herbaceous annual species (see Fig. 9.3). It has linear blades, (2−3 mm wide, 10−15mm long), plant height usually less than 20cm. It has a narrow distribution area, Binchuan, Syanyun, and Mide districts of central Yunnan province. However, recently Tang et al. (2016) reported its distribution in the western Sichuan province. Steward (1930) already listed this species in his list of wild buckwheat species.

As discussed in Ohnishi (2011, 2016), this species is molecularly genetically close to *F. urophyllum*-Dali (the samples of *F. urophyllum* from Dali) in spite of great morphological differences between the two (see Fig. 9.3 and Kawasaki and Ohnishi, 2006).

**FIGURE 9.3** A herbarial plant of *Fagopyrum lineare* based on a plant from Binchuan district, central Yunnan Province.

## Fagopyrum jinshaense

This species was first found in a small village Benzilang, Deqin district, Yunnan province, in 1998, near the Jinsha river. It looks like *Fagopyrum gilessi* and *F. leptopodum* in general morphology, but has distinct inflorescence and leaf blades (see Fig. 9.4, after Ohsako et al., 2002).

Later, it was found that this species distributes in the Jinshajiang river valley and the valleys of its tributaries. It does not grow in the Nu river valley, nor in the Lanzang river valley, only in the Jinshajiang river valley (see Fig. 1 of Ohnishi and Tomiyoshi, 2005). Hence, the name of *F. jinshaense* was given (Ohsako et al., 2002). The plants are erect and are not tall (usually less than 30cm), as seen in Fig. 9.4.

Curiously, *F. jinshaense* is closely related to *F. urophyllum*-Kunming, in molecular taxonomy (Ohnishi, 2011), as *F. lineare* is closely related to *F. urophyllum*-Dali. Both *F. lineare* and *F. jinshaense* differentiated from *F. urophyllum* at a very early time of differentiations of the species of the urophyllum group at different places of Yunnan province. Therefore, even at present, *F. lineare* and *F. jinshaense* have a close genetic relationship to *F. urophyllum*-Dali and *F. urophyllum*-Kunming, respectively.

## Fagopyrum gracilipedoides

*Fagopyrum gracilipedoides* is distributed in a narrow area in the center of distribution of *Fagopyrum* species (Ohnishi, 2012), namely Baoshan village and the surrounding villages in Lijiang city, Yunnan province. As shown in Fig. 9.4, the plants are not erect, rather it is creeping and medium in size (20−30 cm in height). It has similar general morphology as *F. leptopodum* and

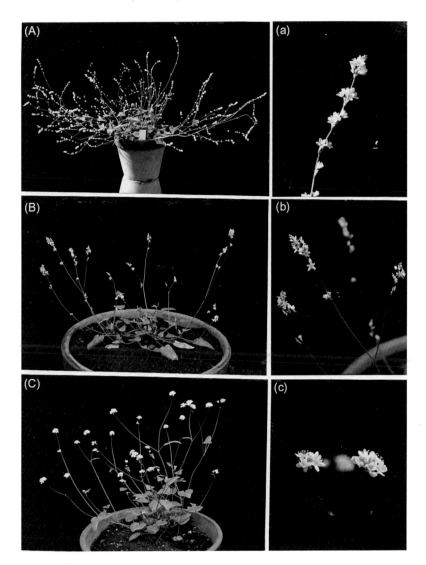

**FIGURE 9.4** (A), (B) and (C) are a plant of *Fagopyrum gracilipedoides*, *Fagopyrum jinshaense* and *Fagopyrum gilesii*, respectively. (a), (b) and (c) are inflorescence of each species.

*F. gracilipes*; however, it is not erect and has specific inflorescence as shown in Fig. 9.4a.

It was clarified that this species is diploid and is genetically closely related with a tetraploid species *Fagopyrum rubufolium* (Ohsako et al., 2002, Nishimoto et al., 2003). It is a curious phenomenon that this middle size (20−30 cm in height) diploid species contributed to the establishment of a very tiny (10−20 cm in height) tetraploid species, *F. rubifolium*. Furthermore, the distribution area of *F. rubifolium*, i.e., the Maerkan district of Aba state of Sichuan province, is far away from the distribution area of *F. gracilipedoides*, Lijiang city, Yunnan province. At present, we do not know why the tetraploid species, *F. rubifolium*, is growing in a narrow area of Maerkang district, Aba state, Sichuan province far away from the distribution area of diploid ancestor, *F. gracilipedoides*, Lijiang city, Yunnan province, China.

### *Fagopyrum capillatum*

As Fig. 9.5 shows this species is tall (more than 50 cm) and erect, having many branches and having a huge number of small flowers on each branch. As its name shows, plants of this species have many fine pubescents on stems and on the surface of the leaf blades. This species was first found in the Wulang river valley of Yongsheng district, Yunnan province at the same time as the discovery of a wild ancestor of cultivated common buckwheat in 1990 (Ohnishi, 1991). Later, this species was found in many places on the north-western corner of the Yunnan province (see Fig. 1 of Ohnishi, 2013).

It is a very curious phenomenon that this tall erect (usually more than 50 cm in height) species is diploid and seems to contribute to the establishment of the small herbaceous tetraploid species *F. gracilipes* (usually 10−30 cm in height). We do not know whether *F. gracilipes* is autotetraploid or allotetraploid. A search for the second species that contributed to tetraploid of *F. gracilipes* should be conducted. Ohsako et al., (2002)'s data showed that the *Fagopyrum* species which is most closely related with *F. gracilipes*

**FIGURE 9.5** *Fagopyrum capillatum.*

**FIGURE 9.6** *Fagopyrum leptopodum*, a plant with many flowers on each branch (left); a plant with less flowers (right).

is *F. capillatum*, and *F. gracilipedoides* is the next. Any other diploid species relates far away from *F. gracilipes*.

### Fagopyrum leptopodum

This medium to small plant species distributes in relatively wide areas of the southern Sichuan province to the northwestern part of the Yunnan province. This species has large morphological variations. The group of plants from the southern Sichuan province—for example Ximeng and Hanyuan districts of the Sichuan province—have larger and more vigorous plants than the typical *F. leptopodum* in the Yunnan province, therefore it was once considered to be a different species from the typical *F. leptopodum* in the Yunnan province. Ohsako and Ohnishi (1998, 2001)'s molecular analysis concluded that both groups of plants from the Sichuan province and the Yunnan province should be treated as one species— *F. leptopodum*. Furthermore, *F. leptopodum* is shown to be very closely related at the molecular level with a perennial species *F. statice*, which is distributed in the eastern Yunnan province (Ohsako and Ohnishi, 2001). This implies that *F. leptopodum* is phylogenetically a derivative species of the ancestral species *F. statice*.

As shown in Fig. 9.6, some plants of *F. leptopodum* have branches with many flowers, but some plants have branches with a few flowers (compare the plants on the left and the right in Fig. 9.6. Both types should be included in species *F. leptopodum*.

### Fagopyrum gilesii

This species was first classified in the genus *Cephalophilon* by Hooker (1886). Later Gross (1913) classified this species in *Fagopyrum*. This species has a characteristic inflorescence, head like small (1 cm) inflorescence (see Fig. 9.4C, 9.4c). This species is distributed in the deeper (northern) parts of the valleys of the Jinshajian, the Lanzanjiang and the Nujiang (see Ohnishi, 2012, 2013). This species is said to be distributed in the Himalayan hills, but

**FIGURE 9.7** A large natural population of *Fagopyrum gilesii* growing in Changbo village of Batang district, Sichuan province.

this author has never seen this species on the Himalayan hills of Bhutan, Nepal, or India. This species was already well known to Steward (1930), and he mentioned it in his list of wild *Fagopyrum* species.

It is phylogenetically closely related to the *F. statice−F. leptopodum* complex, although the distribution areas of *F. gilesii, F. statice*, and *F. leptopodum* are quite different: the deeper parts of the three river area for *F. gilesii,* eastern Yunnan province for *F. statice,* northwestern part of Yunnan province and southwestern part of Sichuan province for *F. leptopodum.*

This species occasionally grows as a large population in conditions of "not-so-wet, not-so-dry" in barren fields where no graminaceae plants are growing (see for example Fig. 9.7).

## ACKNOWLEDGMENTS

This research was supported by the Investigation of Forage Germplasm in Central China (grant number 2017FY100604).

## REFERENCES

Gross, H., 1913. Remarques sur les Polygonees de l'Asie Orientale. Bull. Geogr. Bot. 23, 7−32.

Hooker, J.D., 1886. Flora of British India, vol. 22-61. Reeve & Co, London.

Kawasaki, M., Ohnishi, O., 2006. Two distinct groups of natural populations of *Fagopyrum urophyllum* (Bur. et Franch.) Gross revealed by the nucleotide sequence of non-coding region in chloroplast DNA. Genes, Genet. Syst. 81, 323−332.

Konishi, T., Ohnishi, O., 2007. Close genetic relationship between cultivated and natural populations of common buckwheat in the Sanjiang area is not due to recent gene flow between them — An analysis using microsatellite markers. Genes Genet. Syst. 82, 53−64.

Konishi, T., Yasui, Y., Ohnishi, O., 2005. Original birthplace of cultivated common buckwheat inferred from genetic relationships among cultivated populations and natural populations of wild common buckwheat revealed by AFLP analysis. Genes Genet. Syst. 80, 113−119.

Nishimoto, Y., Ohnishi, O., Hasegawa, M., 2003. Topological incongruence between nuclear and chloroplast DNA trees suggesting hybridization in the urophyllum group of the genus *Fagopyrum* (Polygonaceae). Genes Genet. Syst. 78, 139−153.

Ohnishi, O., 1991. Discovery of the wild ancestor of common buckwheat. Fagopyrum 11, 5−10.

Ohnishi, O., 2004. On the origin of cultivated buckwheat. In: Proceedings of 9th International Symposium Buckwheat at Prague, pp. 16−21.

Ohnishi, O., 2007. Natural populations of the wild ancestor of cultivated common buckwheat, *Fagopyrum esculentum* ssp. *ancestrale* from the Dongyi river valley—Their distribution and allozyme variations. In: Proceedings of 10th International Symposium Buckwheat at Yangling, pp. 13−18.

Ohnishi, O., 2009. On the origin of cultivated common buckwheat based on allozyme analyses of cultivated and wild populations of common buckwheat. Fagopyrum 26, 3−9.

Ohnishi, O., 2010. Distribution and classification of wild buckwheat species. 1. Cymosum group. Fagopyrum 27, 1−8.

Ohnishi, O., 2011. Distribution and classification of wild buckwheat species. 2. Urophyllum group. Fagopyrum 28, 1−8.

Ohnishi, O., 2012. On the distribution of natural populations of wild buckwheat species. Fagopyrum 29, 1−6.

Ohnishi, O., 2013. Distribution of wild buckwheat species and perspective for their utilization. Fagopyrum 30, 9−14.

Ohnishi, O., 2016. Molecular taxonomy of the genus *Fagopyrum*. Chapt. 1 (pp. 1-12). In: Zhou, M., Kreft, I., Woo, S.H., Chrungoo, N., Weislander, G. (Eds.), "Molecular Breeding and Nutritional Aspects of Buckwheat". Elsevier, Amsterdam.

Ohnishi, O., Tomiyoshi, M., 2005. Distribution of cultivated and wild buckwheat species in the Nu river valley of southwestern China. Fagopyrum 22, 1−5.

Ohsako, T., Ohnishi, O., 1998. New *Fagopyrum* species revealed by morphological and molecular analyses. Genes Genet. Syst. 73, 85−94.

Ohsako, T., Ohnishi, O., 2001. Nucleotide sequence variation of the chloroplast *trnK/matK* region in two wild *Fagopyrum* (Polygonaceae) species, *F. leptopodum* and *F. statice*. Genes Genet. Syst. 76, 39−46.

Ohsako, T., Yamane, K., Ohnishi, O., 2002. Two new *Fagopyrum* (Polygonaceae) species, *F. gracilipedoides* and *F. jinshaense* from Yunnan, China. Genes Genet. Syst. 77, 399−408.

Shao, J.R., Zhou, M.L., Zhu, X.M., Wang, D.Z., Bai, D.Q., 2011. *Fagopyrum wenchuanense* and *Fagopyrum qiangcai*, two new species of polygonaceae from Sichuan, China. Novon 21, 256−261.

Steward, A.N., 1930. Polygoneae in eastern Asia, 88. Contributions from the Gray Herbarium at Harvard University, pp. 1−129.

Tang, Y., Shao, J.R., Zhu, X.M., Ding, M.Q., Zhou, M.L., 2016. The investigation and collection of wild buckwheat in Sichuan China. In: Proceedings of 13th International Symposium Buckwheat at Cheongju, pp. 99−105.

Zhou, M.L., Bai, D.Q., Tang, Y., Zhu, X.M., Shao, J.R., 2012. Genetic diversity of four new species related to southwestern Sichuan buckwheats as revealed by karyotype, ISSR and allozymes characterization. Plant Syst. Evol. 298, 751−759.

Zhou, M., Wang, C., Wang, D., Zheng, Y., Li, F., Zhu, X., et al., 2014. Phylogenetic relationship of four new species related to southwestern Sichuan *Fagopyrum* based on morphological and molecular characterization. Biochem. Syst. Ecol. 57, 403−409.

# Chapter Ten

# Annual Self-Compatible Species

**Ohmi Ohnishi[1] and Meiliang Zhou[2]**
*[1]Kyoto University, Kyoto-fu, Japan, [2]Institute of Crop Sciences, Chinese Academy of Agricultural Sciences, Beijing, China*

Of the total 25 *Fagopyrum* species, the annual self-compatible wild species consist of *Fagopyrum tataricum* ssp. *potanini*, *Fagopyrum homotropicum*, *Fagopyrum gracilipes*, *Fagopyrum rubifolium* and *Fagopyrum crispatifolium*.

## ANNUAL SELF-COMPATIBLE SPECIES IN THE CYMOSUM GROUP

### *Fagopyrum tataricum* ssp. *potanini*, the Wild Ancestor of Cultivated Tartary Buckwheat

*Fagopyrum tataricum* ssp. *potanini* Batalin is the wild ancestor of cultivated Tartary buckwheat (Tsuji and Ohnishi, 2000, 2001) and it falls into the category of annual self-compatible species. This subspecies was found by a Russian explorer and botanist G. Potanin in the Gansu province of China at the end of 19th century and registered by Batalin based on a plant grown in St. Petersburg (see Bredtschneider, 1898).

This subspecies is crossable with cultivated Tartary buckwheat. Yet, crossing this wild subspecies with cultivated Tartary buckwheat has not been practiced for improving cultivated Tartary buckwheat, because of the technical difficulty of crossing them.

The natural populations of wild Tartary buckwheat from the Sichuan province are most variable in allozymes, and Amplified Fragment Length Polymorphism markers (Ohnishi, 2000; Tsuji and Ohnishi, 2001). The natural populations of wild Tartary buckwheat from Tibet are very similar to the cultivated populations of Tartary buckwheat, so far as allozyme variations are concerned (Ohnishi, 2000, 2002). Natural populations of wild Tartary buckwheat from the Sichuan and Yunnan provinces have more variations than the natural populations of wild Tartary buckwheat from Tibet (Ohnishi, 2000,

Buckwheat Germplasm in the World. DOI: https://doi.org/10.1016/B978-0-12-811006-5.00010-0
© 2018 Elsevier Inc. All rights reserved.

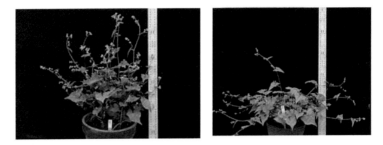

**FIGURE 10.1** The erect type (on the left) and the creeping type (on the right) of *Fagopyrum tataricum* ssp. *potanini.*

**FIGURE 10.2** Wild Tatary buckwheat *Fagopyrum tataricum* ssp. *potanini* creeping on rocky mountain ground.

2002). This implies that the wild Tatary buckwheat samples from the Sichuan and Yunnan provinces have a higher potential to be used as a breeding material for improving cultivated Tartary buckwheat. Ohnishi (2009) considered that genotype similarity between natural populations of the wild Tartary buckwheat from Tibet and cultivated Tartary buckwheat are due to the fact that both cultivated Tartary buckwheat and wild Tartary buckwheat in Tibet are differentiated from the Yunnan-Sichuan type of wild Tartary buckwheat, namely, the wild Tartary buckwheat from Tibet and the cultivated Tartary buckwheat are not in a parent-offspring relationship, rather they are in a relationship of brothers.

At present, there is no attempt to improve cultivated Tartary buckwheat by incorporating the wild ancestor's favorable traits by crossing them.

Wild Tartary buckwheat is as erect as cultivated Tartary buckwheat in some cases (shown in Fig. 10.1, left), but in some cases it is a creeper on the ground (see Fig. 10.1, right, and Fig. 10.2).

A weed type of Tatary buckwheat is also growing in the areas where both wild and cultivated Tatary buckwheat are grown. Tsuji and Ohnishi (2009) showed that the weedy type of Tartary buckwheat is the hybrid between wild and cultivated Tartary buckwheat.

**FIGURE 10.3** *Fagopyrum homotropicum.*

## *Fagopyrum homotropicum*, the Wild Self-Fertile Species, Morphologically Similar to *Fagopyrum esculentum* ssp. *ancestrale*

*Fagopyrum homotropicum* is a self-fertile new species found by Ohnishi (1995). Fig. 10.3 is a plant of *F. homotropicum* grown at Kyoto, Japan. Both diploid and tetraploid forms are distributed in western Sichuan (Aba and Dalingshan states) and northwestern Yunnan province, and eastern Tibet, in all the valleys of the three big rivers, Jinshajiang (the Yangtze river), Lanzangjiang (the Mekong river), and Nujiang (the Salween river) (see Ohnishi, 1995, 2013; Ohnishi and Asano,1999; Tomiyoshi et al., 2012 for the distribution areas of *F. homotropicum*). The distribution area of this species is slightly wider than that of the wild ancestor of common buckwheat. *F. homotropicum* is adapted to a slightly cooler and higher-altitude climate than *Fagopyrum esculentum* ssp. *ancestrale*.

Because of the self-fertile property of this species, it was used to make a new cultivated buckwheat, the "self-fertile common buckwheat." The idea is as follows: crossing *F. homotropicum* with cultivated common buckwheat and making subsequent backcrossings of the hybrids with culti-vated common buckwheat, choosing self-fertile individuals at each genera-tion of backcrossings. After a great enough number of generations, self-fertile common buckwheat may be obtained. Campbell (1995), Wang and Campbell (1998), and Woo et al. (1999) have reported successful interspe-cific hybridization between diploid *F. esculentum* (2n = 16) and diploid *F. homotropicum* (2n = 16). These crosses improved agronomic characters of hybrids of the cross. At the same time, the self-compatibility character has been introgressed from *F. homotropicum* into *F. esculentum*. The percent-age of successful crosses between *F. homotropicum* and cultivated common

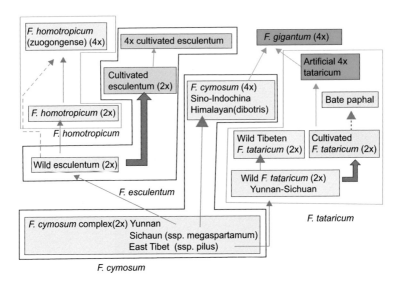

**FIGURE 10.4** Phylogenetic relationships among the species of the cymosum group. Modified after Chen, 2012.

buckwheat is highly dependent on the population (experimental line) of *F. homotropicum* used in the crosses. This may imply that the origin of *F. homotropicum* from *F. esculentum* ssp. *ancestrale* has multiple origins; different *F. homotropicum* lines originated from different lines of *F. esculentum* ssp. *ancestrale* (Ohnishi and Asano, 1999). If *F. homotropicum* and cultivated common buckwheat have the common ancestor of *F. esculentum* ssp. *ancestrale*, then the cross between *F. homotropicum* and cultivated common buckwheat is easy, whereas if *F. homotropicum* and cultivated common buckwheat do not have the common ancestor of *F. esculentum* ssp. *ancestrale*, the cross between them is difficult.

Phylogenetic relationships among the wild and cultivated species of the cymosum group, particularly phylogenetic relationships among *F. esculentum* ssp. *ancestrale*, *F. esculentum* ssp. *esculentum* (cultivated common buckwheat), and *F. homotropicum* shown in Fig. 10.4 (modified after Chen, 2012), explain the phenomena observed. (see also Tomiyoshi, et al., 2012).

## SELF-FERTILE SPECIES OF THE UROPHYLLUM GROUP

Self-fertile species in the urophyllum group are restricted to two tetraploid species of two 2x−4x complex pairs, that is, *F. gracilipes* and *F. rubifolium* (Ohnishi, 2011).

### *Fagopyrum gracilipes*

*Fagopyrum gracilipes* is a well-known example of this category. This tiny herbaceous species is growing primarily in farmer's cultivated fields of any

**FIGURE 10.5** *Fagopyrum gracilipes* in farmer's cultivated field.

summer crops, such as buckwheat, corn, potato, and indeed any vegetables (see Fig. 10.5). Due to the self-compatible character and colonizing ability as a weed of any cultivated species, *F. gracilipes* has extended its distribution area from the center of the distribution of *Fagopyrum*—the northwestern corner of Yunnan province (see Ohnishi, 2012, 2013)—to whole area of southern China and also west, up to the western part of Bhutan, Thimphu, Paro, and Chandebji areas, although this species has never entered into Nepal and India (Hooker, 1886 and Hara, 1966 did not mention this species). Now this species has the second widest area of distribution next to *Fagopyrum cymosum* (Ohnishi, 2012, 2013). *F. gracilipes* does not distribute to the north of the Qinling mountains. This species is not distributed in the Silk Road area or on the Tibetan plateau.

Even in this species, as this author examined a long time ago (Ohnishi, unpublished), both heterostyly plants (5 pin, 1 thrum individuals) and homostyly plants in 24 individuals of *F. gracilipes* are growing together (% of heterostyly plants are higher than the author expected). This data shows that *F. gracilipes* is not perfectly homostyly and is not completely self-fertile.

*Fagopyrum gracilipes* whose seeds have wing is treated as a distinct variety, *F. gracilipes* var. *odontopterum* (Wu, 1984). However, both winged plants and nonwinged plants are growing together in the same fields of cultivated crops. Hence, the author considers that we do not need to treat one form as a distinct variety or subspecies.

*Fagopyrum gracilipes* is a tetraploid species. While *F. cappilatum* is a diploid species which is the most closely related with *F. gracilipes*. Hence, *F. cappilatum* is considered as the species contributed a genome to *F. gracilipes*. All molecular data support this (Ohsako and Ohnishi, 1998, 2000; Nishimoto et al., 2003). There then remains a curious problem—a tall diploid plant species, *Fagopyrum capillatum*, contributed to the establishment of small herbaceous species *F. gracilipes*.

A similar problem also remains for the establishment of small tetraploid species *F. rubifolium*. Is it true that *Fagopyrum gracilipedoides* is a parent of small species *F. rubifolium*?

**FIGURE 10.6** *Fagopyrum rubifolium* in a barren stony field.

## Fagopyrum rubifolium

As a similar result to the *F. capillatum—F. gracilipes* 2x—4x complex it is also true for the tetraploid species *F. rubifolium* of 2x—4x complex *F. gracilipedoides—F. rubifolium* 2x—4x complex. In the case of *F. rubifolium*, *F. rubifolium* grows in a very narrow area in Maerkang district, Aba state of Sichuan province, very far away from the distribution area of the diploid counterpart, *F. gracilipedoides* in the Baoshan village and surrounding villages of Lijiang city of the Yunnan province.

Fagopyrum rubifolium is self-fertile and very small plant species, as small as *F. gracilipes*. However, *F. rubifolium* has a distinct character of red blades (see Fig. 10.6). No other species of *Fagopyrum* than *F. rubifolium* has this character.

Nishimoto et al. (2003) found two distinct sequences of FLO/LFY gene of *F. rubifolium* and two sequences lie in two distinct positions of FLO/LFY phylogenetic tree. They interpreted their result by assuming multiple interspecific hybrids in the urophyllum group from which *F. rubifolium* was born and assuming tetraploid of *F. rubifolium* is allotetraploid. *F. homotropicum* (2x and 4x) and *F. rubifolium* are the new species found by Ohnishi's group at Kyoto University.

## Fagopyrum crispatifolium

Fagopyrum crispatifolium is mainly distributed in Puge, Sichuan, China. It can grow at altitudes of 1850—1900m. It usually grows on vacant lands, orchards, fields, slopes, valleys, barren hills, woodlands, grass, etc. The flowering and fruiting is September—November. The flowers are of self-pollinated. As shown in Fig. 10.7, it is closely related to *F. gracilipes* (Hemsl.) Dammer ex Diels, but differs in that the leaves are alternate, blade papery, broadly ovate, ovate, long ovate. The chromosome number is: 2n = 4x = 32. (Liu et al., 2008).

**FIGURE 10.7** The plant of *Fagopyrum crispatifolium.* (A) clearly shows a characteristic leaf morphology (shrunken leaf) of *Fagopyrum crispatifolium.* (B) shows erect nature of stems of *Fagopyrum crispatifolium* similar to *Fagopyrum gracilipes.*

## ACKNOWLEDGMENTS

This research was supported by the Investigation of Forage Germplasm in Central China (grant number 2017FY100604).

## REFERENCES

Bredtschneider, E., 1898. History of European Botanical Discoveries in China. Press Imp. Acad. Sci., Petersburg.

Campbell, C.G., 1995. Inter-specific hybridization in the genus *Fagopyrum*, Proc. 6th Intl. Symp. Buckwheat at Ina, 255—263.

Chen, Q.F., 2012. Plant Sciences on Genus *Fagopyrum*. ScientificPublisher, Beijing.

Hara, H., 1966. The Flora of East Himalaya. Tokyo University Press, Tokyo.

Hooker, J.D., 1886. Flora of British India. v., 22-61. Reeve & Co, London.

Liu, J.L., Tang, Y., Xia, M.Z., Shao, J.R., Cai, G.Z., Luo, Q., et al., 2008. *Fagopyrum crispatifolium* J. L. Liu, a new species of Polygonaceae from Sichuan, China. J. Syst. Evol. 46, 929—932.

Nishimoto, Y., Ohnishi, O., Hasegawa, M., 2003. Topological incongruence between nuclear and chloroplast DNA trees suggesting hybridization in the genus *Fagopyrum* (Polygonaceae). Genes Genet. Syst. 78, 139—153.

Ohnishi, O., 1995. Discovery of new *Fagopyrum* species and its implication for the studies of evolution of Fagopyrum and of the origin of cultivated buckwheat, Proc. 6th Intl. Symp. Buckwheat at Ina, 175—190.

Ohnishi, O., 2000. Geographical distribution of allozymes in natural populations of wild Tartary buckwheat. Fagopyrum 17, 29—34.

Ohnishi, O., 2002. Wild buckwheat species in the border area of Sichuan, Yunnan and Tibet and allozyme diversity of wild Tartary buckwheat in this area. Fagoyrum 19, 3—9.

Ohnishi, O., 2009. On the origin of cultivated common buckwheat based on allozyme analyses of cultivated and wild populations of common buckwheat. Fagopyrum 26, 3—9.

Ohnishi, O., 2011. Distribution and classification of wild buckwheat species. 2. Urophyllum group. Fagopyrum 28, 1—8.

Ohnishi, O., 2012. On the distribution of natural populations of wild buckwheat species. Fagopyrum 29, 1—6.

Ohnishi, O., 2013. Distribution of wild buckwheat species and perspective for their utilization. Fagopyrum 30, 9—14.

Ohnishi, O., Asano, N., 1999. Genetic diversity of *Fagopyrum homotropicum*, a wild species related to common buckwheat. Genet. Resour. Crop Evol. 46, 389—398.

Ohsako, T., Ohnishi, O., 1998. New *Fagopyrum* species revealed by morphological and molecular analyses. Genes Genet. Syst. 73, 85−94.

Ohsako, T., Ohnishi, O., 2000. Intra- and inter-specific phylogeny of wild *Fagopyrum* (Polygonaceae) species based on nucleotide sequences of noncoding regions in chloroplast DNA. Am. J. Bot. 87, 53−582.

Tomiyoshi, M., Yasui, Y., Ohsako, T., Li, C.Y., Ohnishi, O., 2012. Phylogenetic analysis of *AGAMOUS* sequences reveals the origin of the diploid and tetraploid forms of self-pollinating wild buckwheat, *Fagopyrum homotropicum* Ohnishi. Breed. Sci. 62, 241−247.

Tsuji, K., Ohnishi, O., 2000. Origin of cultivated Tartary buckwheat (*Fagopyrum tataricum* Gaert.) revealed by RAPD analysis. Genetic Resour. and Crop Evol. 47, 431−438.

Tsuji, K., Ohnishi, O., 2001. Phylogenetic relationships among wild and cultivated Tartary buckwheat (*Fagopyrum tataricum* Gaert.) populations revealed by AFLP analyses. Genes Genet. Syst. 76, 47−52.

Tsuji, K., Ohnishi, O., 2009. Morphological characters in cultivated, wild and weedy Tatary buckwheat. Fagopyrum 26, 11−19.

Wang, Y.J., Campbell, C., 1998. Inter-specific hybridization in buckwheat among *F. esculentum, F. homptroicum,* and *F. tataricum,* Proc. 7th Intl. Symp. Buckwheat at Winnipeg, 1−13.

Woo, S.H., Adachi, T., Jong, S.K., Campbell, C., 1999. Inheritance of self-compatibility and flower morphology in an inter-specific buckwheat hybrid. Can. J. Plant Sci. 79, 483−490.

Wu, C.Y., 1984. Index Florae Yunnanensis. People's Publishing House, Yunnan, China.

# FURTHER READING

Ohnishi, O., 2016. Molecular taxonomy of the genus *Fagopyrum*. Chapt. 1. In: Zhou, M., Kreft, I., Woo, S.H., Chrungoo, N., Weislander, G. (Eds.), Molecular Breeding and Nutritional Aspects of Buckwheat. Elsevier, Amsterdam, pp. 1−12.

# Chapter Eleven

# Annual Partially Self-Fertile Species

**Meiliang Zhou[1] and Ohmi Ohnishi[2]**

[1]*Institute of Crop Sciences, Chinese Academy of Agricultural Sciences, Beijing, China,* [2]*Kyoto University, Kyoto-fu, Japan*

Typical wild buckwheat species of annual partially self-fertile species are six new species found in the upper Min river valley, in the Sichuan province, China. These species consist of *Fagopyrum pleioramosum*, *Fagopyrum callianthum*, and *Fagopyrum macrocarpum*, discovered by Ohnishi's group, and *Fagopyrum luojishanense*, *Fagopyrum pugense*, and *Fagopyrum qiangcai*, discovered by Zhou's group.

The Min river is a tributary of Changjiang (the Yangtze river), and it runs through the central part of the Sichuan basin. The upper streams of the Min river run through the mountain regions of the Aba state of the Sichuan province. Three new wild species of *Fagopyrum*—*F. pleioramosum*, *F. callianthum*, and *F. macrocarpum*, were found by Ohnishi (1995, 1998) in the upper Min river valley. They are similar in general morphology, and they are closely related in molecular taxonomy (Ohnishi and Matsuoka, 1996; Yasui and Ohnishi, 1998a, b; Ohsako and Ohnishi, 1998, 2000, 2002; Nishimoto et al., 2003). Later, another three new wild species of *Fagopyrum*—*F. luojishanense*, *F. pugense*, and *F. qiangcai*, were also found in the Sichuan province by Zhou and his colleagues (Tang et al., 2010; Shao et al., 2011; Hou ct al., 2015).

Figs. 11.1–11.6 show plants of each species in fields—that is, on roadsides, on the margins of cultivated fields, and on mountain slopes.

*Fagopyrum pleioramosum* (Fig. 11.1) primarily grows on roadsides and the margins of cultivated fields in the Wenchuan district of Aba state. *F. callianthum* (Fig. 11.2) is growing mainly on mountain slopes in Lixian, Maoxian, and Wenchuan districts of the Aba state of the Sichuan province, and *F. macrocarpum* grows at the margins of cultivated fields in Wenchuan, Lixian, Maoxian, Heishui, and Maerkang districts of Aba state, Sichuan

**89**

Buckwheat Germplasm in the World. DOI: https://doi.org/10.1016/B978-0-12-811006-5.00011-2
© 2018 Elsevier Inc. All rights reserved.

**FIGURE 11.1** *Fagopyrum pleioramosum* growing on roadside in Wenchuan district of Aba state, Sichuan province.

**FIGURE 11.2** *Fagopyrum callianthum* growing on mountain slope in Lixian district of Aba state, Sichuan province.

**FIGURE 11.3** *Fagopyrum macrocarpum* growing at the margin of a cultivated field in Lixian district of Aba state, Sichuan province.

**FIGURE 11.4** The plant of *Fagopyrum pugense*. (A) Whole plant; (B) Inflorescences; (C) Flowers; (D) Fruits.

**FIGURE 11.5** The plant of *Fagopyrum qiangcai*. (A) Whole plant; (B) Inflorescences; (C) Flowers.

**FIGURE 11.6** The plant of *Fagopyrum luojishanense*. (A) and (B) Whole plant; (C) Flowers and inflorescences; (D) Fruit.

province. All these species are heteromorphic, that is, both pin and thrum forms of the flowers grow. However, possible self-fertilization of the three species was confirmed (Yasui and Ohnishi, 1998b; Nishimoto et al., 2003), namely plants of these three species set seeds even if we covered flowers with insects, although the percentage of seeds set is decreased. Precise experiments to get a percentage of flowers to set seeds among the flowers pollinated by pollen grains from own flowers should be conducted, and the final conclusion on the partially self-fertile nature of these three species should be drawn based on the experiment. Three species from the upper Min river valley are closely related to each other at the molecular level, and three species are isolated from the other wild species of *Fagopyrum* in distribution area and in phylogenetic positions in phylogenetic trees (all molecular data support this conclusion, see Yasui and Ohnishi, 1998a, b; Ohsako and Ohnishi, 1998, 2000, 2002; Nishimoto et al., 2003). The recently identified three new species, *F. luojishanense*, *F. pugense*, and *F. qiangcai*, may be closely related with one of *F. pleioramosum*, *F. callianthum*, and *F. macrocarpum*, although more molecular data is still needed to demonstrate this hypothesis.

# FAGOPYRUM PLEIORAMOSUM

*Fagopyrum pleioramosum* was first found in Yanmeng village of Wenchuan district, in the valley of the Min river, Sichuan province. This species is creeping on weeds on the roadsides or in the margins of cultivated fields. As its name shows, this species has many long (more than 1 m long) branches (Fig. 11.1). This species has many small flowers on cymes of the terminus of branches. This species is abundant on roadsides or margins of cultivated fields in Wenchuan district, Sichuan province, but it becomes rare in the neighboring districts of Lixian, Maoxian, and the Heishui districts of Aba state of Sichuan province.

# FAGOPYRUM MACROCARPUM

*Fagopyrum macrocarpum* was first found on the margins of a cultivated field in Putuo village of Lixian district, Aba state, Sichuan province. It looks like *F. pleioramosum* in general morphology. However, it is critically different from *F. pleioramosum* in the seed size of the species. As its name shows, *F. macrocarpum* has definitely larger seeds than *F. pleioramosum*. *F. macrocarpum* also has many branches, but branches are not so long as *F. pleioramosum*. These branches have a lesser number of flowers on cymes. *F. macrocarpum* extended its distribution area to Maerkang district beyond the high mountains. *F. macrocarpum* is a popular weed of the apple orchards in Maoxian district, Sichuan province, where this species is highly polymorphic in the color of its flowers—white or pink, and in color of its seed grains—brown and dark pink.

# FAGOPYRUM CALLIANTHUM

*Fagopyrum callianthum* was first found on a mountain slope in Guanbao village in Lixian district, Aba state. As its name shows, *F. callianthum* has the most beautiful flowers (see Fig. 11.2) among the wild species of *Fagopyrum*. The flowers are larger than the flowers of common buckwheat. *F. callianthum* has a suitable number of flowers/plant for ornamental flowers. However, this species is not tall enough for ornamental flowers (usually shorter than 30 cm). As for the pot flower, this species is not fit for that purpose because petals will fall down on the table every day.

   *Fagopyrum callianthum* is morphologically slightly different in general morphology from the other two species, *F. pleioramosum* and *F. macrocarpum* (compare Fig. 11.2, to Figs. 11.1 and 11.3). All molecular taxonomic studies showed that three species are closely related each other at the molecular level, and far away from other wild *Fagopyrum* species (Yasui and Ohnishi, 1998a, b; Ohsako and Ohnishi, 1998, 2000, 2002; Nishimoto et al., 2003). This probably implies that the ancestor of the three species diffused from the center of *Fagopyrum* species toward the upper Min river valley.

After arriving at the upper Min river valley, three species were differentiated from the ancestor—*F. callianthum* first, then *F. pleioramosum* and *F. macrocarpum* (Ohsako and Ohnishi, 2000; Nishimoto et al., 2003).

## FAGOPYRUM PUGENSE

*Fagopyrum pugense* is distributed among counties such as Puge, Yanyuan, Huili, Huidong, Butuo, Zhaojue, Meigu, Mianning, Luding, and Kangding, etc., which are located in the southwest of Sichuan, China. It can grow at altitudes of 1300−3100 m. It usually grows on vacant lands, orchards, fields, slopes, valleys, barren hillsides, woodlands, grass, etc., is usually accompanied by *Fagraea gracilipes* and its variant (*F. gracilipes* var. *odontopterum*). The flowering and fruiting is July−October. The flowers are of partially self-pollinated. As shown in Fig. 11.4, it is closely related to *F. gracilipes* (Hemsl.) Dammer ex Diels, but differs in that the plants have thicker stems and branches with densely erect villose pubescence, numerous nodes and short internodes, ovate to cordate leaves that are minutely rugose with small pustules, and small achenes (Tang et al., 2010).

## FAGOPYRUM QIANGCAI

*Fagopyrum qiangcai* is distributed in Wenchuan County, which is located in the southwest of Sichuan, China. It can grow at altitudes of 1210 1240 m. It usually grows on the grass of slopes or valleys, but a few also grow in the fields, by the roadsides, or the sides of the valley. It is highly resistant to drought and barrenness. The flowering and fruiting is August−November. The flowers are of partially self-pollinated. As shown in Fig. 11.5, it is closely related to *Fagopyrum esculentum* Moench based on its leafy base, triangular leaves, and terminal or axillary racemes. This species differs in having congested nodes at the base of the plant, triangular to oval terminal leaves with bright red veins, dense inflorescences, white punctate adaxial leaf surfaces, and an articulate pedicel. It is diploid, $2n = 2x = 16$, with a karyotype of 12 metacentric and four submetacentric chromosomes (Shao et al., 2011).

## FAGOPYRUM LUOJISHANENSE

*Fagopyrum luojishanense* is distributed in counties such as Puge, Yanyuan, Butuo, Huili, Huidong, Meigu, Mianning, Luding, and Kangding, etc., which are located in the southwest of Sichuan, China. It can grow at altitudes of 1200−3100 m. It usually grows on barren hillsides, woodlands, ditches, roadsides, the edges of fields, corners, ridges of ground, orchards, the ground around houses, or the croplands. Flowering is July−October, while the fruiting is August−November. The flowers are partially self-pollinated. As shown in Fig. 11.6, this species is similar to *F. gracilipes* (Hemsl.) Dammer ex Diels, and is distinguished by its leaf blades that are ovate, narrowly ovate, or triangular, winged seeds, and inflorescences that are either axillary or terminal. Its fruit surface is densely and irregularly warty, while fruit ornamentation is sparser in *F. gracilipes* (Hemsl.) Dammer ex Diels. This species is a diploid, with 2n = 2x = 16 and a karyotype of 16 metacentric chromosomes, while *F. gracilipes* is tetraploid, 4n = 4x = 32, with a karyotype of 30 metacentric and 2 submetacentric chromosomes (Hou et al., 2015).

All six species mentioned in this chapter are newly found by Ohnishi's group and Zhou's group in Sichuan, China. These species probably are closely related to each other, but we need further research to confirm their relationship (also see Zhou et al., 2012, 2014; Ohnishi, 2016).

## ACKNOWLEDGMENTS
This research was supported by the Investigation of Forage Germplasm in Central China (grant number 2017FY100604).

## REFERENCES
Hou, L.L., Zhou, M.L., Zhang, Q., Qi, L.P., Yang, X.B., Tang, Y., et al., 2015. *Fagopyrum luojishanense*, a new species of Polygonaceae from Sichuan, China. Novon 24, 22−26.

Nishimoto, Y., Ohnishi, O., Hasegawa, M., 2003. Topological incongruence between nuclear and chloroplast DNA trees suggesting hybridization in the urophyllum group of the genus *Fagopyrum* (Polygonaceae). Genes Genet. Syst. 78, 139−153.

Ohnishi, O., 1995. Discovery of new *Fagopyrum* species and its implication for the studies of evolution of *Fagopyrum* and of the origin of cultivated buckwheat, Proc. 6th Intl. Symp. Buckwheat at Ina, 175−190.

Ohnishi, O., 1998. Search for the wild ancestor of buckwheat. I. Description of new *Fagopyrum* (Polygoneae) species and their distribution in China and the Himalayan hills. Fagopyrum 15, 18−28.

Ohnishi, O., 2016. Molecular taxonomy of the genus *Fagopyrum*. In: Zhou, M., Kreft, I., Woo, S.H., Chrungoo, N., Weislander, G. (Eds.), Molecular Breeding and Nutritional Aspect of Buckwheat. Elsevier, Amsterdam, pp. 1−12. Chapt. 1.

Ohnishi, O., Matsuoka, Y., 1996. Search for the wild ancestor of buckwheat. II. Taxonomy of *Fagopyrum* (Polygonaceae) species based on morphology, isozymes and cpDNA variability. Genes Genet. Syst. 71, 383−390.

Ohsako, T., Ohnishi, O., 1998. New *Fagopyrum* species revealed by morphological and molecular analyses. Genes Genet. Syst. 73, 85−94.

Ohsako, T., Ohnishi, O., 2000. Intra- and inter-specific phylogeny of wild *Fagopyrum* (Polygonaceae) species based on nucleotide sequences of noncoding regions in chloroplast DNA. Amer. J. Botany 87, 573−582.

Ohsako, T., Ohnishi, O., 2002. Two new *Fagopyrum* (Polygonaceae) species, *F. gracilipedoides* and *F. jinshense* from Yunnan, China. Genes Genet. Syst. 77, 399−408.

Shao, J.R., Zhou, M.L., Zhu, X.M., Wang, D.Z., Bai, D.Q., 2011. *Fagopyrum wenchuanense* and Fagopyrum *qingcai*, two new species of Polygonaceae from Sichuan. China Novon 21, 256−261.

Tang, Y., Zhou, M.L., Bai, D.Q., Shao, J.R., Zhu, X.M., Wang, D.Z., et al., 2010. *Fagopyrum pugense* (Polygonaceae), a new species from Sichuan, China. Novon 20, 239−242.

Yasui, Y., Ohnishi, O., 1998a. Interspecific relationships in *Fagopyrum* (Polygonaceae) revealed by the nucleotide sequences of the *rbcL* and *accD* genes and their intergenic region. Am. J. Botany 85, 1134−1142.

Yasui, Y., Ohnishi, O., 1998b. Phylogenetic relationships among *Fagopyrum* species revealed by the nucleotide sequences of the ITS region of the nuclear rRNA gene. Genes Genet. Syst. 3, 201−210.

Zhou, M., Wang, C., Wang, D., Zheng, Y., Li, F., Zhu, X., et al., 2014. Phylogenetic relationship of four new species related to southwestern Sichuan *Fagopyrum* based on morphological and molecular characterization. Biochem. Syst. Ecol. 57, 403−409.

Zhou, M.L., Bai, D.Q., Tang, Y., Zhu, X.M., Shao, J.R., 2012. Genetic diversity of four new species related to southwestern Sichuan buckwheat as revealed by karyotype, ISSR and allozyme characterization. Plant Syst. Evol. 298, 51−55.

# Chapter Twelve

# Useful Genetic Resources Among the Wild Species of Buckwheat

**Ohmi Ohnishi**

*Kyoto University, Kyoto-fu, Japan*

## INTRODUCTION

Analyses of genetic resources for future breeding and making new cultivars for farmers have not been developed well in buckwheat. Farmers are still growing mostly local landraces which have been inherited from farmers' ancestors. Therefore, the evaluation of local landraces of common buckwheat and Tartary buckwheat is the first task for buckwheat breeders.

The author is not familiar with genetic resources of cultivated common buckwheat and of cultivated Tartary buckwheat. The author can, however, contribute to the analyses of genetic germplasm for future buckwheat breeding through the analyses of genetic resources among the wild buckwheat species as a specialist on the wild species of buckwheat.

This article will briefly summarize the possibility of the wild species as genetic resources for the future breeding of common and Tartary buckwheat.

As is well-known, the wild buckwheat species can be divided into two groups, (Ohnishi and Matsuoka, 1996; Ohnishi, 2010, 2011, 2016). One is the big seed group, the cymosum group, including two cultivated buckwheat, *Fagopyrum esculentum* and *Fagopyrum tataricum*. The wild species of this group are frequently crossable with cultivated buckwheat. Hence, several species of this group have been attempted to be used for breeding of buckwheat. The other group is the small seed group, the urophyllum group, which consists of more than 10 species. However, no species of this group is crossable with cultivated buckwheat, hence having never been used for breeding of

**97**

Buckwheat Germplasm in the World. DOI: https://doi.org/10.1016/B978-0-12-811006-5.00012-4
© 2018 Elsevier Inc. All rights reserved.

cultivated buckwheat. As a consequence, this article will mostly treat the wild species of the cymosum group.

## CLASSIFICATION OF THE WILD *FAGOPYRUM* SPECIES

As described by Ohnishi (2010, 2011, 2016), wild *Fagopyrum* species can be subdivided into two major groups, the cymosum group and the urophyllum group, as first suggested by Ohnishi and Matsuoka (1996). The cymosum group is the group of large seed species, including two cultivated species, *F. esculentum* and *F. tataricum*. This group includes pure wild species (*Fagopyrum cymosum, Fagopyrum homotropicum* Ohnishi = *F. esculentum* ssp. *homotropicum* Chen, 2012) and wild subspecies (*F. esculentum* ssp. *ancestrale* Ohnishi, *F. tataricum* ssp. *potanini* Batalin) which may be useful genetic germplasm for the future improvement of cultivated buckwheat. All the species in the urophyllum group have small seeds. Hence, at present, the use of the species in the urophyllum group as a breeding material seem to have no merit so far as the yield of cultivated buckwheat varieties are concerned. The impossibility of crossing the species of the urophyllum group and the species of the cymosum group leads to no use of the species of the urophyllum group for breeding of cultivated buckwheat. Future research may find great merit in the use of the wild species in the urophyllum group, e.g., disease resistance or cold tolerance, etc., or perhaps quite different merit such as use as an ornamental flower.

## USEFUL GENETIC GERMPLASM AMONG THE WILD SPECIES IN THE CYMOSUM GROUP

### Fagopyrum esculentum ssp. ancestrale

*Fagopyrum esculentum* ssp. *ancestrale* Ohnishi is the wild ancestor of the cultivated common buckwheat. It is a heterostylous self-incompatible species as cultivated common buckwheat. Natural populations of this subspecies are growing in southwestern China, i.e., Yunnan and Sichuan provinces, and eastern Tibet (see Figs. 12.1 and 12.2 for natural growing of *F. esculentum* ssp. *ancestrale* in the Dongyi river valley, see also Ohnishi, 2007).

This subspecies is easily crossable with the cultivated common buckwheat, hence this subspecies has potential as a genetic resource for future breeding of cultivated common buckwheat. At present, more than 30 populations of this subspecies have been collected from Yunnan and Sichuan provinces and eastern Tibet (see Fig. 12.1, of Ohnishi, 2013). Some of them were analyzed for allozyme, AFLP, and SSR variability (Ohnishi, 2009; Konishi et al., 2005; Konishi and Ohnishi, 2007). However, those studies are primarily for genetic relationships among the populations of the wild ancestor.

Table 12.1 shows genetic variability of the populations of the wild ancestor, measured by % of polymorphic loci, the average number of alleles/locus, the average heterozygosity. The data in Table 12.1 was

**FIGURE 12.1** *Fagopyrum esculentum* ssp. *ancestrale* is growing on mountain slopes covering an entire mountain at Zhouke, the Dongyi river valley.

**FIGURE 12.2** *Fagopyrum esculentum* ssp. *ancestrale* is growing on roadsides along the Dongyi river (the Dongyi river runs along the bottom of the valley).

obtained by studying 18 isozyme loci of 14 enzymes using starch gel electrophoresis (see Ohnishi and Nishimoto, 1988, for the procedures of electrophoresis and the enzymes studied).

As shown in Table 12.1, cultivated populations are more polymorphic (higher percent of polymorphic loci, higher number of alleles/locus and higher average heterozygosity), because of a larger population size in cultivated populations. Among the wild ancestral populations, the populations from the Dongyi river valley, i.e., the Dongyi and Yijie populations in Table 12.1, have as much variation as cultivated populations have. Hence, the search for useful traits in the ancestor populations in the Dongyi river valley should be practiced and the incorporation of the wild species' favorable traits (e.g., disease resistance) into cultivars should be tried.

**TABLE 12.1** Percentage of Polymorphic Loci (P), No. of Alleles/Locus, and Average Heterozygosity (He) of Cultivated and Wild Populations of Common Buckwheat

| Wild Ancestor Populations Gst = 0.173 | | | | Cultivated Populations Gst = 0.020 | | | |
|---|---|---|---|---|---|---|---|
| Population | P | A | He | Population | P | A | He |
| Jinan | 25.0 | 1.45 | 0.113 | Zhouba | 45.0 | 1.95 | 0.138 |
| Yongsheng | 35.0 | 1.45 | 0.134 | Zhebalong | 50.0 | 2.00 | 0.144 |
| Yanyuan | 45.0 | 1.90 | 0.123 | Batang | 50.0 | 1.95 | 0.149 |
| Boke | 50.0 | 1.90 | 0.117 | Yanjing | 50.0 | 2.10 | 0.139 |
| Dongyi | 60.0 | 2.10 | 0.178 | Deqin | 45.0 | 2.05 | 0.154 |
| Yijie | 60.0 | 2.05 | 0.181 | Derong | 50.0 | 2.00 | 0.160 |
| Nidong | 40.0 | 1.70 | 0.098 | Xiancheng | 50.0 | 2.05 | 0.149 |
| Adong | 40.0 | 1.70 | 0.107 | Meirl | 50.0 | 2.00 | 0.153 |
| Yanjing | 50.0 | 2.05 | 0.130 | Boke | 50.0 | 1.95 | 0.154 |
| Xihe | 40.0 | 1.85 | 0.113 | Muli | 50.0 | 1.95 | 0.149 |
| Zheke | 40.0 | 1.70 | 0.121 | Yanyuan | 45.0 | 1.85 | 0.141 |
| | | | | Yongsheng | 45.0 | 1.85 | 0.141 |
| Average | 44.1 | 1.80 | 0.129 | Average | 48.3 | 1.97 | 0.149 |

Ohnishi (2009) showed that the natural populations of the ancestral subspecies can be divided into three groups. One is a group of the populations from the San Jiang area (Nidong, Adong, Yanjing, Xihe, and Zheke populations in Table 12.1), which are most closely related to cultivated populations of common buckwheat. Hence, Ohnishi (2009) can conclude that the original birthplace of cultivated common buckwheat is the Sanjiang area. The second group is the populations from the Dongyi river valley of Sichuan province and from the Nyiru river valley of Yunnan province (Dongyi and Yijie populations in Table 12.1), which are genetically most variable, but are genetically far away from cultivated populations of common buckwheat. Populations from Muli and Yanyuan districts of Sichuan province and some populations from Deqin district of Yunnan province (Jinan, Yongsheng, Yanyuan, and Boke populations in Table 12.1) consist of the third group which is characterized by the fact that populations of this group have the same amount of genetic variability as the populations from the Sanjiang area and have intermediate genetic relationships with cultivated populations of common buckwheat between the previous two groups (see Ohnishi, 2009 for more detail). Thus, the natural populations of the wild ancestor from the Dongyi river valley or from the Nyiru river valley should be searched for useful genetic resources for breeding better cultivars, because they are rich in genetic variation.

However, so far as the author knows, the wild ancestor has never been used for improving cultivars of common buckwheat. Evaluation of the populations of the subspecies have never practiced systematically, except for allozyme variability of natural populations in the search of the original birthplace of cultivated common buckwheat (see Ohnishi, 2009).

## *F. tataricum* ssp. *potanini,* the Wild Ancestor of Cultivated Tartary Buckwheat

*Fagopyrum tataricum* ssp. *potanini* Batalin is the wild ancestor of cultivated Tartary buckwheat. It was found by a Russian explorer and botanist G. Potanin in Gansu province, China (see Bredtschneider, 1898). It is a homostylous self-fertilizing species, just as cultivated Tartary buckwheat, and it is crossable with cultivated Tartary buckwheat. Yet, crossing this wild subspecies with cultivated Tartary buckwheat has never been practiced because of the technical difficulty of crossing them.

The natural populations of wild Tartary buckwheat from Tibet are similar to the cultivated populations of Tartary buckwheat, so far as allozyme variations are concerned (Ohnishi, 2000, 2002). Natural populations of wild Tartary buckwheat from Sichuan and Yunnan provinces have more variations than the natural populations from Tibet (Ohnishi, 2000, 2002). This is clearly shown in Fig. 12.3, where the variations of allozyme frequency at the polymorphic locus, *Pgm-2* are figured. This implies that the Tatary buckwheat

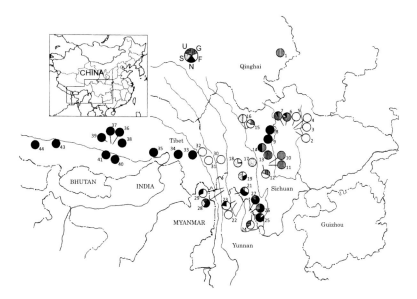

**FIGURE 12.3** Geographical distribution of the alleles at the polymorphic locus, *Pgm-2.* See Ohnishi (2002) for the identification of populations.

samples from Sichuan and Yunnan provinces have a higher potential to be used as a breeding material for improving cultivated Tartary buckwheat. At present, there is no attempt to improve cultivated Tartary buckwheat by incorporating the wild ancestor's favorable traits by crossing them.

Tartary buckwheat, in general, has a higher yield and a higher content of flavonoids important for functional food than common buckwheat. However, almost all cultivars of Tartary buckwheat have thick shell and deep grooves, and these characters make dehulling of Tartary buckwheat difficult. Recently, rice Tartary buckwheat was of interest to buckwheat breeders, because of its trait of easy dehulling. Only a few cultivars of rice Tartary buckwheat are cultivated in southern China, and in the Himalayan hills also on only small scale, because of their low yield.

No wild plant of rice Tartary buckwheat is known at present. This probably implies that rice Tartary buckwheat was differentiated from cultivated Tartary buckwheat after the domestication of cultivated Tartary buckwheat. Recently, Chen (2016) got hybrid rice Tartary buckwheat which has a higher yield and a good adaptability to hot climate in the progeny of rice Tartary buckwheat crossed with ordinary Tartary buckwheat. Similar attempts should be tried for better ordinary Tartary buckwheat and better rice Tartary buckwheat.

## *Fagopyrum homotropicum,* the Wild Self-Fertile Species

*Fagopyrum homotropicum* is a homostylous self-fertile new species found by Ohnishi (1995). There are both diploid and tetraploid forms distributed in western Sichuan (Aba and Dalingshan states) and northwestern Yunnan province, and eastern Tibet (see Ohnishi, 1995, 2013, Ohnishi and Asano, 1999, for the distribution areas of *F. homotropicum*). The distribution area of this species is slightly wider than that of the wild ancestor of common buckwheat. *F. homotropicum* is adapted to slightly a cooler and higher altitude climate than *F. esculentum* ssp. *ancestrale*.

Because of self-fertile property of this species, this species was used to make a new cultivated buckwheat, "self-fertile common buckwheat." The idea is as follows: crossing *F. homotropicum* with cultivated common buckwheat, and making subsequent backcrossing of the hybrids with cultivated common buckwheat, choosing self-fertile individuals at each generation of backcrossing. After a great enough number of generations, self-fertile common buckwheat may be obtained. In Canada, Campbell (1995), Wang and Campbell (1998), and Woo et al. (1999), and in Japan, Matsui et al. (2003, 2008), have reported successful inter-specific hybridization between diploid *F. esculentum* ($2n = 16$) and diploid *F. homotropicum* ($2n = 16$). These crosses improved agronomic characters of descendants of hybrids of the cross. At the same time, the self-compatibility character has been introgressed from *F. homotropicum* into *F. esculentum*. The percentage of successful crosses between *F. homotropicum* and *F. esculentum* is highly dependent on the population (the experimental linc) of *F. homotropicum* used in the crosses. This may imply that the origin of *F. homotropicum* from *F. esculentum* ssp. *ancestrale* has multiple origins, different *F. homotropicum* lines originated from different lines of *F. esculentum* ssp. *ancestrale*.

In spite of successive backcrosses of more than five generations, economically sufficiently good self-fertile common buckwheat lines have not yet obtained, because unfavorable wild characters which were inherited from the wild *F. homotropicum* used at the beginning of the cross still remain in self-fertile common buckwheat. Wang and Campbell (1998) and Wang et al. (2001) developed new hybrids between *F. esculentum* ($2n = 16$) and tetraploid *F. homotropicum* ($2n = 32$). In this case, sterility of descendants of hybrids caused by chromosome unbalance, in spite of the rescued hybrids by chromosome doubling, resulted in poor limited results. Choice of *F. homotropicum* lines in the experiment, selection of favorable self-fertile individuals in successive backcross generations, and continuation of long-enough generations of backcrosses are the main issues to be overcome for obtaining useful self-fertile common buckwheat.

### *Fagopyrum cymosum*, Wild Perennial Buckwheat

*Fagopyrum cymosum* (see Fig. 12.4) is a heterostylous self-incompatible perennial species. Once this species was considered as the wild ancestor of cultivated common buckwheat. However, this species is now believed to be more closely related with *F. tataricum* rather than *F. esculentum* based on knowledge from molecular analyses (Kishima et al., 1995, Ohnishi, 2010, 2016). This species has both diploid and tetraploid forms (see Ohnishi, 2016) and has the most wide distribution area among the wild buckwheat species, from southern China to the Himalayan hills (Ohnishi, 2013). This species has many morphological variants, among them *Fagopyrum pilus* Chen and

**FIGURE 12.4** *Fagopyrum cymosum* (4x) (*Fagopyrum dibotris*) growing at the margin of a cultivated field in Nepal.

*Fagopyrum megaspartanum* Chen were once registered as a new species (Chen, 1999), but now Prof. Chen himself (Chen, 2012) calls them "the cymosum complex" including variants related with *F. cymosum*, e.g., *F. cymosum, F. pilus, F. megaspartanum, Fagopyrum divotris,* and others.

Now *F. cymosum* is very important as a parent of interspecific crosses among the species of the cymosum group of *Fagopyrum.* Woo et al. (2016) showed ease and difficulty of interspecific crosses among species of the cymosum group. Recently, Suvorova (2016) reviewed the results of inter-specific crosses among species belonging to the cymosum group. *Fagopyrum cymosum* is the key species for interspecific crosses among species of the cymosum group.

*Fagopyrum cymosum* is most closely related with the cultivated Tartary buckwheat as described above. Hence it is natural that the first inter-specific hybrid in the genus *Fagopyrum* was an artificial allopolyploid *Fagopyrum giganteum* Krotov obtained as a result of a conventional crossing between tetraploid *F. tataricum* and *F. cymosum* (Krotov and Golubeva, 1973). The allopolyploid nature of the new hybrid was confirmed by Krotov (1975). *F. gigantum* was annual, and morphologically looks like *F. tataricum. F. gigantum* has a good fertility, and is morphologically stable, but does not have any economic profit. Hence, *F. gigantum* is interesting only as a new interspecific hybrid. Recently, Chen (2016) made inter-specific crosses between *F. cymosum* and *F. tataricum,* and searched for fertile inter-specific hybrids that are morphologically similar to *F. cymosum.* He obtained 20 lines at $F_6$ generation (Ren and Chen, 2016), which have woody high plants with large achenes, like perennial Tartary buckwheat. Ren and Chen (2016) named this plant *F. tatari-cymosum* Chen as a new species.

In conclusion of reviews of interspecific hybrids among species in the cymosum group, Suvorova (2016) described that the development of a biotechnological method of embryo rescue technique, on the one hand, and a discovery of a new species *F. homotropicum* (Ohnishi, 1995), on the other hand, extended the possibilities of the interspecific hybridization of buckwheat. In fact, interspecific hybrids were successfully obtained (for examples see Ujihara et al. 1990, and Suvorova et al., 1994 for *F. esculentum* crossed with *F. cymosum*; Suvorova, 2001, 2010 for trispecific hybrids of *F. esculentum, F. cymosum,* and *F. homotropicum*). However, most of the hybrids were aborted in later generations, and no economically useful inter-specific hybrids have been obtained.

## *Fagopyrum cymosum* ssp. *pillus* = *Fagopyrum pillus* Chen

*Fagopyrum cymosum* ssp. *pillus* (or *F. pillus* Chen) was found in eastern Tibet by Chen (1999) and Tsuji et al. (1999), independently. This subspecies has a unique morphology, having achenes similar to those of wild Tartary buckwheat, and also has a unique distribution area (see Fig. 12.5). In other aspects, it has similar morphology with *F. cymosum.* Molecular taxonomic results showed that this new subspecies lies within the scope of *F. cymosum,*

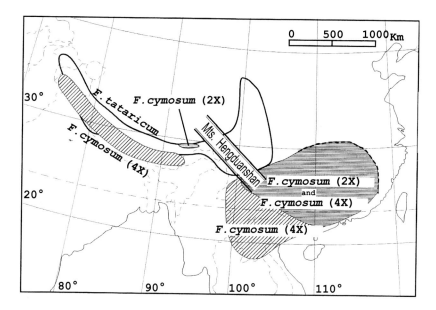

**FIGURE 12.5** Geographic distribution of diploid and tetraploid *Fagopyrum cymosum* and wild *Fagopyrum tataricum* in southern China and the Himalayan hills. *F. cymosum* (2x) and *F. cymosum* (4x) in the west of Mts. Hengduanshan are *Fagopyrum pillus* and *Fagopyrum dibotris*, respectively. After Yamane, 2003.

shifting slightly in the direction of *F. tataricum* (Yamane et al., 2003). Hence, this new subspecies can be used as a bridge species in the cross of *F. tataricum* with *F. cymosum*. However, it is not clear that interspecific hybrids and their descendants show more favorable characters as breeding materials than *F. gigantum*.

## *Fagopyrum cymosum* Tetraploid Form in China and *Fagopyrum dibotris* Hara in the Himalayan Hills

The natural populations of *F. cymosum* distributed in the Himalayan hills are entirely tetraploid, and they are morphologically similar with tetraploid populations in southern China and Thailand. Hara (1966) classified *F. cymosum*-like plants on the Himalayan hills as new species *F. dibotris* Hara. Buckwheat scientists in India and Nepal followed him and have used the name *F. dibotris*.

Although *F. dibotris* and *F. cymosum*-4x in southern China have completely different geographical distributions (see Fig. 12.5), they are not so different that they should be treated as different species. In fact, Yamane (2003) showed that tetraploid *F. cymosum* is autotetraploid, consisting of the combinations of two from three types of the intron of *Adh* gene, S, L, and L'. She also showed that the difference between the east type and the west type (originated and now growing in east and west of the Hengduanshan mountains, that is, in southern China and in the Himalayan hills, respectively) can be recognized both in cpDNA and nDNA, but the differences are not so large, hence *F. dibotris*, the west type, should be treated as the same species as *F. cymosum* tetraploid form in China, the east type.

*Fagopyrum cymosum* 4x in the Himalayan hills, i.e., *F. dibotris*, propagates both with seeds and by runners and is well adapted to the climate in the Himalayan hills, and grows thick there. Farmers in the Himalayan hills use *F. dibotris* as cattle fodder as well as a medicinal plant.

## Genetic Resources Among the Species in the Uropyllum Group

Wild species in the urophyllum group can be divided into the following four groups (Ohnishi, 2011): (1) *Fagopyrum urophyllum* and associated herbaceous species, *Fagopyrum lineare* and *Fagopyrum jinshaense*; (2) Diploid-Tetraploid complex, *Fagopyrum capillatum* (2x)-*Fagopyrum gracilipes* (4x), *Fagopyrum gracilipedoides* (2x)-*Fagopyrum rubifolium* (4x); (3) *Fagopyrum statice-Fagopyrum leptopodum* complex; (4) Three new species in the upper Min river valley.

All these species have small achens covered with remaining perianths (Ohnishi and Matsuoka, 1996; Ohnishi, 2011). All species cannot be crossed with the cultivated species *F. esculentum* and *F. tataricum*. Hence, so far as the author knows, no trial has been conducted to make inter-specific hybrids between any species in the urophyllum group and the two cultivated species, *F. esculentum* and *F. tataricum*. Although the wild species in the urophyllum group have no hope of contributing to improved cultivated buckwheat, each species may have a role to contribute to our understanding of the evolutionary processes in *Fagopyrum*.

A description of the author's opinions on the role of each species in future studies follows below.

### FAGOPYRUM UROPHYLLUM AND ASSOCIATED SPECIES, FAGOPYRUM LINEARE AND FAGOPYRUM JINSHAENSE

*Fagopyrum urophyllum* is known as a woody perennial species. This species can be divided into two species, *F. uropyllum*-Dali and *F. urophyllum*-Kunming by the help of molecular classification (Kawasaki and Ohnishi, 2006). Two forms are not so different in morphological character, but they are so different at the molecular level that they should be treated as distinct species. Furthermore, both *F. urophyllum*-Dali and *F. urophyllum*-Kunming have an herbaceous annual species closely related at the molecular level, but quite different in morphology. *Fagopyrum urophyllum*-Dali has *F. lineare*, and *F. urophyllum*-Kunming has *F. jinshaense* as an associated species (closely related to each of *F. urophyllum* at the molecular level, but quite different in morphological characters).

*Fagopyrum urophyllum* has been considered as a primitive species having a long history of evolution (Ohnishi, 2011). The fact mentioned above may imply that the ancestral species of *F. lineare* and *F. jinshaense* was differentiated from *F. urophyllum* at the very early stages of evolution of *F. urophyllum*. *Fagopyrum urophyllum*-Dali and *F. urophyllum*-Kunming then evolved independently with the evolution of their associated herbaceous annuals. This speculation can be tested experimentally by analyzing suitable genes of these species and by comparing DNA sequences of each species. Solving the issue does not contribute to improving cultivated buckwheat, however it does offer a great contribution to our understanding of the evolutionary processes of the species in *Fagopyrum*.

## 2X−4X COMPLEX OF *Fagopyrum capillatum* AND OF *Fagopyrum gracilipedoides*

Each of the diploid species *F. capillatum* and *F. gracilipedoides* has genetically the closely related tetraploid species, *F. gracilipes* and *F. rubifolium*, respectively. As discussed in Ohnishi (2016), tetraploid species with similar a genome as diploid species are classified in the same species in the cymosum group, but classified in different species in the urophyllum group. Two 2x−4x pairs are genetically closely related in a phylogenetic tree as Fig. 12.2 of Ohnishi (2011) shows. What is meant by this observation? At present, since we do not know whether tetraploid species *F. gracilipes* and *F. rubifolium* are auto-tetraploid or allo-tetraploid, the author's speculation based on these observations is not a sophisticated one, however we may draw a conclusion that chromosome doubling may not occur randomly, but rather occur in specific lineages of phylogeny. There remains a fact which is difficult to be understood. Why do such tall diploid plants, *F. capillatum* and *F. gracilipedoides*, become such small plants, *F. gracilipes* and *F. rubifolium*, respectively, when diploid plants become tetraploid?

In *Fagopyrum*, the plants and seeds of tetraploids seem smaller than those of diploids. This is also true for *F. cymosum*, a species of the cymosum group.

## *Fagopyrum statice−Fagopyrum leptopodum* COMPLEX

Ohsako and Ohnishi (2001) analyzed in detail the species in this group.

Perennial *F. statice*-Yilang (*F. statice* from Yilang, Yunnan province) is the primitive type of this group and *F. statice*-Yuanmao is a derivative from *F. statice*-Yilang, judging from their distribution areas. *F. statice*-Yuanmao distributes only in a small area in the district of Yuanmao, in central Yunnan province. *Fagopyrum leptopodum* has a rather wider area of distribution—the northwestern Yunnan province and the southern Sichuan province. *F. gilesii* is morphologically similar to *F. leptopodum*, except in its cymes, but has a quite different distribution area from that of *F. statice* and *F. leptopodum* (see Fig. 12.6).

## THREE NEW SPECIES IN THE UPPER MIN RIVER VALLEY

These three species are geographically isolated from other *Fagopyrum* species and they also situated at a quite different position in the phylogenetic tree (see Fig. 12.6 and Ohnishi, 2011). This is a reasonable result, and the quite different position in distribution implies a quite different position in the

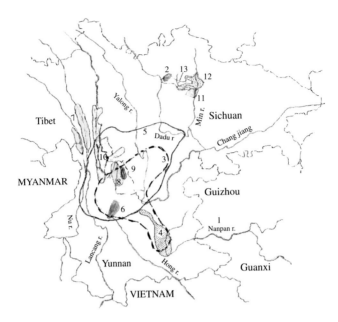

**FIGURE 12.6** Distribution of wild buckwheat species of the urophyllum group.
1. *Fagopyrum gracilipes* distributes almost everywhere on this map. 2. *Fagopyrum rubifolium*. 3. *Fagopyrum urophyllum*. 4. *Fagopyrum statice* 5. *Fagopyrum leptopodum*. 6. *Fagopyrum lineare*. 7. *Fagopyrum gilesii*. 8. *Fagopyrum capillatum*. 9. *Fagopyrum gracilipedoides*. 10. *Fagopyrum jinshaense*. 11. *Fagopyrum pleioramosum*. 12. *Fagopyrum callianthum*. 13. *Fagopyrum macrocarpum*

phylogenetic tree. This means that a diffusion of the ancestor of these three species may occur at the very early stages of evolution of the genus *Fagopyrum*. More detailed analyses on the phylogenetic tree of these species and on the distribution areas may lead to the assessment of the time of diffusion from the center of distribution and to the estimation of the speed of diffusion.

## THE NEWLY FOUND SPECIES BY CHINESE SCIENTISTS

There are several new species claimed by Chinese scientists, however, their taxonomic positions are entirely unclear (see Ohnishi, 2016). They are *Fagopyrum caudatum* (Ye and Guo, 1992), *Fagopyrum crispatofolium* (Liu et al., 2008), *Fagopyrum pugense* (Tang et al., 2010), *Fagopyrum wenchuanense* and *Fagopyrum qiangcai* (Shao et al., 2011), *Fagopyrum luojishanense* (Hou et al., 2015), and *Fagopyrum hailuogouense* (Zhou et al., 2015).

Among the new species found by Chinese scientists, *Fagopyrum zwogongense* is only the species in the cymosum group. According to Chen (1999) and Chen (2012)'s description, it would seem that *Fagopyrum zuogongense* is the tetraploid form of *F. homotropicum*. *F. zuogongense*, a homostyly self-fertile species, looks like *F. esculentum* ssp. *ancestrale* in morphology. The distribution area of *F. homotropicum* covers eastern Tibet including Zuogong district (see Ohnishi, 2013)—where *F. zuogongense* was found by Chen (1999). From these descriptions one can conclude that *F. zugongense* is a tetraploid form of *F. homotropicum*. Dr. Meiliang Zhou and his colleagues recently discovered four new wild species, *F. pugense, F. wenchuanense, F. qiangcai*, and *F. crispatofolium*. Assessment of the phylogenetic positions of these new species showed that *F. wenchuanense* belongs to the cymosum group, while *F. qiangcai, F. pugense*, and *F. crispatifolium* are classified into the urophyllum group (Zhou et al., 2014). However, further molecular studies will provide a better understanding of the evolutionary mechanisms and genetic relationships of cultivated and recently discovered wild *Fagopyrum* species.

Fig. 12.6 showed the distribution of wild buckwheat species of the urophyllum group around the center of the distributions of wild buckwheat species—the area of northwest corner of Yunnan province and the southwest corner of Sichuan province. Comparing this figure and the table for the distributions of wild species in Sichuan (Tang et al., 2016), the author found several discrepancies in the distributions of such wild species as *F. lineare* and *F. esculentum* ssp. *ancestrale*. Efforts to make a complete distribution map of the wild buckwheat species must be made by buckwheat scientists in the next generation. By doing so, the wild buckwheat species which cannot contribute to increased yields of cultivated buckwheat make great contributions to our deep understanding on the evolutionary processes of buckwheat species, including cultivated buckwheat *F. esculentum* and *F. tataricum*.

## REFERENCES

Bredtschneider, E., 1898. History of European Botanical Discovery in China. Press Imp. Acad. Sci., Petersburg.

Campbell, C.G., 1995. Inter-specific hybridization in the genus *Fagopyrum*. Proc. 6th Intl. Symp. Buckwheat at Ina, 255−263.

Chen, Q.F., 1999. A study of resources of *Fagopyrum* (Polygonaceae) native to China. Bot. J. Linnean Soc. 130, 53−64.

Chen, Q.F. (Ed.), 2012. Plant Sciences on Genus *Fagopyrum*. China Sci. Press, Beijing.

Chen, Q.F., 2016. Recent progresses on interspecific crossing of genus *Fagopyrum* Mill. Proc. 13th Intl. Symp. Buckwheat at Cheongju, pp. 285−298.

Hara, H., 1966. The Flora of East Himalaya. Tokyo University Press, Tokyo.

Hou, L.L., Zhou, M.L., Zhang, Q., Qi, L.P., Yang, X.B., Tang, Y., et al., 2015. *Fagopyrum luojishanense*, a new species of Polygonaceae from Sichuan. China Novon. 24, 22−26.

Kawasaki, M., Ohnishi, O., 2006. Two distinct groups of natural populations of *Fagopyrum urophyllum* (Bur. et Franch.) gross revealed by the nucleotide sequence of noncoding region in chloroplast DNA. Genes Genet. Syst. 81, 323−332.

Kishima, Y., Ogura, K., Mizukami, K., Mikami, T., Adaci, T., 1995. Chloroplast DNA analysis in buckwheat species: phylogenetic relationships, origin of the reproductive systems and extended inverted repeats. Plant Sci. 108, 173−179.

Konishi, T., Ohnishi, O., 2007. Close genetic relationship between cultivated and natural populations of common buckwheat in the Sanjiang area is not due to recent gene flow between them—an analysis using microsatellite markers. Genes Genet. Syst. 82, 53−64.

Konishi, T., Yasui, Y., Ohnishi, O., 2005. Original birthplace of cultivated common buckwheat inferred from genetic relationships among cultivated populations and natural populations of wild common buckwheat revealed by AFLP analysis. Genes Genet. Syst. 80, 113–119.

Krotov, A.S., 1975. Buckwheat–*Fagopyrum* Mill. Cultivated flora of USSR, Kolos. Lenningrad, pp. 7–118. (in Russian).

Krotov, A.S., Golubeva, E.A., 1973. Cytological studies on an interspecific hybrid *F. tataricum* × *F. cymosum*. Bull. Appl. Bot. Genet. Plant Breed. 51 (1), 256–260 (in Russian).

Liu, J.L., Tang, Y., Xia, Z.M., Shao, J.R., Cai, G.Z., Luo, Q., et al., 2008. *Fagopyrum crispatofolium* J. L. Liu, a new species of Polygonaceae from Sichuan, China. J. Syst. Evol. 46, 929–932.

Matsui, K., Tetsuka, T., Nishio, T., Hara, T., 2003. Heteromorphic incompatibility retained in self-compatible plants produced by a cross between common and wild buckwheat. New Phytol. 159, 701–708.

Matsui, K., Tetsuka, T., Hara, T., Morishita, T., 2008. Breeding and characterization of a new self-compatible common buckwheat (*Fagopyrum esculentum*) parental line, 'Buckwheat Norin-PL-1'. Bull. Natl. Agric. Res. Center Kyushu Okinawa Region 49, 11–17.

Ohnishi, O., 1995. Discovery of new *Fagopyrum* species and its implication for the studies of evolution of *Fagopyrum* and of the origin of cultivated buckwheat. Proc. 6th Intl. Symp. Buckwheat at Ina, 175–190.

Ohnishi, O., 2000. Geographical distribution of allozymes in natural populations of wild Tartary buckwheat. Fagopyrum 17, 29–34.

Ohnishi, O., 2002. Wild buckwheat species in the border area of Sichuan, Yunnan and Tibet and allozyme diversity of wild Tartary buckwheat in this area. Fagoyrum 19, 3–9.

Ohnishi, O., 2007. Natural populations of the wild ancestor of cultivated common buckwheat, *Fagopyrum esculentum* ssp. *ancestrale* from the Dongyi river valley—Their distribution and allozyme variations. Proc. 10th Intl. Symp. Buckwheat at Yangling, 13–18.

Ohnishi, O., 2009. On the origin of cultivated common buckwheat based on allozyme analyses of cultivated and wild populations of common buckwheat. Fagopyrum 26, 3–9.

Ohnishi, O., 2010. Distribution and classification of wild buckwheat species. 1. Cymosum group. Fagopyrum 27, 1–8.

Ohnishi, O., 2011. Distribution and classification of wild buckwheat species. 2. Urophyllum group. Fagopyrum 28, 1–8.

Ohnishi, O., 2013. Distribution of wild buckwheat species and perspective for their utilization. Fagopyrum 30, 9–14.

Ohnishi, O., 2016. Molecular taxonomy of the genus *Fagopyrum*. Chapter 1. In: Zhou, M., Kreft, I., Woo, S.H., Chrungoo, N., Weislander, G. (Eds.), Molecular Breeding and Nutritional Aspects of Buckwheat. Elsevier, Amsterdam, pp. 1–12.

Ohnishi, O., Asano, N., 1999. Genetic diversity of *Fagopyrum homotropicum*, a wild species related to common buckwheat. Genet. Resour. Crop Evol. 46, 389–398.

Ohnishi, O., Matsuoka, Y., 1996. Search for the wild ancestor of buckwheat. II. Taxonomy of *Fagopyrum* (Polygonaceae) species based on morphology, isozymes and cpDNA variability. Genes Genet. Syst. 72, 383–390.

Ohnishi, O., Nishimoto, T., 1988. Population genetics of cultivated common buckwheat, *Fagopyrum esculentum* Moench. V. Further studies on allozyme variability in the Indian and Nepali Himalaya. Jpn. J. Genet. 63, 51–66.

Ohsako, T., Ohnishi, O., 2001. Nucleotide sequence variation of the chloroplast *trnK/matK* region in two wild *Fagopyrum* (Polygonaceae) species, *F. leptopodum* and *F. statice*. Genes Genet. Syst. 76, 39–46.

Ren, C.Z., Chen, Q.F., 2016. Buckwheat Breed Records in China. China Agricultural Science and Technology Press, Beijing.

Shao, J.R., Zhou, M.L., Zhu, X.M., Wang, D.Z., Bai, D.Q., 2011. *Fagopyrum wenchuanense* and *Fagopyrumqin qincai*, two new species of Polygonaceae from Sichuan, China. Novon 21, 256−261.

Suvorova, G.N., 2001. The problem of interspecific crosses of *Fagopyrum esculentum* Moench x *Fagopyrum cymosum* Meisn. Proc. 8th Intl. Symp. Buckwheat at Chunchon, 311−318.

Suvorova, G.N., 2010. Perspectives of interspecific buckwheat hybridization. Proc. 11th Intl. Symp. Buckwheat at Orel, pp. 295−299.

Suvorova, G., 2016. Interspecific crosses in buckwheat breeding. Chapter 7. In: Zhou, M., Kreft, I., Woo, S.H., Chrungoo, N., Weislander, G. (Eds.), Molecular Breeding and Nutritional Aspects of Buckwheat. Elsevier, Amsterdam, pp. 87−97.

Suvorova, G.N., Fesenko, N.N., Kostrubin, M.M., 1994. Obtaining of interspecific buckwheat hybrid (*Fagopyrum esculentum* Moench. x *F. cymosum* Meisn.). Fagopyrum 14, 13−16.

Tang, Y., Zhou, M.L., Bai, D.Q., Shao, J.R., Zhu, X.M., Wang, D.Z., et al., 2010. *Fagopyrum pugense* (Polygonaceae), a new species from Sichuan, China. Novon 20, 239−242.

Tang, Y., Shao, J., Zhu, X., Ding, M., Zhou, M., 2016. The investigation and collection of wild buckwheat in Sichuan China. Proc. 13th Intl. Symp. Buckwheat at Chengju, 99−105.

Tsuji, K., Yasui, Y., Ohnishi, O., 1999. Search for *Fagopyrum* species in eastern Tibet. Fagopyrum 16, 1−6.

Ujihara, A., Nakamura, Y., Minami, M., 1990. Interspecific hybridization in genus *Fagopyrum*−properties of hybrid (*F. esculentum* Moench. x *F. cymosum* Meisn.) through ovule culture. GammaField Symposium 29, 45−53.

Yamane, K., 2003. Evolutionary processes of polyploidization and genetic differentiation in perennial buckwheat (*Fagopyrum cymosum* Meisn.) revealed by molecular markers. Ph.D. thesis at Kyoto University, Kyoto.

Yamane, K., Yasui, Y., Ohnishi, O., 2003. Interspecific cpDNA variations of diploid and tetraploid perennial buckwheat, *Fagopyrum cymosum* (Polygonaceae). Amer. J. Bot. 90, 339−346.

Ye, N.G., Guo, G.Q., 1992. Classification, origin and evolution of genus *Fagopyrum* in China. Proc. 5th Intl. Symp. Buckwheat at Taiyuan, 19−28.

Wang,Y.J., Campbell, C., 1998. Interspecific hybridization in buckwheat among *Fagopyrum esculentum, F. homotropicum* and *F. tataricum*. Proc. 7th Intl. Symp. Buckwheat at Winnipeg, 1−13.

Wang, Y.J., Scarth, R., Campbell, C., 2001. Chromosome variations in interspecific hybrids between, *F. esculentum* (2n = 2x = 16) and *F. homotropicum* (2n = 4x = 32). Proc. 8th Intl. Symp. Buckwheat at Chunchon, 301−307.

Woo, S.H., Adachi, T., Jong, S.K., Campbell, C.G., 1999. Inheritance of self-compatibility and flower morphology in an inter-specific buckwheat hybrid. Can. J. Plant Sci. 79, 483−490.

Woo, S.H., Adachi, T., Park, C.H., Campbell, C., 2016. Possibilities of interspecific hybridization in the genus *Fagopyrum*. Proc. 13th Intl. Symp. Buckwheat at Cheongju, 267−273.

Zhou, M., Wang, C., Wang, D., Zheng, Y., Li, F., Zhu, X., et al., 2014. Phylogenetic relationship of four new species related to southwestern Sichuan *Fagopyrum* based on morphological and molecular characterization. Biochem. Syst. Ecol. 57, 403−409.

Zhou, M.L., Zhang, Q., Zheng, Y.D., Tang, Y., Li, F.L., Zhu, X.M., et al., 2015. *Fagopyrum hailuogouense* (Polygonaceae), one new species from Sichuan, China. Novon. 24, 222−224.

# FURTHER READING

Ohnishi, O., 2012. On the distribution of natural populations of wild buckwheat species. Fagopyrum 29, 1−6.

# Chapter Thirteen

# Utilization of Wild Buckwheat Species

**Meiliang Zhou[1], Yu Tang[2], Xianyu Deng[3], Chen Ruan[3], Mengqi Ding[4], Jirong Shao[4], Yixiong Tang[5], and Yanmin Wu[5]**

[1]Institute of Crop Sciences, Chinese Academy of Agricultural Sciences, Beijing, China, [2]Sichuan Tourism University, Chengdu, China, [3]Kunming University of Science and Technology, Kunming, Yunnan, China, [4]Sichuan Agricultural University, Yaan, China, [5]Biotechnology Research Institute, Chinese Academy of Agricultural Sciences, Beijing, China

The roots, stems, leaves, flowers, and fruits of wild buckwheat are rich in proteins, amino acids, carbohydrates, minerals, flavonoids, and other nutritional and medical composition, which have the efficacy of clearing heat and detoxicating, nourishing the spleen and eliminating dampness, inhibiting the invasion of tumor cells and preventing and curing cardiovascular disease and diabetes. Therefore wild buckwheat is an important resource plant that has great value for development and utilization.

## THE NUTRITIONAL VALUES

### The Protein Contents

Tang et al. (2011) measured the protein contents of three wild buckwheat species (*Fagopyrum cymosum*, *Fagopyrum gracilipes*, and *Fagopyrum urophyllum*), and two cultivars (*Fagopyrum esculentum* and *Fagopyrum tataricum*). The protein content of *F. cymosum* is the highest among the five buckwheat species, and can reach 141 mg/g, while *F. urophyllum* comes second at 129 mg/g. There's not much difference (in protein content) between *F. gracilipes* and *F. esculentum*, which come third (118 mg/g) and fourth (115 mg/g) respectively. The protein content of *F. gracilipes* is the lowest at 106 mg/g. The protein contents of wild buckwheat are higher than those of buckwheat cultivars. Zhang et al. (1999) analyzed the proteins in two grain samples of *F. cymosum* that come from Hunan and Guizhou provinces, China, and the

**113**

Buckwheat Germplasm in the World. DOI: https://doi.org/10.1016/B978-0-12-811006-5.00013-6
© 2018 Elsevier Inc. All rights reserved.

protein contents in both the samples were higher than 126 mg/g, reaching the protein content level of quality cereal crops. Zhao et al. (2002) analyzed the protein contents in three grain samples of *F. cymosum* that come from Sichuan, Yunnan, and Guizhou, whose protein contents were 131 mg/g, 128 mg/g, and 125 mg/g respectively and all of which were higher than that of the buckwheat cultivars *F. tataricum* (115 mg/g)and *F. esculentum* (98 mg/g). Lu et al. (1996) analyzed the protein contents in roots, stems, and leaves of *F. cymosum*, showing the contents of crude protein in roots, stems, and leaves are 42 mg/g, 51 mg/g, and 87 mg/g, respectively. These researches fully illustrate that the protein contents of wild buckwheat (especially that of *F. cymosum*) are not only higher than those of other cereal crops, but also than those of buckwheat cultivars.

Moreover, Zhang et al. (1999) also analyzed the protein fractions in grains of *F. cymosum*, and the results ranking in descending order are as follows: the content of globulin (70.9 mg/g) is the highest, accounting for 56.3% of the total protein content; the content of albumin is 23.8 mg/g, which accounts for 18.9% of the total protein content; the content of glutelin is 12.3 mg/g, which accounts for 9.8% of the total protein content; and the content of prolamine is 4.7 mg/g, which accounts for 3.7% of the total protein content. In addition, the content of insoluble protein in grains of *F. cymosum* is 14.4 mg/g, which accounts for 11.4% of the total protein content. The protein fraction of *F. cymosum* is similar to that of buckwheat cultivars, both of which have no gluten, have low viscosity, and are hard to form into elastic and malleable dough.

## The Contents and Compositions of Amino Acids

Tang et al. (2011) reported the contents and compositions of amino acids in grains of three wild buckwheat species (*F. cymosum, F. gracilipes* and *F. urophyllum*) and two cultivars (*F. esculentum* and *F. tataricum*) (Table 13.1). The five kinds of buckwheat grains contain 18 kinds of amino acids, while the total content of amino acids in *F. cymosum* is the highest, and can reach 13.72%. *Fagopyrum urophyllum* comes second, whose total content of amino acids is 12.24%. There's not much difference (in total content of amino acids) between *F. esculentum* and *F. gracilipes*, which come third (11.06%) and fourth (10.99%), respectively, and that of *F. tataricum* is the lowest, at 10.21%. What is the most noteworthy is that all the contents of eight essential amino acids for the human body (threonine, valine, methionine, isoleucine, leucine, phenylalanine, lysine, and tryptophan) of wild buckwheat are higher than those of buckwheat cultivars, and the results ranking in descending order are as follows: *F. cymosum* (4.64%), *F. urophyllum* (4.15%), *F. gracilipes* (3.78%), *F. esculentum* (3.73%) and *F. tataricum* (3.37%).

Zhao et al. (2002) reported the amino acids' compositions and contents in three grain samples of *F. cymosum* from Sichuan, Yunnan, and Guizhou, whose total contents of 18 amino acids are 11.51%, 11.17%, and 11.0% respectively, and the average value is 11.23%, which is higher than those of

**TABLE 13.1** Comparison of Contents (%) of 18 Types of Amino Acids in Seeds of 5 Buckwheat Species

| | Species | | | | |
|---|---|---|---|---|---|
| | Fagopyrum esculentum | Fagopyrum tataricum | Fagopyrum cymosum | Fagopyrum gracilipes | Fagopyrum urophyllum |
| Threonine | 0.48 | 0.41 | 0.59 | 0.52 | 0.58 |
| Valine | 0.54 | 0.46 | 0.73 | 0.53 | 0.60 |
| Methionine | 0.18 | 0.29 | 0.16 | 0.17 | 0.16 |
| Isoleucine | 0.41 | 0.34 | 0.55 | 0.46 | 0.51 |
| Leucine | 0.78 | 0.72 | 0.95 | 0.77 | 0.84 |
| Phenylalanine | 0.52 | 0.46 | 0.67 | 0.56 | 0.60 |
| Lysine | 0.73 | 0.61 | 0.88 | 0.68 | 0.76 |
| Tryptophan | 0.09 | 0.08 | 0.11 | 0.09 | 0.10 |
| Aspartic acid | 1.12 | 0.98 | 1.49 | 1.06 | 1.19 |
| Serine | 0.58 | 0.35 | 0.74 | 0.61 | 0.70 |
| Glutamic acid | 1.92 | 2.01 | 2.39 | 1.83 | 2.04 |
| Glycine | 0.69 | 0.48 | 0.83 | 0.66 | 0.73 |
| Alanine | 0.57 | 0.51 | 0.66 | 0.57 | 0.63 |
| Cystine | 0.24 | 0.29 | 0.27 | 0.25 | 0.30 |
| Tyrosine | 0.31 | 0.28 | 0.38 | 0.32 | 0.35 |
| Histidine | 0.30 | 0.28 | 0.39 | 0.31 | 0.35 |
| Arginine | 1.11 | 1.24 | 1.33 | 1.13 | 1.28 |
| Proline | 0.48 | 0.42 | 0.60 | 0.47 | 0.52 |
| Total amount of amino acids | 11.06 | 10.21 | 13.72 | 10.99 | 12.24 |

*F. tataricum* and *F. esculentum* (10.69% and 10.52%). Moreover, the contents of most amino acids in these three grain samples are higher than those of *F. tataricum* and *F. esculentum*.

Lu et al. (1996) reported the compositions and contents of combined amino acids in the roots, stems, leaves, and flowers, etc., different organs of *F. cymosum* (Table 13.2).

From Table 13.2 it can be seen that while the roots of *F. cymosum* contain 17 kinds of amino acids, the stems, leaves, and flowers (of *F. cymosum*) contain 16 kinds of amino acids, but not cystine. The total contents of amino acids in roots, stems, leaves and flowers are 4.05%, 1.93%, 11.23%, and 9.86% respectively, whose results, ranking in descending order, are as follows: leaves > (greater than) flowers > roots > stems.

**TABLE 13.2** The Contents of Amino Acids in Different Organs of *Fagopyrum cymosum* (%)

| The Amino Acid Compositions | Roots | Stems | Leaves | Flowers |
|---|---|---|---|---|
| Aspartic acid | 0.18 | 0.17 | 1.19 | 0.95 |
| Threonine | 0.07 | 0.08 | 0.59 | 0.31 |
| Serine | 0.09 | 0.10 | 0.60 | 0.33 |
| Glutamic acid | 0.31 | 0.32 | 0.87 | 1.14 |
| Glycine | 0.09 | 0.10 | 0.70 | 0.56 |
| Alanine | 0.12 | 0.11 | 0.83 | 0.73 |
| Cystine | 0.04 | | | |
| Valine | 0.13 | 0.12 | 0.79 | 0.76 |
| Methionine | 0.07 | 0.06 | 0.18 | 0.19 |
| Isoleucine | 0.10 | 0.09 | 0.63 | 0.60 |
| Leucine | 0.16 | 0.15 | 1.16 | 0.98 |
| Tyrosine | 0.04 | 0.04 | 0.51 | 0.42 |
| Phenylalanine | 0.15 | 0.09 | 0.70 | 0.58 |
| Lysine | 0.16 | 0.13 | 0.78 | 0.77 |
| Histidine | 1.07 | 0.05 | 0.31 | 0.29 |
| Arginine | 1.13 | 0.11 | 0.69 | 0.65 |
| Proline | 0.14 | 0.21 | 0.70 | 0.60 |
| Total contents | 4.05 | 1.93 | 11.23 | 9.86 |

## The Minerals and Vitamins

Lu et al. (1996) reported the compositions and contents of various minerals in the roots, stems, leaves, and flowers of *F. cymosum* (Table 13.3). The roots, stems, leaves, and flowers of *F. cymosum* contain K, Fe, Cu, Zn, I, Mn, P, Ni, S, Ca, Ti, Mo, and Cr, etc. 24 kinds of minerals, among which the contents of K, P, Ca, Mg, Fe, and S etc. inorganic minerals are pretty high. The trace elements that are essential for the life activities of human beings and animals, such as Fe, Cu, Zn, Co, Mn, Ni, Cr, Mo, Se, I, and B, are in highest concentration in the leaves and flowers of *F. cymosum* and come second in the stems and roots.

Zhang et al. (1999) reported the mineral contents in the grains of *F. cymosum*. They measured the contents of Cu, Zn, Mn, Fe, Co, Cd, Cr, and Se, etc. 8 kinds of minerals, whose contents are 0.38 mg/100 g, 1.70 mg/100 g, 0.69 mg/100 g, 8.43 mg/100 g, 0.01 mg/100 g, 0.004 mg/100 g, 0.008 mg/100 g and 0.025 mg/100 g respectively.

**TABLE 13.3** The Mineral Contents in Different Organs of *Fagopyrum cymosum* (µg/g)

| Elements | Roots | Stems | Leaves | Flowers |
|---|---|---|---|---|
| K | 6869.97 | 13547.47 | 22799.97 | 14009.97 |
| P | 2953.81 | 1571.55 | 3052.06 | 4701.81 |
| Ca | 1063.14 | 6823.55 | 14893.00 | 7168.55 |
| Mg | 899.27 | 1431.52 | 5209.52 | 2732.77 |
| Fe | 191.38 | 86.75 | 1045.03 | 424.08 |
| S | 552.99 | 561.49 | 1181.49 | 1697.24 |
| Na | 299.81 | 33.59 | 100.86 | 40.06 |
| Al | 245.48 | 38.73 | 925.33 | 246.16 |
| Zn | 5.91 | 5.88 | 25.39 | 26.04 |
| Mn | 29.83 | 10.45 | 53.49 | 58.86 |
| B | 12.34 | 13.45 | 16.14 | 20.14 |
| Ba | 11.13 | 31.35 | 60.42 | 13.08 |
| I | 6.32 | 7.09 | 31.43 | 15.80 |
| As | 0.12 | 0.16 | 0.50 | 0.28 |
| Cd | 0.08 | 0.07 | 0.32 | 0.18 |
| Cr | 0.45 | 0.31 | 1.67 | 4.43 |
| Co | 0.10 | 0.11 | 0.47 | 0.27 |
| Cu | 4.14 | 2.19 | 7.28 | 8.98 |
| Hg | 0.05 | 0.09 | 0.27 | 0.16 |
| Mo | 1.46 | 0.80 | 1.83 | 1.47 |
| Ni | 0.44 | 0.25 | 2.25 | 2.00 |
| Pb | 0.23 | 0.25 | 2.18 | 0.95 |
| Se | 0.09 | 0.23 | 0.54 | 0.31 |
| Ti | 1.04 | 1.39 | 21.46 | 0.35 |

Zhang et al. (1999) also measured the contents of $VB_1$, $VB_2$, and VE, etc. in the grains of *F. cymosum*. The content of $VB_1$ (in the grains of *F. cymosum*) is 0.40 mg/100 g, the content of $VB_2$ is 0.25 mg/100 g, the content of VE is more than 1.66 mg/100 g and the content of niacin is 3.85 mg/100 g that are higher than that of *F. esculentum* (3.11 mg/100 g), *F. tataricum* (3.42 mg/100 g) or wheat flour (2.0 mg/100 g). Zhao et al. (2002) reported the compositions and contents of vitamins in the three grain samples of *F. cymosum* from Sichuan, Yunnan, and Guizhou, among which the average content of $VB_1$ is 0.45 mg/100 g, the average content of $VB_2$ is 0.29 mg/ 100 g, the average content of Vpp is 3.86 mg/100 g, and the average content of Vp is 1.39 mg/100 g. While the (average) contents of $VB_1$ and Vpp (in *F. cymosum*) are higher than those of *F. tataricum* (0.39 mg/100 g, 2.44 mg/ 100 g) and *F. esculentum* (0.31 mg/100 g, 2.67 mg/100 g), the (average) contents of $VB_2$ and Vp (in *F. cymosum*) are lower than those of *F. tataricum* (0.37 mg/100 g, 1.51 mg/100 g), but well above than those of *F. esculentum* (0.24 mg/100 g, 0.14 mg/100 g).

# THE MAIN PHYTOCHEMICAL COMPOSITION AND PHARMACOLOGICAL FUNCTIONS
## The Contents of Flavonoids and Its Main Compositions

Flavonoids, a group of polyphenolic compounds consisting of a 15-carbon basic skeleton (C6−C3−C6), found widely in plants and the human diet, are potent antitumor, antioxidant and microcirculation improvers (Cook and Samman, 1996). Tang et al. (2011) reported the flavonoids contents in different organs of three kinds of wild buckwheat (*F. cymosum, F. gracilipes* and *F. urophyllum*) and two kinds of buckwheat cultivars (*F. esculentum* and *F. tataricum*) (Table 13.4). From the table, it can be seen that the flavonoids contents differ considerably in different organs of these buckwheat species. Among the different organs, the flavonoids content of leaves is the highest, that of flowers comes second, followed by that of fruits, and lastly that of stems. In the same organ, the flavonoids contents differ considerably in different buckwheat species. In terms of leaves, the flavonoids content of *F. tataricum* is the highest, followed by those of *F. cymosum* and *F. gracilipes*, and that of *F. urophyllum* is lower than other species. When it comes to the flavonoids contents in stems, that of *F. gracilipes* is the highest, that of *F. cymosum* comes second, that of *F. tataricum* comes third, and that of *F. urophyllum* is the lowest. As for the flowers, that of *F. cymosum* is the highest, followed by those of *F. gracilipes* and *F. urophyllum*. In terms of fruits, that of *F. cymosum* is the highest, that of *F. tataricum* comes second,

**TABLE 13.4** The Comparison of Flavonoids Contents in Different Organs of Several Buckwheat Species (mg/g)

| Organs | Fagopyrum cymosum | Fagopyrum gracilipes | Fagopyrum urophyllum | Fagopyrum esculentum | Fagopyrum tataricum |
|---|---|---|---|---|---|
| Leaves | 59.1 | 42.7 | 26.9 | 0.26 | 81.2 |
| Stems | 5.4 | 6.3 | 1.6 | – | 4.5 |
| Flowers | 69.9 | 13.8 | 13.0 | – | – |
| Fruits | 17.5 | 6.3 | 11.2 | 3.1 | 12.9 |

that of *F. urophyllum* comes third, that of *F. gracilipes* comes fourth, and that of *F. esculentum* is the lowest.

Flavonoids have been proven to be the major active compounds in buckwheat, and the kinds and contents of flavonoids vary in different parts of buckwheat. For example, while six kinds of flavonoids (rutin, quercetin, orientin, vitexin, isovitexin, and isoorientin) were found in the hulls of *F. esculentum*, only rutin and isovitexin were found in its seeds. Tang et al. (2011) reported the rutin contents in the seeds of *F. cymosum*, *F. gracilipes*, *F. urophyllum*, *F. esculentum* and *F. tataricum*. Among the seeds of these five buckwheat species, the rutin content of *F. cymosum* is the highest, which is 18.5 mg/g; that of *F. tataricum* comes second (13.2 mg/g); that of *F. urophyllum* comes third (10.6 mg/g); that of *F. gracilipes* comes fourth (7.8 mg/g); and that of *F. esculentum* is the lowest (2.6 mg/g),which is only 14.1% of that of *F. cymosum*. Since the discovery of rutin in *F. esculentum* in the 20th century, more than 30 flavonoids have been isolated and identified from these *Fagopyrum* buckwheat, such as 3-methylquercetin, 3,5-dimethylquercetin, quercetin-3-O-rutinoside-3′-O-β-glucopyranoside, quercitrin, kaempferol, lute, 3′,4′-methylenedioxy-7-hydroxy-6-isopentenyl flavone, hesperidin, rhamnetin, 3-methyl-gossypetin-8-O-β-D-glucopyranoside, and quercetin-3-O-(2″-O-p-hydroxy-coumaroyl)-glucoside from *F. cymosum* (Wang et al., 2005). In addition, catechins (a derivative of flavanols) and condensed tannins (proanthocyanidins mainly) were also found in buckwheat. While (−)-epicatechin, (−)-epicatechin-3-O-p-hydroxybenzoate, (−)-epicatechin-3-O-(3,4-di-O-methyl)-gallate, and (+)-catechin-7-O-glucoside were found in *F. esculentum* (Watanabe, 1998), (+)-catechin and (−)-epicatechin were reported from *F. cymosum* (Wang et al., 2005).

Tang et al. (2013) measured the contents of epicatechin in various organs of *F. cymosum*, and the results show that epicatechin is mainly concentrated in rhizomes (the content is 0.88 mg/g), but the content of epicatechin is very low in flowers and seeds, and even entirely zero in leaves and stems. Wang et al. (2005) reported four kinds of major condensed tannins and dimers of catechin, derivatives including procyanidin B-1, procyanidin B-2, 3,3-di-O-galloyl-procyanidin B-2, and 3-O-galloyl-procyanidin B-2, which were isolated from the rhizomes of *F. cymosum* and display significant radical-scavenging activities. Especially, 3,3-di-O-galloyl-procyanidin B-2 and 3-O-galloyl-procyanidin B-2 are the most active ones due to their abundance of phenolic hydroxyl groups (Wang et al., 2005).

## Phenolics

Phenolic compounds are secondary metabolites derived from the pentose phosphate, shikimate, and phenylpropanoid pathways in plants (Reena et al., 2004). This class of compounds exhibits a wide range of physiological properties (antioxidant, antitumor, antibacterial activities, etc.), and is ubiquitous in plants (Balasundram et al., 2006). The major phenolic constituents in the buckwheat include phenylpropanoids and derivatives of hydroxybenzoic and hydroxycinnamic acid. The phenolics in *F. cymosum* include diboside A, lapathoside A, L-benzene, benzoic acid, gallic acid, p-hydroxybenzoic acid, vanillic acid, protocatechuic acid methyl ester, 6-O-galloyl-D-glucose, 3,4-dihydroxybenzaldehyde, and trans-p-hydroxy cinnamic methyl ester.

## Triterpenoids

A few triterpenoids have been reported from buckwheat. Glutinone and glutinol were isolated from the rhizomes of *F. cymosum*. In addition, ursolic acid was isolated from *F. cymosum* (Wang et al., 2005).

## Other Compounds

Hecogenin was isolated from *F. cymosum* (Liu et al., 1983). Succinic acid, emodin, emodin-8-O-β-D-glucopyranoside, and 7-hydroxycoumarin were isolated from the rhizomes of *F. cymosum* (Wang et al., 2005; Wu et al., 2008; Bao et al., 2003; Tian and Xu, 1997).

## The Pharmacological Functions

Modern pharmacological studies revealed that the above-mentioned five *Fagopyrum* species possessed versatile bioactivities, including antitumor, antioxidant, antiinflammatory, antiaging, hepatoprotective, hypoglycemic, antiallergic, antifatigue activities, etc. (Kim et al., 2003; Chan, 2003; Gao and Meng, 1993; Choi et al., 2007; Hafeez et al., 2009). Several types of bioactive phenolics including flavonoids, condensed tannins, phenylpropanoids and phenol derivatives were isolated from *Fagopyrum* species. Flavonoids in *Fagopyrum* buckwheat exhibited remarkable antioxidant and cardio-cerebral vascular protective effects (Cook and Samman, 1996; Wang et al., 2005; Watanabe, 1998), and thus these buckwheat were considered as valuable dietary supplements. The condensed tannins, isolated from the rhizomes of *F. cymosum*, showed excellent antitumor and antioxidant effects (Wang et al., 2005; Takuo et al., 1985; Nobuko et al., 1985; Barbehenna and Constabel, 2011; Bernadetta and Zuzana, 2005). In vitro experiments revealed that a commercial extract of *F. cymosum* obtained from the International Herbal Pharmaceuticals Inc. (Whitestone, New York, NY, USA), exhibited broad-spectrum cytotoxicity that significantly inhibited the growth of cancer cells from lung (H460), liver (HepG2), colon (HCT116), leukocytes (K562), and bone (U2OS) with concentrations that cause 50% inhibition of cell growth (G50) approximately in the range of $25-40\,\mu g/mL$, whereas cancer cells derived from prostate (DU145), cervix (HeLa-S3), ovary (OVCAR-3), and brain (T98G) were not sensitive to *F. cymosum*. A synergistic inhibition effect of *F. cymosum* and daunomycin was also observed in human lung cancer cells (H460) (Chan, 2003; Gao and Meng, 1993). Flavonoids of *F. cymosum* contained mainly Hyperoside, epicatechin, protocatechuic acid, procyanidin B2, quercetin, red clover flavonoid, luteolin, 7,4′-dimethyl ether, isorhamnetin, etc., in which epicatechin has a significant antitumor, anticancer, antibacterial, and antiinflammatory effect (Chai et al., 2004).

Many phenolic compounds, including flavonoids, tannins, phenolic acids, coumarins, lignans, stilbenes, and curcuminoids, have been reported to possess potent antioxidant activity (Cai et al., 2006). The wide use of buckwheat as a medicinal food largely benefits from the abundance of phenolic compounds (Bernadetta and Zuzana, 2005). Rutin was early on found to be rich in *F. esculentum* and showed significant antioxidant activity. At the concentration of 0.05 mg/mL, ascorbic acid (Vc), butylated hydroxytoluene (BHT), and rutin exhibited 92.8%, 58.8%, and 90.4% inhibition against 1,1-diphenyl-2-picryl-hydrazyl (DPPH) radical, respectively, and rutin also showed effective inhibition on lipid peroxidation (Yang et al., 2008). According to this, the contents of rutin and the total flavonoids in buckwheat play an important role in antioxidant activity (Jiang et al., 2007).

In addition, the extract of *F. cymosum* roots (Fag) significantly inhibited the acetic acid-induced writhing in mice, and reduced the peritoneal permeability and the exudation of Evans blue in mice, indicating that Fag possessed antinociceptive and antiinflammatory effects (Liu et al., 2012).

## REASONABLE USE OF WILD BUCKWHEATS

*F. cymosum*, *F. gracilipes*, and *F.urophyllum*, and other wild buckwheat have not only wide distributions, abundant resources, but also their protein contents, total amounts of essential amino acids for human beings (such as threonine, valine, methionine, isoleucine, leucine, phenylalanine, lysine, and tryptophan, and the amounts of medicinal components, are all higher than those of *F. tataricum* cultivar and *F. esculentum* cultivar. Hence, wild buckwheat has a greater value of development and utilization. So the wild buckwheat, with abundant resources as well as greater value for development, should be used reasonably (such as in the research and development of products) to comprehensively utilize the grains and plants of wild buckwheats to produce food or medicine products with higher added value and better healthcare and therapeutic function, as well as quality silage for developing stock breeding.

### Health Foods and Drinks

The grains of wild buckwheat are not only rich in nutrients, but also rich in medicinal components. Hence, in order to make various nutritional health foods and drinks, the grains can be made full use of into other flour products, such as tea, wheat sprouts, cakes, bread, biscuits, vermicelli, and snack powder, etc., as well as rice wines.

Nowadays, *F. cymosum*, one of the wild buckwheats, is used to develop related flour products, such as *F. cymosum* vermicelli, plain cakes, toast, and nutritious snack powder, etc. *F. gracilipes* is made into *F. gracilipes* sprouts; the grains of *F. cymosum* are made into *F. cymosum* tea, etc. (Tang et al.,

2011, 2013). The flour products that are made of wild buckwheat not only can be used as nutritional health foods for middle aged and elderly people, diabetic patients, and hypercholesterolemic patients—just as the *F. tataricum* cultivar is—but also have a smooth mouth feeling that is without the bitter taste of *F. tataricum* products. The sprout of wild buckwheat is a kind of new vegetable product, which has a soft and somewhat crisp texture with a pleasant scent, a high protein content, a reasonable composition of amino acids, a good amount of the unsaturated fatty acids, a high flavonoid (a kind of bioactive substance) content, and a high vitamin content. As a seasonable vegetable, the sprout of wild buckwheat can be used to make soup, and for frying or for cold dishes. Due to the high content of flavonoids, the sprout of *F. gracilipes* could be further processed into sprout powder and used as an additive to foods or drinks to enhance their healthy functions. The tea soup of *F. cymosum* is clear with a brilliant pale yellow color, the taste is delicious and the smell shows a special toasty aroma. The texture is nice and the content of flavonoids is rich, so this tea is suitable for long-term use in order to keep fit.

## Medical Value

*F. cymosum* has a long history of medical use. In china, the root, leaf, stem, flower, and grain of *F. cymosum* all can be used as medicinal materials, and this is very common in the south of China, especially Yunnan Province. According to the ancient medical books in China, *F. cymosum* had a cold nature and a sour-bitter taste, with the function of *F. cymosum* roots including heat cooling, detoxification, clearing the lungs and eliminating phlegm, removing swelling and empyema, resolving hard lumps, and regulating painful menstruation. This medicine is helpful in relation to the diseases of amygdalitis, pneumonia, dysentery, abnormal menstruation, lumbago, and over exertion. The root powder, for external use, can cure injury caused by insects, snakes, dogs, fallings, fractures, and contusions, and also has good effect on ulcers and carbuncles.

The root, leaf, stem, flower, and grain of *F. cymosum* are all rich in bioflavonoids including rutin, quercetin, etc. Hence, *F. cymosum* could be used as a raw material to extract bioflavonoids, and be further processed as soft capsule or purified to rutin or quercetin or other medicinal material. The extracts from different parts of *F. cymosum* all have antiinflammatory functions, extracts from the roots of *F. cymosum* (the main components are secondary products of flavonoids) in particular have provided significant efficiencies in anticancer, inhibition of tumor growth and metastasis, antisepsis, and antiinflammation, so that it has become a major component of several anticancer drugs and cancer prevention drugs. Nowadays, the root, stem, leaf, and flower are made into mixtures, tablets, or compounds and are being widely used clinically in China (such as for Jinqiao compounds, Jinqiao tablets, and Weimaining capsules). The major component of the Weimaining capsule is the ethanol extracts of root and stem of *F. cymosum*, and it has a certain effect on lung cancer—the total effective rate reaches 71%. The side effects of radiotherapy and chemotherapy can be also relieved (Zhang et al., 1994; Lin and Li, 2003).

## Feeding Value

It is provided that the stem, leaf, flower, and grain of the wild buckwheat such as *F. cymosum*, *F. gracilipes* (and its variety *F. graclipes var. odontopterum), F. urophyllum*, and *F. leptopodum* are all high in protein, amino acids, carbohydrates, and inorganic nutrients, so the plant itself and its processed products can meet the feeding requirements for livestock and poultry. Chinese people have already utilized cultivated and wild buckwheat to feed animals for years. The crude protein levels of *F. cymosum* are similar to those of witloof, and the crude fiber contents are similar to those of alfalfa, so that *F. cymosum* are considered as a high-protein forage and fit for the feeding of horses, cows, pigs, and chickens (Yin et al., 2006). *Fagopyrum cymosum* can propagate both sexually using seeds or vegetatively using branches, so it is very easy to expand the plant scales. *Fagopyrum cymosum* has great regenerative capacity and is much branched. The palatability is good, too, so it can be used as a kind of excellent greenfeed. *Fagopyrum cymosum* also has the advantage of mowing-resistance, leanness-resistance, high productivity, and less plant diseases and insect pests. This plant is easy to culture and easy to manage, so it has already become popular as a high-quality forage in many areas of China.

## ACKNOWLEDGMENTS

This research was supported by the Investigation of Forage Germplasm in Central China (grant number 2017FY100604).

## REFERENCES

Balasundram, N., Sundram, K., Samman, S., 2006. Phenolic compounds in plants and agri-industrial by-products: Antioxidant activity, occurrence, and potential uses. Food. Chem. 99, 191−203.

Bao, T.N., Peng, S.L., Zhou, Z.Z., Li, B.G., 2003. Chemical constituents of Fagopytum tataricum (Linn) gaertn. Nat. Prod. Res. Dev. 15, 24−26.

Barbehenna, R.V., Constabel, C.P., 2011. Tannins in plant-herbivore interactions. Phytochemistry 72, 1551−1565.

Bernadetta, K., Zuzana, M., 2005. Prophylactic components of buckwheat. Food Res. Int. 38, 561−568.

Cai, Y.Z., Sun, M., Xing, J., Luo, Q., Corke, H., 2006. Structure-radical scavenging activity relationships of phenolic compounds from traditional chinese medicinal plants. Life Sci. 78, 2872−2888.

Chai,Y., Feng,B.L., Hu,Y.G., 2004. Analysis on the variation of rutin content in different buckwheat genotypes. In: Proceedings of the Ninth International Symposium on Buckwheat; Prague, Czech Republic 18−22 August, pp. 688−691.

Chan, P.K., 2003. Inhibition of tumor growth in vitro by the extract of Fagopyrum cymosum (fago-c). Life Sci. 72, 1851−1858.

Choi, I., Seog, H., Park, Y., Kim, Y., Choi, H., 2007. Suppressive effects of germinated buckwheat on development of fatty liver in mice fed with high-fat diet. Phytomedicine 14, 563−567.

Cook, N.C., Samman, S., 1996. Flavonoids-chemistry, metabolism, cardio protective effects, and dietary sources. J. Nutr. Biochem. 7, 66−76.

Gao, Z., Meng, F., 1993. Effect of Fagopyrum cymosum rootin on clonal formation of four human tumor cells. China J. Chin. Mater. Med. 18, 498−500.

Hafeez, B.B., Adhami, V.M., Asim, M., Siddiqui, I.A., Bhat, K.M., Zhong, W., et al., 2009. Targeted knockdown of Notch1 inhibits invasion of human prostate cancer cells concomitant with inhibition of matrix metalloproteinase-9 and urokinase plasminogen activator. Clin. Cancer. Res. 15, 452−459.

Jiang, P., Burczynski, F., Campbell, C., Pierce, G., Austria, J.A., Briggs, C.J., 2007. Rutin and flavonoid contents in three buckwheat species Fagopyrum esculentum, F. tataricum, and F. homotropicum and their protective effects against lipid peroxidation. Food Res. Int 40, 356−364.

Kim, C.D., Lee, W.K., No, K.O., Park, S.K., Lee, M.H., Lim, S.R., et al., 2003. Anti-allergic action of buckwheat (Fagopyrum esculentum Moench) grain extract. Int. Immunopharmacol. 3, 129−136.

Lin, H.S., Li, G.S., 2003. Clinical research on Wei Mai Ning capsule in treating nsclc. Cancer Res. Clin. 15 (6), 368−370.

Liu, L., Sun, Z., Lu, Y., Shao, M., Yan, J., Chen, G., et al., 2012. The pharmacodynamic study on the effects of antinociception and anti-inflammation of the extracts of Fagopyrum cymosum (Trev.) Meisn. Med. Inf. 25, 49−50.

Liu, Y., Fang, Q., Zhang, X., Feng, X., Zhang, L., He, X., 1983. Study on the effective compounds of Fagopyrum dibotrys. Acta Pharm. Sin. 18, 545−547.

Lu, G.L., Zhang, Y.L., Zhao, B.H., Quan, L.H., Huo, D.J., 1993. Yield and active ingredi ent content of Fagopyrum cymosum by introduction and cultivation. Chin. J. Vet. Drug 29 (4), 19−22.

Lu, G.L., Zhang, Y.L., Li, Y., Lu, G., He, R.J., Bai, Y.J., 1996. Study on the nutritional components of Fagopyrum cymosum. Chin. J. Vet. Drug 30 (1), 19−21.

Nobuko, K., Masao, H., Tsuneo, N., Makoto, N., Takashi, Y., Takuo, O., 1985. Inhibitory effect of tannins on reverse transcriptase from RNA tumor virus. J. Nat. Prod. 48, 614−621.

Reena, R., Lin, Y.T., Shetty, K., 2004. Phenolics, their antioxidant and antimicrobial activity in dark germinated fenugreek sprouts in response to peptide and phytochemical elicitors. Asia. Pac. J. Clin. Nutr. 13, 295−307.

Takuo, O., Kazuko, M., Tsutomu, H., 1985. Relationship of the structures of tannins to the binding activities with hemoglobin and methylene blue. Chem. Pharm. Bull 33, 1424−1433.

Tang, Y., Sun, J.X., Peng, D.C., Liu, J.L., Shao, J.R., 2011. Study on the nutrients and medical elements in wild buckwheat. J. Sichuan Higher Inst. Cuisine 6, 28−31.

Tang, Y., Jia, H.F., Sun, J.X., Zhong, Z.H., Shao, J.R., 2013. Analysis of Active Ingredients in Fagopyrum cymosum. Fagopyrum 30, 45−49.

Tian, L., Xu, L.Z., 1997. Studies on the chemical constituents of the aerial parts of Fagopyrum dibotrys. China J. Chin. Mater. Med. 22, 743−745.

Wang, K.J., Zhang, Y.J., Yang, C.R., 2005. Antioxidant phenolic constituents from Fagopyrum dibotrys. J. Ethnopharmacol. 99, 259−264.

Wang, L.B., Shao, M., Gao, H.Y., Wu, B., Wu, L.J., 2005. Study on bateriostasis of Fagopyrum cymosum meisn. Chin. J. Microecol. 17, 330−331.

Watanabe, M., 1998. Catechins as antioxidants from buckwheat (Fagopyrum esculentum moench) groats. J. Agric. Food. Chem. 46, 839−845.

Wu, H.Z., Zhou, J.Y., Pan, H.L., 2008. Study on chemical constituents of Fagopyrum dibotrys (D. Don) Hara. Chin. J. Hosp. Pharm 28, 21−26.

Yang, J., Guo, J., Yuan, J., 2008. In vitro antioxidant properties of rutin. LWT—Food Sci. Technol. 41, 1060−1066.

Yin, D.X., Tang, H.B., Luo, H.J., Zhen, R.Q., Guo, K.X., Zhen, R., 2006. Preliminary study on comparing the wild forage grass *Fagopyrum cymosum* with extended forage grasses in Guizhou province. Grassland Sci. 23 (7), 45−48.

Zhang, W.J., Li, X.C., Liu, Y.Q., Yao, R.C., Nonaka, G.I., Yang, C.R., 1994. Phenolic constituents from *Fagopyrum dibotry*s. Acta Bot. Yunnanica 16 (4), 354−356.

Zhang, Z., Wang, Z.H., Lin, R.F., Wang, R.Z., 1999. Analysis of nutritional components in the seeds of *Fagopyrum cymosum* (Trevir.)Meissn. Acta Nutrimenta Sinica 21 (4), 480−482.

Zhao, G., Tang, Y., Wang, A.H., 2002. Research on the nutrient constituents and medicinal values of *Fagopyrum cymosu*m seeds. Chin. Wild Plant Resour. 21 (5), 39−41.

## FURTHER READING

Lu, G.L., Zhang, Y.L., Zhao, B.H., Quan, L.H., Huo, D.J., 1995. Yield and active ingredient content of *Fagopyrum cymosum* by introduction and cultivation. Chin. J. Vet. Drug 29 (4), 19−22.

Shao, M., Yang, Y.H., Gao, H.Y., Wu, B., Wang, L.B., Wu, L.J., 2005. Studies on the chemical constituents of *Fagopyrum dibotrys*. J. Shenyang Pharm. Univ. 22, 100−102.

Sun, T., Ho, C.T., 2005. Antioxidant activities of buckwheat extracts. Food. Chem. 90, 743−749.

Xu, G.F., Cui, M.S., Kuang, C.Y., Xu, C., Xue, S.M., 2011. Characteristics and cultivation of *Fagopyrum cymosum*. J. Forage Fee 5 (3), 53−54.

# Chapter Fourteen

# Buckwheat Genetic Resources in Central Europe

**Zlata Luthar**

*University of Ljubljana, Ljubljana, Slovenia*

## INTRODUCTION

Biodiversity has a very broad meaning that goes beyond the national interests of each country and is defined by the international Convention on Biological Diversity (1992). The Convention defines biodiversity as the variability among living organisms—which includes terrestrial, marine, and other aquatic ecosystems—and the ecological complexes of which they are a part. A living organism is understood in the broadest sense and includes all forms, from biological material to humankind. The diversity within a species itself, the diversity between species, and the diversity of ecosystems are covered or defined. The basis of diversity are genes and their allelic forms. The presence of a certain diversity depends on the interaction between genes and alleles, as well as environment, and human impact or activity. Humans have always been the driving force in exploiting natural diversity by raising, for their own needs, varieties of plants, breeds of animals, and strains of microorganisms. The one-sided cultivation of productive varieties, breeds, and strains that satisfy human needs, and the abandonment of the cultivation of less fertile and—at a given moment—uninteresting populations through the establishment of large estates and the change in the mode of production, have brought about a dramatic reduction in plant diversity, the so-called plant genetic erosion. This is currently faster than animal genetic erosion and has affected all those species which are directly or indirectly linked to agricultural use. Erosion of genetic resources can also lead to loss of valuable genetic material which has not been used yet. This means losing a lot of interesting characteristics. If we want to further increase the production and at the same time

**127**

Buckwheat Germplasm in the World. DOI: https://doi.org/10.1016/B978-0-12-811006-5.00014-8
© 2018 Elsevier Inc. All rights reserved.

improve the quality of food, we should protect and use the genetic resources effectively, as they are crucial for agricultural development. For this reason, more attention is being paid to the conservation and sustainable use of genetic resources.

The concept of sustainable development is closely linked to the Presidency of the United Nations G. H. Brundtland Commission on Environment and Development. The report of the Commission from 1987 states: We are already today borrowing environmental capital from future generations without the intention of ever returning it to them. For this reason, the report represents the beginning of the concept of sustainable development. This means a development that considers the needs of the present without hindering future generations from being able to use natural resources. The requirement of sustainable development should be followed by all countries. Sustainable development is a long-term concept of environmental, economic and social regulations which should be independent of the current political policies and national borders. One of the priority objectives of sustainable development is to protect genetic or biotic diversity.

The countries which signed the International Treaty on Plant Genetic Resources for Food and Agriculture (ITPGRFA), adopted by The Food and Agriculture Organisation of the United Nations (FAO) in November 2001, accepted the responsibility and confirmed the awareness of the importance of genetic resources for global food security and of the risk of reducing the genetic diversity and causing genetic erosion. The objective of this treaty is the conservation and sustainable utilization of plant genetic resources, and the equitable distribution of the benefits arising from their use in agriculture and food security. Activities to achieve these objectives can make a significant contribution to the conservation of landscape, species, and genetic diversity. Where they are threatened due to natural variations and human interventions, genetic resources can be collected, inventoried, and stored.

The Convention on Biological Diversity (1992) and the Nagoya Protocol (2011) address the conservation and sustainable use of genetic resources as a comprehensive and important part of nature conservation. They also enable that the benefits arising from the utilization of genetic resources are shared fairly and honestly. On the other hand, agreements and conventions governing the rights of patents enable their exploitation, which may also excessively reduce or threaten genetic or biological diversity. Therefore, many international associations and organizations, such as the United Nations Environmental Programme (UNEP), FAO, the World Intellectual Property Organisation (WIPO), the World Trade Organisation (WTO), the Consultative Group on International Agricultural Research (CGIAR), and others, seek to strike a balance, preserving human welfare, simultaneously harmonizing regimes of use and study of genetic resources and protecting intellectual property rights.

The beginnings of a systematic study of plant species diversity and genetic resources date back to the period after World War I. Among the first who began to systematically study areas with a lot of genetic diversity, as well as to collect cultivated plant species and their wild relatives, was N. I. Vavilov (1887–1943). On the expeditions he led between 1916 and 1943, he and his colleagues found out from which areas of the world various agricultural crops originated and conducted basic inventories, and by virtue of these determined the eight genetic centers. His work represents the first documented research of genetic centers and the study of the origin of individual crops around the world, even outside genetic centers. Through his approach towards collecting, describing, analyzing, and archiving, he established the principles which still govern the organization and operation of gene banks. Vavilov was convinced that the global gene pool was inexhaustible (Vavilov, 1926).

In the aftermath of World War II, it became increasingly clear that the diversity of genetic resources is very vulnerable and closely associated with varying degrees of genetic erosion. Collecting, characterizing, and evaluating genetic resources, mainly agricultural crops, and establishing gene banks at national and international levels, began in the sixties, following the pioneering work of Vavilov.

The first organizations that perceived the danger of genetic erosion and started activities to prevent it were FAO, founded in 1945, and the European Association for Plant Breeding (EUCARPIA), founded in 1956. In 1974 FAO established the International Board for Plant Genetic Resources (IBPGR), which was in 1991 renamed the International Plant Genetic Resources Institute (IPGRI), and in 2006 became Bioversity International. Under the auspices of FAO are also the centers of CGIAR, which promote agriculture and place much emphasis on the conservation of genetic resources at local, regional, and international levels, and link their practical value with breeding. The International Buckwheat Research Association (IBRA) was founded in 1980 in Ljubljana, Slovenia. IBRA points out the importance of collecting, storing, and researching common and Tartary buckwheat and related wild species through symposia (organized in Europe, Asia, and Canada), and through publishing the *Fagopyrum Journal*.

Genetic resources can be maintained in different ways. In situ means preservation of free living or agricultural species in their natural environment or where they have developed. The main objective of the in situ conservation of species through appropriate mechanisms is to ensure conservation of sufficiently large and viable populations in suitable natural habitats. Such conservation is in principle not excluded from evolutionary processes. Ex situ means conservation of genetic resources outside their natural habitats, primarily in the collections of the so-called gene banks. This method of conservation is important for agricultural genetic resources, because due to the current economic benefits their selection varies and quickly tapers. Ex situ conservation, such as botanical and zoological gardens, museums, and gene banks, is important for rare and threatened species and where in situ protection is no longer possible. The value and usefulness of genetic resources is in their polymorphism that is present between specimens within populations, as well as between populations and species. Rich genetic diversity is also invaluable for all who use genetic resources for any purpose. Special attention should be paid to free living organisms, which may be commercially interesting for agriculture, fisheries, forestry, health, pharmacy, food industry, environmental protection, etc. Access to genetic resources and their management will be of great and universal importance also in future.

## THE PURPOSE OF COLLECTING AND THE METHODS OF CONSERVING BUCKWHEAT GENETIC RESOURCES IN EUROPE

If 90 years ago it was still thought that the global gene pool was inexhaustible (Vavilov, 1926), it becomes now increasingly clear that the global picture has changed significantly and that even remote parts of the world undergo major changes in agricultural production. Of the approximately 5000 species farmed today, only about 100 are significant, and imported varieties displace local populations. This applies especially to the native populations of buckwheat in Europe, for they are almost nowhere any longer grown. This does not only hold true for Europe—according to IPGRI (now Bioversity International), buckwheat is ranked among the most threatened plant species. IPGRI in 1993 suggested that it should be treated as one of those plants which have a decisive advantage in being collected and stored (Annual Report, 1993).

Due to the reduction of buckwheat production in Europe and the emergence of bred varieties, five countries decided to collect the existing populations and preserve them. In their gene banks they store, besides other plant species, also buckwheat accessions. Two collections of buckwheat are stored in Germany, while France, Poland, the Czech Republic, and Slovenia have one each. France, Germany (Braunschweig, Gatersleben), and Slovenia keep primarily accessions of common and Tartary buckwheat, as well as some cultivars. Poland keeps only cultivars and the Czech Republic mainly accessions from other gene banks (Table 14.1).

### Reasons for Collecting Slovenian Populations of Common and Tartary Buckwheat and Tasks of the Gene Bank

Slovenia already began abandoning the cultivation of Tartary buckwheat before World War II. After 1960, the decline was significant, as it was believed to have been suitable only for fodder. Growers also abandoned domestic populations of common buckwheat and replaced them with imported domestic and foreign varieties and populations whose seeds were available in sufficient quantities. As the remainder of the domestic populations did not mate with the then-growing material, Prof. Ivan Kreft after 1975 collected most of Slovenia's preserved buckwheat genetic resources, which are now kept in the Slovenian plant gene bank at the Biotechnical Faculty, University of Ljubljana.

**TABLE 14.1** Buckwheat Gene Banks in Europe and the Number of Stored Accessions of Common and Tartary Buckwheat

| No. of Common Buckwheat | No. of Tartary Buckwheat | No. of Cultivars | Storage Conditions |
|---|---|---|---|
| France, Station d'Amélioration des Plantes, Institut National de la Recherche Agronomique (INRA), Domaine de la Motte, BP 29, 35650 Le Rheu | | | |
| 90 | 80 | 20 | Medium and long term |
| Poland, Plant Breeding and Acclimatization Institute, 05-870 Blonie Radzików near Warsaw | | | |
| | | 41 | Long term |
| | | 18 breeding or inbred lines | |
| Institute of Crop Science, Federal Research Centre for Agriculture, Bundesallee 50, 38116 Braunschweig | | | |
| 73 from Bhutan, Canada, Germany, Japan, Mexico, Nepal, Zimbabwe, former USSR, and former Yugoslavia | 8 | 24 from Poland, Japan, former USSR and Czechoslovakia | Long term |
| Institute for Plant Genetics and Crop Plant Research Genebank, Corrensstrasse 3, 06466 Gatersleben | | | |
| | 84 | 13 | Long term |
| Czech Republic, Crop Research Institute, Department of Gene Bank, Dmovská 507, 16106 Prague 6-Ruzyně | | | |
| 160 accessions of common and tartary, most of which from foreign gene bank | | | |
| Slovenia, Biotechnical Faculty, University of Ljubljana, Jamnikarjeva 101, 1000 Ljubljana | | | |
| 361 | 26 | | Medium term |

The accessions seed of common and Tartary buckwheat is kept ex situ in cold storage, packed in glass jars with lids or vacuumed in aluminum bags. Before storing, the seed is cleaned, dried to about 8% moisture and stored at 4°C. Buckwheat in these conditions maintains good germination for up to 30 years or longer. After this period, the accessions are renewed or reproduced, but it is recommendable to do this already earlier—depending on the preserved germination.

The basic tasks and work at the gene bank is focused on the appropriate medium-term storage of accessions, which allows for maintaining proper germination, regeneration, and multiplication of seeds, as well as for describing and evaluating the data collected by international descriptors (Descriptors for buckwheat, 1994).

## Determining Germination of Stored Accessions

The attempt of germination included the accessions of common and Tartary buckwheat from gene banks collected or obtained in Slovenia between 1977 and 1987. The common buckwheat accessions were divided into groups with brown seeds (24 accessions, Table 14.2) and a group of gray seeds (24 accessions, Table 14.3), while the Tartary buckwheat accessions were all in one group (10 accessions, Table 14.4). Germination was checked at the same accessions in 1990 and 2012 to determine the vitality of seeds and loss of the germination ability during the 22-year period. All tested accessions were stored, from the acquisition, under the same ex situ conditions.

The accessions, when enough seeds were available, had $4 \times 50$ seeds (those with few seeds $2 \times 50$) and were placed on moist filter paper in Petri dishes with a diameter of $90 \times 15$ mm at 20°C. After $7-8$ days we counted germinable and ungerminable seeds, calculated the percentage of germination, and classified the results into three germinating categories, depending on the priority of recovery or the reproduction of seeds (Table 14.5). In 1990, the average germination of 24 accessions with brown seeds was 87.9%, in 2012 it was 74.7%. During the 22-year storage it dropped on average by 13.2% and in the range of 5% (accession 104) to 27% (accession 97). In the first germination class with the best germination ability between 80% and 100% there were 18 accessions in 1990, 22 years later there were 13. In the second germination class with the germination ability between 50% and 79% there were in 1990 only six accessions, in 2012 there were 9. In 1990 in the third germination class with the germination rate 49% or less, there were no accessions, in 2012 in this class appeared two accessions labeled 97 and 106 (Tables 14.2 and 14.5).

In 1990, the average germination of 24 accessions with gray seeds was 85.8%, in 2012 it was 70%. During the 12-year storage it dropped on average by 15.8% and in the range of 2% (accessions 69) to 40% (accessions 43). In the first germination class with the best germination ability between

**TABLE 14.2** The Germination Percentage of Buckwheat Accessions With Brown Seeds in 1990 and 2012 Held in Slovenian Gene Bank

| ID of Accession | Year of Crop | Collecting Location in Slovenia | %Germination in 1990 | 2012 | The Difference in the 12-Year Period (%) |
|---|---|---|---|---|---|
| 10 | 1978 | Radovljica | 96 | 82 | 14 |
| 11 | 1978 | Slovenj Gradec | 98 | 87 | 11 |
| 27 | 1978 | Slovenj Gradec | 78 | 54 | 24 |
| 36 | 1078 | Slovenj Gradec | 94 | 89 | 5 |
| 36A | 1978 | Slovenj Gradec | 74 | 66 | 8 |
| 12 | 1980 | Gorenja vas | 71 | 52 | 19 |
| 26 | 1980 | Cerklje | 91 | 82 | 9 |
| 87A | 1980 | Kleče | 80 | 67 | 13 |
| 29 | 1981 | Goričko | 97 | 86 | 11 |
| 133 | 1981 | Škofja Loka | 97 | 82 | 15 |
| 97 | 1982 | Podgorje, Slov. Gradec | 76 | 49 | 27 |
| 72 | 1983 | Rut nad Tominom | 89 | 70 | 19 |
| 108 | 1983 | Ravne | 89 | 84 | 5 |
| 109 | 1983 | Javorje Ruše | 91 | 82 | 9 |
| 113 | 1983 | Holmec | 92 | 80 | 12 |
| 150 | 1983 | Podgorje | 99 | 89 | 10 |
| 151 | 1983 | Cerkno | 85 | 71 | 14 |
| 110 | 1984 | Dolga Brda, Ravne | 97 | 90 | 7 |
| 111 | 1984 | Hamunov vrh, Ravne | 98 | 87 | 11 |
| 149A | 1984 | Pristava | 89 | 71 | 18 |
| 104 | 1985 | Rut nad Tolminom | 97 | 92 | 5 |
| 105 | 1985 | Jeprca | 90 | 78 | 12 |
| 106 | 1985 | Kranj | 68 | 44 | 24 |
| 157 | 1985 | Kovor, Tržič | 73 | 58 | 15 |
| Average | | | 87.9 | 74.7 | 13.2 |

80% and 100% in 1990 were 21 accessions (3 more in comparison with the brown seeds group), 12 years later there were 9 (4 less than by the accessions with brown seeds). In the second germination class with the germination ability between 50% and 79% were in 1990 only three accessions, in

**TABLE 14.3** The Germination Percentage of Buckwheat Accessions with Gray Seeds in 1990 and 2012 Held in Slovenian Gene Bank

| ID of Accession | Year of Crop | Collecting Location in Slovenia | %Germination in 1990 | %Germination in 2012 | The Difference in the 12-Year Period (%) |
|---|---|---|---|---|---|
| 1 | 1977 | Vrhtrebnje | 87 | 62 | 25 |
| 2 | 1977 | Vrhtrebnje | 85 | 46 | 39 |
| 3 | 1978 | Vrhtrebnje | 84 | 62 | 22 |
| 18 | 1978 | Lahinje, Cerkno | 59 | 38 | 21 |
| 30 | 1980 | Žirje, Sežana | 94 | 82 | 12 |
| 31T | 1980 | Sežana | 82 | 68 | 14 |
| 32P | 1980 | Povir, Sežana | 82 | 76 | 6 |
| 119 | 1980 | Žirje, Sežana | 63 | 44 | 19 |
| 33 | 1981 | Sevnica | 98 | 86 | 12 |
| 34 | 1981 | Sevnica | 85 | 54 | 31 |
| 43 | 1981 | Vrhtrebnje | 82 | 42 | 40 |
| 77 | 1982 | Vujščina, Kostanjevica | 95 | 84 | 11 |
| 91A | 1982 | Vrh na Višnjo Goro | 58 | 42 | 16 |
| 96 | 1982 | Brusnice, Novo Mesto | 97 | 86 | 11 |
| 69 | 1983 | Humarje, Kanal | 96 | 94 | 2 |
| 69A | 1983 | Kanal | 98 | 93 | 5 |
| 70 | 1983 | Krajna Vas | 87 | 78 | 9 |
| 71 | 1983 | Lipa pri Komnu | 88 | 82 | 6 |
| 74 | 1983 | Temenica, Kras | 92 | 81 | 11 |
| 76 | 1983 | Sela na Krasu | 89 | 76 | 13 |
| 78 | 1983 | Zagrad pri Škocjanu | 93 | 84 | 9 |
| 134 | 1983 | Vrh nad Višnjo Goro | 87 | 71 | 16 |
| 114 | 1984 | Vrh nad Višnjo Goro | 86 | 72 | 14 |
| 152 | 1984 | Podhosta, Dolenjske Toplice | 93 | 78 | 15 |
| Average | | | 85.8 | 70.0 | 15.8 |

2012 there were 10. The 1990 third germination class with the germination rate of 49% or less had no accessions, while in 2012 this class produced five accessions, which is three more than the accessions with brown seeds (Tables 14.3 and 14.5).

**TABLE 14.4** The Germination Percentage of Tartary Buckwheat Accessions in 1990 and 2012 Held in Slovenian Gene Bank

| ID of Accession | Year of Crop | Collecting Location in Slovenia | %Germination in 1990 | 2012 | The Difference in the 12-Year Period (%) |
|---|---|---|---|---|---|
| $46_1$ | 1977 | Radohova Vas 1 | 63 | 34 | 29 |
| 63 | 1977 | Radohova Vas | 57 | 40 | 17 |
| 64 | 1977 | Dolina Krme 1 | 50 | 14 | 36 |
| 64A | 1977 | Dolina Krme 2 | 52 | 26 | 26 |
| 65 | 1977 | Dolenjska | 56 | 41 | 15 |
| 137 | 1977 | Sevnica | 54 | 39 | 15 |
| 66 | 1981 | Radohova Vas | 52 | 37 | 15 |
| 194 | 1984 | Straška Vas | 92 | 76 | 16 |
| 115 | 1985 | Žirovski Vrh | 82 | 62 | 20 |
| 116 | 1987 | Sv. Miklavž nad Litijo | 86 | 63 | 23 |
| Average | | | 64.4 | 43.2 | 21.2 |

**TABLE 14.5** Number of Common Buckwheat Accessions with Brown and Gray Seeds and Tartary Buckwheat in Three Germinating Classes in 1990 and 2012 Held in Slovenian Gene Bank

| Germination Class | | The Number of Buckwheat Accessions in | | | | | |
|---|---|---|---|---|---|---|---|
| | | 1990 | | | 2012 | | |
| | | Common | | | Common | | |
| ID | % | Brown | Gray | Tartary | Brown | Gray | Tartary |
| 1 | 80–100 | 18 | 21 | 3 | 13 | 9 | 0 |
| 2 | 50–79 | 6 | 3 | 7 | 9 | 10 | 3 |
| 3 | <49 | | | | 2 | 5 | 7 |
| Total | | 24 | 24 | 10 | 24 | 24 | 10 |

In comparison with the common buckwheat, the Tartary buckwheat accessions had on average lower germination. In 1990 it was 64.5%, in 2012 only 43.2%. During the 12-year storage it dropped on average by as far as 21.2%. In 1990, the lowest 50% germination was at accession labeled 64, all others had a germination of 52% and more. In the first germination class with the best germination ability between 80thers had a germination of 52% and more. In the first germination class with the best germination ability it was 64.5%, in 2012 only 43.2%. After 12 years, due to a loss of germination, all accessions fell into a lower germination class. In the third germination class as many as seven accessions were ranked. Among them were five accessions with 17% or less of germination (Tables 14.4 and 14.5).

The accessions of buckwheat which in 2012 ranked in the first germinating class have very good germination ability and over the next 10 years need not be restored or reproduced. The populations of the second germinating class have a relatively well-preserved germination ability, but it is necessary to reproduce them within 5−8 years. The populations in the third germinating class require this within 5 years, some already within 3 years or, as a priority, within 1 year. Such accessions with brown seeds are 2 (97 and 106), with gray seeds 5 (2, 18, 119, 43, and 91 A) and with Tartary seeds 7 (461, 63, 64, 64 A, 65, 137, and 66) (Tables 14.2 and 14.5).

Our germination experiments found out that common buckwheat can at the current storage conditions retain for a relatively long time a high percentage rate of germination. During the 12-year storage the accessions germination of common buckwheat dropped by 14.5%, while in the same period the germination rate of Tartary buckwheat dropped by 21.2%. The Tartary buckwheat accessions in 1990 and 2012 had on average a lower germination rate than the common buckwheat accessions. It is possible that they had a lower germination than the accessions of common brown and gray buckwheat already in the period of collection, prior to storing, but we cannot confirm this. For some plant species germination after years of storage is not falling linearly, but it may, after a period, drop very rapidly. We found differences or fluctuations in germination between accessions also in common and Tartary buckwheat.

Even Nagel et al. (2009) and Rechman-Arif et al. (2012) reported the intraspecific variation of wheat and barley, and that the accessions of both genera in the first 5 years of storage at 0₀C maintained a very high and uniform germination. After 5 years of storage, the intraspecific differences between samples in germination emerged, after 20 years of storage these differences were already noticeable, and after 34 years these differences varied in wheat from 0% to 87% and in barley from 40% to 90%. They emphasized that significant differences in germination had not occurred immediately, but only after medium- or long-term storing. Differences in germination within species of long-term stored samples under the same conditions whose seeds were produced in the same year and underwent the same technical processes of preparation for storage are attributed to genetic variation and to the fact that this is a genetically determined trait.

Börner et al. (2012) also reported inter- and intra-species variability in regard to preserving a satisfactory level of vitality and germination among the long-term stored accessions. Nagel and Börner (2010) quote the results of maintained germination of 18 species which were stored for more than 26 years at about 20°C and 50% humidity. Peas and beans maintained good germination in the aforementioned circumstances for more than 21 years, chives only for 5 years and lettuce for 7 years. They point out that germination is, after many years of storing, influenced by several factors, including the composition of the seed. Therefore, they classified plants in regard to the preservation of long-term germination into those with more proteins (legumes), carbohydrates (cereals), and fat (oilseeds). They found that among them there were also large interspecies differences in maintaining germination ability. The germination between different samples of peas after the 15-year storage ranged from 0% to 95%.

## Multiplication of Accessions

Common buckwheat is a cross-pollinated plant, pollinated by the insects, especially by bees. To prevent unwanted pollination between accessions and to ensure the equality of the genotype of restored seed with the original accession, it is necessary to reproduce accessions in isolation. This may be provided through spatial isolation with nets or the appropriate distance of 1 km or more, depending on the terrain and natural barriers. We can also, in a larger crop of tetraploid (4n) buckwheat, reproduce (within a reasonable distance) several smaller diploid (2n) accessions. The latter method of multiplication has been used for the recovery of stored accessions since 1997.

We sow diploid accessions of common buckwheat in a plot sized $2-4$ m$^2$, within a larger crop of tetraploid buckwheat. Diploid buckwheat with 16 chromosomes and tetraploid buckwheat with 32 chromosomes do not crossbreed with each other. The distance between plots, as recommended by Adhikari and Campbell (1998), should be 12 m or more. This zone of tetraploid isolation prevents insects, the main pollinators among them being bees, to transfer pollen between diploid accessions.

Due to a relatively high coefficient of multiplication in the case of buckwheat, the preferred plot size is about $2-4$ m$^2$. Such an area requires very little basic or original seed and in favorable climatic conditions the produced quantity is sufficient for restoring the basic collection and providing several samples of 50 seeds for exchange with other users.

The seeds are sown in five rows 20 cm apart. This distance is slightly greater than required by optimum agricultural technology. There are several reasons for bigger row spacing: plants have more room for growth and development, it provides greater transparency when describing and sampling in the field and causes less damage to plants. In our case—a larger plot—a distance of at least 20 m is required. This allows us easier and less expensive work and saves time, as if there were more scattered isolations at the distance of 1 km or more. The findings of Adhikari and Campbell (1998) confirm the suitability of this method and indicate 4.8% undesired (foreign) pollination at the distance of 12 m. In order to determine the appropriateness of the method in our conditions with the distance of 20 m, we carried out an experiment of crossbreeding in 1997 and 1998, when we started the systematic reproduction of accessions from the gene bank. Within about 1.5 ha of 4n crop we sowed 40 plots of 4 m$^2$ with diploid accessions. Thirty-seven plots were sown with accessions from the gene bank, one with the standard variety "Darja" and two with the determinant variety "Darina," which was used for testing potential cross-pollination between plants of individual plots.

The expression of determinant growth habitus is affected by the recessive gene or allele ($d$). In the case of 2n plants, four determinant forms of growth or habitus needed two recessive genes or alleles ($dd$). The accessions from the gene bank involved in the attempt of restoration and reproduction were largely indeterminate growth forms ($DD$). Sometimes individual plants with determinant growth appeared among the populations of buckwheat with gray seeds, which were also involved in the experiment, and because this trait is inherited dominant/recessive at a ratio of 3:1, there were about 25% of such plants. The genotype of these indeterminate plants can be homozygous ($DD$) or heterozygous ($Dd$) at a ratio of 1:2. The phenotype of both is the same—indeterminate growth form. If a determinate plant ($dd$) crossbreeds with a homozygous indeterminate one ($DD$), this already reflects the next year in the phenotype of progenies—the growth form. The progenies of crossbred plants will no longer be determinate, but indeterminate heterozygotes ($Dd$). If a determinate plant ($dd$) is crossbred with a heterozygous plant ($Dd$), half of the progenies are heterozygotes and half recessive homozygotes.

Part of the seeds of the variety "Darina" produced in 1997 were in 1998 randomly seeded on two plots inside the 4n crop. After the counting of phenotypes within each plot, we found that the percentage of the indeterminate plants, or such that arose after the crossbreeding with indeterminate plants from plots where we reproduced other accessions, was very small, on average 3.5%. If we take into consideration the possibility of the occurrence of a low percentage of determinant plants on plots where we reproduced accessions, in which occasionally occurred the recessive gene ($d$) (in 1997 there were four such accessions), and the possibility of interbreeding of these determinate plants with the variety "Darina," the emergence of the determinate plants which are the result of this breed-crossing could reduce the percentage of indeterminate plants. However, such coincidences are negligibly small, though still possible. Since the 3.5% of cross-pollination is very little, we, on the basis of this experiment, opted for this manner of seed reproduction, as it maintains a satisfactory identicalness with the base, the stored accession. At the same time, the reproduction of a larger number of accessions in one location in the same year is associated with lower costs as compared with reproduction in different locations.

Tartary buckwheat is a self-pollinating plant and does not require such extensive isolation. Multiplication is technically less demanding but has a slightly longer growing season than common buckwheat or the seeds ripen very unevenly, even more unevenly than those of common buckwheat. Years without early autumn hoarfrost and a season with minimal rain or moisture during ripening are favorable for multiplication of these accessions.

## Describing the Accessions Through Descriptors

During the growing season, we described the accessions with 43 recommended international descriptors (Descriptors for buckwheat, 1994). The follow-up period or describing starts with the phase of germination or emergence and ends with full maturity or plant harvesting. This period is also an indicator of earliness. In between we describe the vegetative part of the plants—stems and leaves, and the generative parts—flowers and seeds. A stem is described or monitored with 10 descriptors: the manner of growth—habitus, determination, plant height, branching of the plant, the number of internodes, length, color and diameter of the main shoot, the thickness of the stem tissue and sensitivity to lying. Leaves are described with 11 descriptors: color, leaf edges, leaf veins and petiole, the number of leaves on the main shoot, petiole length, length, the width and shape of lamina, the weight of fresh leaves and air-dried leaves. Flowers or inflorescence are described by 10 descriptors: the number of days from germination to flowering, the number of compound mixed inflorescences (cymes), the compactness or compaction of inflorescences, inflorescence length, the branching of inflorescences, the color of inflorescence peduncle, the number of flowers in a cluster and the peak of cyme, flower color morphology or the shape of flowers and flower atrophies. The seed is described with 10 descriptors: the number of seeds in a cluster and the peak of cyme, the color of the seeds, testa and hull, the form and surface of seeds, the appearance of seeds, seed length and width, the average harvest per plant, and weight of 1000 seeds.

## Evaluation of the Data Collected on Accessions

The differences and similarities between the 10 restored and described accessions and the standard variety "Darja" with agronomically important descriptors that affect the amount of seeds crop are shown in Table 14.6. It roughly follows the first five accessions from a group of light seeds—light gray without streaks, and a darker gray with dark streaks. The remaining five accessions come from a group with brown seeds—from light to dark brown with darker streaks. Statistically significant differences and equality within and between the two groups were calculated and tested by the analysis of variance and Duncan's test at risk ($P < 0.05$) with the program Statgraphics and are represented by the letters that follow the average and the percentage for each accession. The values marked with a different letter are statistically significantly different, while the ones with the same letters are not statistically significantly different (Table 14.6).

Common buckwheat can have strong or weak branching and usually all side shoots form inflorescence, and under favorable conditions even seed. The intensity of branching depends on the genotype and density of the crop and nutrient availability. Full inflorescence of one shoot is called a composite mixed inflorescence (cyme), whose essence is a cluster and ends with a cyme (Cawoy et al., 2009). Within composite mixed inflorescences (cyme)

**TABLE 14.6** Average Values and Percentages of Characteristics That Affect the Size of the Crop in the 10 Accessions of Buckwheat and Variety "Darja"

| Accession ID | No. Flowers in the Cluster and the Peak Cyme | %Seeds in Cluster and Peak Cyme | %Hulls in Cluster and Peak Cyme | Length of Seeds (mm) | Width of Seeds (mm) | The Weight of Seeds per Plant (g) | Mass of 1000 Seeds (g) |
|---|---|---|---|---|---|---|---|
| **Light seeds—Light and dark gray with streaks** | | | | | | | |
| 114 | 21.7 abcd | 24.6 a | 23.3 ab | 5.7 d | 3.3 c | 0.6 c | 20.7 gh |
| 119 | 23.5 abcd | 7.5 d | 21.4 ab | 6.0 abc | 3.6 b | 0.1 f | 21.7 efg |
| 134 | 24.7 ab | 7.3 d | 29.4 a | 5.4 ef | 3.0 d | 0.1 f | 21.2 fgh |
| 149 | 23.3 abcd | 18.5 ab | 19.5 ab | 5.3 f | 3.3 c | 0.3 d | 22.3 def |
| 152 | 19.5 bcd | 14.5 bcd | 27.2 a | 5.6 de | 3.2 cd | 0.2 e | 20.3 h |
| **Dark seeds—Light and dark brown with streaks** | | | | | | | |
| 113 | 25.3 a | 9.5 bcd | 15.9 b | 5.9 bc | 3.5 b | 0.9 b | 21.4 efgh |
| 135 | 24.4 ab | 15.2 bcd | 23.0 ab | 6.1 ab | 3.8 a | 0.7 bc | 23.9 c |
| 150 | 18.6 d | 17.6 abc | 20.5 ab | 5.2 f | 3.1 cd | 0.5 cd | 22.6 cde |
| 151 | 18.8 cd | 14.0 bcd | 23.1 ab | 5.8 cd | 3.8 a | 0.6 c | 25.6 b |
| 157 | 23.5 abcd | 8.6 cd | 18.8 ab | 6.2 a | 3.9 a | 0.6 c | 23.1 cd |
| **Control** | | | | | | | |
| "Darja" | 26.1 a | 10.4 bcd | 23.5 ab | 5.8 cd | 3.8 a | 1.6 a | 27.2 a |

Averages and percentages marked with different letters are statistically significantly different between accessions, tested by Duncan's test at risk ($P < 0.05$).

of indeterminate buckwheat are more inflorescences, twigs of clusters, which end with a peak cyme. Determinant buckwheat has an inflorescence (cyme), a cluster which ends with the final cyme (Descriptors for buckwheat, 1994).

It is evident from Table 14.6 that the accessions included in restoration set up a relatively large number of flowers. Each inflorescence had on average a twig or cluster with the corresponding peak cyme and from 19 to 26 flowers, depending on in which shoot it was located (main or side). This is quite a lot, if we take into consideration that the length of the inflorescence was on average 0.5−2.5 cm, depending on the branching of inflorescences and its location on the plant. Only 7.3%−24.6% of flowers formed seeds after flowering. On average, there were 15.9%−29.4% empty seeds. At least half or more than half of the flowers (46%−77%) aborted or collapsed in the first days after fertilization in the first phase of creating the seeds. This may have been caused by poor pollination or fertilization, a major problem is probably an incorrect concept formulation of seeds and an inflow of assimilates in forming seeds, as many of these gentle forms perished unnoticed. The studied accessions, given the number of flowers, did not differ significantly, except the accessions labeled 113, 134, and 135, which stood out from the average with 24.8 flowers in a cluster with the corresponding cyme. The accessions with brown seeds had slightly larger conception—hulls, which is manifested by the final weight of the seeds—at the proper inflow of assimilates. The inflow and filling of seeds with assimilates in these was probably better and more intense (Table 14.6). Even though the accessions with brown seeds needed more assimilates for full seed, these accessions did not create more empty seeds in comparison to the gray accessions. There were no significant differences in the percentage of empty seeds between the groups of gray and brown accessions. Three groups formulated a maximum of 29.4% of empty seeds in a cluster with the corresponding cyme in the accessions, it was labeled 134, and in the same group with 27.2% of empty seeds in the accessions 152, both from the group with gray seeds. All other values overlapped or there were no significant differences between them, except for the accessions 113 from the group of dark brown seeds that had at least 15.9% of empty seeds. The mass of seeds per plant was higher in the accessions with brown seeds and varied from 0.5 to 0.9 g, while the accessions with gray seeds had lower average weight of 0.1−0.6 g per plant. The variety "Darja" had the highest average weight of seeds, at 1.6 g per plant. Four of the five accessions with brown seeds had a mass of 1000 seeds greater than 22.6 g, while only one accession from the group of gray seeds had 22.3 g, the other four had lower weight. The filled brown seeds were also bigger than the gray ones. Three accessions labeled 135, 151, 157, and "Darja" had the widest and longest triangular seeds. The accessions 113, 135, and 150 had triangular and ovoid seeds. The accessions with gray seeds had slightly shorter seeds, three of five accessions had shorter and narrower seeds, the predominant form was ovoid. In the accessions 114 and

149 appeared ovoid and conical forms. Generally, the accessions with gray seeds have shorter and narrower, conical, or ovoid seeds, while the accessions with brown seeds have longer and wider triangular seeds.

The variety "Darja," which was included in the description as a control, had some properties, primarily those that affect the size of the crop, which differed significantly from the tested accessions. Breeder's work on the variety "Darja," apart from other improved characteristics, reflects mainly in the mass of seeds, which was statistically significantly greater (27.2 g weight in 1000 seeds) from the weight of seeds of other accessions. Also, the mass of seeds produced per plant was with 1.6 g significantly larger than in other accessions. It was in the group with most voluptuous flowering. It filled 10.4% of seeds, 23.5% of seeds were empty, which ranked it in the middle class. Despite these shortcomings, which were at the time of breeding not eliminated, the crops were more stable and more satisfactory than those of the accessions. On average, they reached 0.8 t/ha, in favorable growing conditions also 1.0−1.2 t/ha. Describing accessions from the gene bank is important, as only in this way do we get in-depth information about what is stored and which accessions or their characteristics are interesting for breeding of improved genotypes—varieties.

## CONCLUSION

Narrowing of biodiversity may result in partial and even complete genetic erosion. It is present in agriculture and in the natural environment. In such cases, the solution to biodiversity conservation is in systematic recording, collecting, and storing of specific genetic resources in in situ and ex situ collections or gene banks. As per estimates by FAO (2010), genetic erosion in agriculture is very high. Its cause is in the promotion of economically important varieties and breeds suitable for food, work, clothing, sports, etc. Genetic erosion in the natural environment is generated by habitat loss due to human activities and climate changes. Due to some of the above reasons, some areas of Europe experienced genetic erosion of buckwheat after 1960. This is the reason why a part of the endangered accessions is being collected and maintained in six gene banks across Europe. Two collections of buckwheat are in Germany, while France, Poland, the Czech Republic, and Slovenia each has one collection. The Slovenian Plant Gene Bank keeps accessions of common and Tartary buckwheat. The majority of accessions of common buckwheat are diploid (2n), some of them are tetraploid (4n). Tetraploid accessions are different from diploid, they have larger leaves with wavy edges, slightly larger flowers with large rounded petals, and darker seeds. Only one of the tetraploid accessions has gray seeds. The collected diploid accessions can be, considering the previous descriptions, roughly divided into two groups. The accessions with gray seeds, which have a finer gray seed, from light to dark gray, often with darker streaks and predominantly white flowers, only in some accessions occur individual plants with slightly pink flowers. Some plants in these accessions have the recessive gene (*d*) for determinant growth form. They are adapted to lowland and hilly soil and climate conditions, as well as to areas without frequent early autumn frosts and fogs. In the second group are the accessions with a slightly thicker dark seed—from light to dark brown, dark streaks are often present in the basic color of seeds. The basic color of flowers is pale to deep pink, some flowers can be also slightly red. They are suitable for elevated, hilly locations with a 7–10 day shorter growing season. Accessions of Tartary buckwheat are distinguished from each other by the color and size of seeds, plant height, branching, leaf size, and earliness.

## ACKNOWLEDGMENT

The gene bank of buckwheat was supported by the Slovenian Ministry of Agriculture, Forestry and Food, project PP 142910.

## REFERENCES

Adhikari, K.N., Campbell, C.G., 1998. Natural outcrossing in common buckwheat. Euphytica 102, 233–237.

Annual Report, 1993. International Plant Genetic Resources Institute (IPGRI), Rome., p. 47.

Börner, A., Nagel, M., Rehman-Arif, M.A., Allam, M., Lohwasser, U., 2012. Ex situ genebank collections – important tools for plant gnetics and breeding. In: Bedo, Z., Lang, L. (Eds.), Proceedings of the 19th EUCARPIA General Congress. Budapest, Hungary, pp. 69–72.

Brundtland, G.H., 1987. Our common future. Report of the World Commission on Environment and Development, pp. 1–187. http://www.channelingreality.com/Documents/Brundtland_Searchable.pdf.

Cawoy, V., Ledent, J.F., Kimet, J.M., Jacquemart, A.L., 2009. Floral biology of common buckwheat. Eur. J. Plant Sci. Biotechnol. (special issue 1) 1–9. Global Science Books.

Convention on Biological Diversity, 1992. United Nations, pp. 1–30. https://www.cbd.int/doc/legal/cbd-en.pdf.

Descriptors for buckwheat (*Fagopyrum* spp.), 1994. International Plant Genetic Resources Institute, Rome, Italy. pp. 1–48. http://www.bioversityinternational.org/e-library/publications/detail/descriptors-for-buckwheat-fagopyrum-spp/.

FAO, 2010. Themi second report on the state of the world's plant genetic resources for food and agriculture. Commission on Genetic Resources for Food and Agriculture, Food and Agriculture Organization of the United Nations, Rome, pp. 1–399. http://www.fao.org/docrep/013/i1500e/i1500e.pdf.

FAO, 2001. International Treaty on Plant Genetic Resources for Food and Agriculture (ITPGRFA). pp. 1–68. http://www.fao.org/3/a-i0510c.pdf.

Nagel, M., Vogel, H., Landjeva, S., Buck-Sorlin, G., Lohwasser, U., Scholz, U., et al., 2009. Seed conservation in *ex situ* genebanks – genetic studies on longevity in barley. Euphytica 170, 5–14.

Nagel, M., Börner, A., 2010. The longevity of crop seeds stored under ambient conditions. Seed Sci. Res. 2, 1–20.

Nagoya Protocol on Access to Genetic Resources and the Fair and Equitable Sharing of Benefits Arising from their Utilization to the Convention on Biological Diversity, 2011. Secretariat of the Convention on Biological Diversity, pp. 1–25. https://www.cbd.int/abs/doc/protocol/nagoya-protocol-en.pdf.

Rechman-Arif, M.A., Nagel, M., Neumann, K., Kobilijski, B., Lohwasser, U., Börner, A., 2012. Genetic studies of seed longevity in hexaploid wheat using segregation and assocition mapping approaches. Euphytica 186, 1–13.

Vavilov, N.I., 1926. Studies on the origin of cultivated plants. Bull. Appl. Bot. 2, 1–16.

Chapter | Fifteen

# Existing Variability and Future Prospects of Buckwheat Germplasm in the Indian Himalayan Tract

**Ravinder N. Gohil and Geeta Sharma**
*University of Jammu, Jammu, Jammu and Kashmir, India*

## INTRODUCTION

Unlike in some other parts of the world, buckwheat (*Fagopyrum* spp.) is not a favored staple crop in the Indian subcontinent because Indians are mostly either rice or wheat eaters. Not consumed regularly on large scale, its cultivation, consumption, and trade are both irregular and limited. While in some hilly tracts it is an important crop, in the rest of the country it is one of the preferred substitutes during some auspicious periods only when true cereals are not taken. Besides as a food constituent in the hilly areas, almost all parts of buckwheat are used. Its flour is used for making Indian bread (chapattis), small leaves and shoots as leafy vegetables, and fermented grains for making alcoholic drink (Gohil, 1984). The post-harvest remnants of buckwheat are plowed back into soil, forming excellent organic matter that improves the soil texture in the field. According to Joshi and Paroda (1991), buckwheat helps in binding soil particles, thereby checking erosion during the rainy season. Since the Himalayan tract is continuous from northeast to northwest India with a large part of it in Nepal, some information available about buckwheat in the latter region is also included in the present chapter.

**145**

Buckwheat Germplasm in the World. DOI: https://doi.org/10.1016/B978-0-12-811006-5.00015-X
© 2018 Elsevier Inc. All rights reserved.

## DISTRIBUTION

In India, the genus *Fagopyrum*, commonly called *ogal, phapar, kuttu, bresha, mittahe, titae, dyat, dro, bro*, is represented by six annuals, namely *Fagopyrum esculentum* Moench (Common buckwheat), *Fagopyrum sagittatum* Gilib. (Coarse buckwheat), *Fagopyrum kashmirianum* Munshi (Kashmir buckwheat), *Fagopyrum tataricum* Gaertn (Tartary buckwheat), *Fagopyrum giganteum*, and *Fagopyrum emarginatum*, and one perennial, *Fagopyrum cymosum* Meissn. species (Munshi, 1982, 1983; Gohil et al., 1983; Joshi, 1999). *F. tataricum* is represented by two subspecies viz. *F. tataricum* ssp. *annum* and *F. tataricum* ssp. *potanini* (Ohnishi, 1989, 1991, 1992), and a variety, *F. tataricum* var. *himalianum*. While *F. kashmirianum* is morphologically akin to *F. tataricum* var. *himalianum* (Gohil and Rathar, 1981; Farooq and Tahir, 1987), *F. emarginatum* has been treated as conspecific with *F. esculentum* (Arora and Engels, 1992). Taxonomic status of IC-13145 accession, maintained in the National Bureau of Plant Genetic Resources (NBPGR), Phagli, Shimla, is controversial. Though Joshi and Paroda (1991) raised it to the status of a new species, "*F. himalianum*," International Plant Genetic Resources Institute (IPGRI), relegated it to the varietal level (*F. himalianum* var. *himalianum*) (http://www.ipgri.cgiar.org/publication). Campbell (1997), however, considered it as a race of *F. esculentum*.

Buckwheat exhibits rich diversity throughout the Indian Himalayas with preponderance in northwestern over the northeastern region (Farooq and Tahir, 1987; Rana, 2004). While in the former region, five species (*F. esculentum, F. sagittatum, F. tataricum* ssp. *potanini, F. kashmirianum*, and *F. cymosum*) are available, in central and eastern areas, three (*F. esculentum, F. tataricum* ssp. *annum*, and *F. cymosum*) are prevalent (Ohnishi, 1989, 1991, 1992; Arora et al., 1995). This disparity in the availability of species of *Fagopyrum* in the two regions is most likely because of the common border between the western Himalayan region and Tibet that must have led to migration of some taxa across the Himalayas along several trade links through this area from southern China and Tibet.

Depending upon climatic conditions associated with altitude varying between 400 m and 2400 m, common and Tartary buckwheat are widely cultivated throughout the Indian Himalayas. While the former is adapted to low, mid, and high hills ranging from 400 to 2400 m, the latter is mainly confined to high hills at 1500−2400 m or even higher altitudes. Common buckwheat is grown in plains during winter and Tartary buckwheat is cultivated as a spring crop at higher altitudes because of its frost resistant characteristic. Barley, wheat, potato, amaranths, millet, and members of the Brassicaceae family are companion crops of buckwheat in high and mid hills and rice, soyabean, finger millet, and maize in the lower hills (Arora et al., 1995).

In Jammu and Kashmir, the northwestern state of India, four annual taxa are widely grown at high altitudes like Sonamarg, Baktour, Kupwara, Machil, Dawar, Nilnag, Gogjipathar Sind Valley, Ladakh and Zanskar, Kargil and Drass sectors, and Gurez valley. Till recently most of these areas were isolated and inaccessible with poor irrigation facilities and limited growing periods (Rathar, 1982; Tahir and Farooq, 1983; Gohil, 1984). Along with common and Tartary buckwheat, *F. kashmirianum* and *F. sagittatum* are also grown in varying frequencies and patterns (Rathar, 1982). While pure cultivation of *F. esculentum* or *F. sagittatum* has been noticed at some places, at others, mixed cultivation of all the four species is on record (Rathar, 1982; Joshi, 1999). In Leh and Kargil districts of Ladakh, relatively warmer areas permitting double-cropping, common buckwheat is grown as the second crop after harvesting barley (Ahmad and Raj, 2012).

Himachal Pradesh, Uttrakhand, Sikkim, Arunachal Pradesh, and higher elevations of West Bengal, Meghalaya, Assam, Manipur, and Nagaland are other Himalayan regions where buckwheat is cultivated (Sharma and Chatterjee, 1960; Joshi, 1999; Hore and Rathi, 2002). Buckwheat is also cultivated in some parts of the Nilgiris and Palani hills in the southern parts of India (Joshi, 1999).

## DIAGNOSTIC CHARACTERISTICS

Seed morphology is a major diagnostic tool in identifying the cultivated species of buckwheat. The seeds of *F. esculentum* are brown, ovate, smooth, 3-ridged with zebra linings (Gupta, 2001). On the other hand, *F. sagittatum*, *F. tataricum*, and *F. kashmirianum* set gray colored, oblong-ovate, rough surfaced, and notched seeds which either lack projections (*F. sagittatum*) or have a number of projections (*F. tataricum* and *F. kashmirianum*), with two projections on extreme of the ridge being larger (*F. tataricum*), or have a single projection at the base of each ridge which is partially curved downwards (*F. kashmirianum*). These taxa also differ in floral traits and the breeding system. While white/pink showy flowers of *F. esculentum* are heterostylous, with clear-cut differentiation into pin and thrum types, green dull flowers of the remaining three (*F. sagittatum*, *F. tataricum*, and *F. kashmirianum*) are predominately homostylous (Tahir and Farooq, 1987a; Gupta, 2001). Tahir and Farooq (1987a) reported dimorphic stigma and pollen grains in common buckwheat; pin type flowers had knobbed stigma and smaller pollen grains, whereas thrum type flowers had papillate stigma and larger pollen grains. Subsequent bagging experiments on these species revealed that *F. esculentum*, a predominantly self-incompatible taxon, is pollinated mainly by insects such as honey bees (*Apis indica*) and wasps (*Vespa* sp.). Forced inbreeding results in depression due to the expression of recessive deleterious mutations (Gohil et al., 1983; Tahir and Farooq, 1987a). The other three species are self-compatible.

## DIVERSITY
### Morphology and Growth Parameters

Since the 17th century, buckwheat has been cultivated in isolation in the Indian Himalayas along large stretches of land under low input conditions.

Considerable environmental heterogeneity in these regions of the Himalayas has led to enormous diversity in this crop in terms of landraces (Campbell, 1997). Most of these races, selected by the farmers as per the needs, traditions, and preferences of the local populace, have not yet mixed with other populations mainly because of geographical isolation. For instance, in Kargil district of Jammu and Kashmir, two variants of common buckwheat differing in size and color are grown. One is a yellow-colored and small sized "brosuk" while the other is a black and larger "gyamrus" (Ahmad and Raj, 2012). Eleven collections of *F. sagittatum* raised from the seed procured from 2 to 3 fields of four villages of Kargil district were analyzed for morphological descriptors. Three collections each from different fields of Titismik, Andu, Chanigund, and two from Hunakipuyum exhibited wide variations in plant height, number of branches per plant, time of maturity, yield, and number of kernels per plant. Variations existed not only among the collections but equally so and particularly in those made from the same village. Variation of this nature and magnitude is likely to arise on account of lack of seed exchange even between adjacent hamlets. In view of this, Gohil and Misri (1986) proposed that these individual collections be treated as different germlines.

Random seed samples collected from different regions of Jammu and Kashmir; Sonamarg, Nilnag, Gurez, Machil, and Gogjipathar (Kashmir) by Tahir and Farooq (1983) and Skurbuchan and Baima (Ladakh Province), Udhampur and Sunderbani (Jammu Province) by Gupta (2001) were subjected to critical morphological analysis. Results revealed that none of the collections was pure for any species, instead each comprised seeds of common, coarse, Tartary and Kashmir buckwheat. While all the samples worked out by Gupta (2001) had maximum number of seeds of *F. esculentum* (85.95%−99.62%), those by Tahir and Farooq (1983) showed a frequent predominance of *F. sagittatum* and to a lesser extent of *F. esculentum*. Morphologically also the three common buckwheat populations from Skurbuchan, Baima, and Sunderbani varied in plant height, internode length, number, and size of stomata and floral characters (Gupta, 2001). Amongst these, the plants of Skurbuchan had maximum height, longest internodes, largest but least stomata, and maximum proportion of pink versus white flowers (72:28). Further detailed studies on the floral morphology, especially that of essential organs by Gupta (2001), revealed some interesting features which have a direct bearing upon the breeding system of the species. These are (1) presence of some homostylous flowers in addition to normal heterostylous ones in a few plants; (2) existence of pistillode in flowers of some pin and thrum type plants; (3) formation of stigma at a level lower than the anther lobes in some flowers of pin eyed plants. Additionally, some variations were also observed in the standard floral structure. Against the normal flowers characterized by five tepals, eight stamens and a pistil with three ridged ovary, three styles and three stigmas, the variants have either of these features; 4−8 tepals, 7−12 stamens, a 4-ridged ovary, 4 styles, 4 stigmas, and 2 pistils. Gupta (2001) has recorded some unusual flowers, that is, with five and six staminodes in Skurbuchan and Baima populations of *F. sagittatum*. In five staminode flowers, all the five staminodes constitute the outer whorl of stamens. In those with six staminodes, all the five stamens of outer whorl are modified into staminodes with the sixth staminode present in the inner whorl along with two normal stamens. These populations also show in situ seed germination resulting in overall reduced yield in common buckwheat despite its exhibiting less shattering as compared to other species.

Joshi (1999) evaluated 408 accessions from the Himalayan region for 31 descriptors and found marked variation in plant height, number of branches, leaf number and size, days taken to flower and mature, number of seeds per cyme, weight of 100 seeds, and yield per plant. Phenotypic and genotypic variances were the maximum for leaf length, branches per plant, and seed weight. Leaf width followed by 100 seed weight had the strongest positive effect on the yield.

Similarly, 30 populations of *F. esculentum* and 16 of *F. tataricum* from the Himalayan ranges of Himachal Pradesh, Jammu and Kashmir and Uttaranchal were studied for 23 morphological descriptors and most of these traits were found to show a wide range of variations (Senthilkumaran et al., 2008). Analysis of variance indicated highly significant differences for all the quantitative traits for the landraces of common buckwheat. However, those of Tartary buckwheat do not differ significantly in leaf and petiole lengths and number of branches and inflorescences per plant. In so far as yield is concerned, none of the accessions performed better than the local control varieties VL7 and Shimla B1 of *F. esculentum* and *F. tartaricum*, respectively. Distinct regional variations for some important quantitative traits were noted in the landraces of these two species. While common buckwheat accessions from Uttaranchal bloomed early and showed indeterminate growth, two accessions from Jammu and Kashmir were relatively determinate in nature. Yield potential of populations from Himachal Pradesh was greater. Similarly, *F. tataricum* accessions from Himachal Pradesh were early flowering/maturing, whereas those from Uttaranchal were determinate in growth habit. At population level, the latter were more variable than the former.

Rout and Chrungoo (2007) analyzed morphological variability among 22 accessions of *F. esculentum*, *F. tartaricum*, and *F. cymosum* from the Indian Himalayas for color, coat striations of seeds, blade shape, color of stem, leaf margin, and flowers. Common buckwheat formed smooth-surfaced seeds, with coat striation patterns varying with each population, whereas all accessions/varieties of *tataricum* and *cymosum* have seeds with a rough coat and without any striations. Variations were also noticed in leaf and floral traits. While common buckwheat has a cordate and sagittate leaf blade, leaves of Tartary buckwheat are exclusively sagittate. So far as flower color is concerned, *F. esculentum* and *F. cymosum* produced white/pink and green flowers, respectively. Rout and Chrungoo (2007) also observed that common buckwheat plants from different Himalayan regions produced pink and white flowers and winged and/non-winged grains. Plants from Himachal Pradesh (Western Himalayas) bore white flowers and winged plus non-winged grains, those from Uttaranchal (Central Himalayas) formed pink flowers and non-winged seeds (except VL7) and the rest from Arunachal Pradesh (Eastern Himalayas) produced white flowers and non-winged grains.

Tremendous diversity has also been reported in landraces of Nepalese buckwheat by Baniya et al. (1992, 1995). These workers found significant variation in days to flowering (26−45), maturity (67−98), plant height (25−116.4 cm), numbers of branches (1.4−14), and leaves (2−18.4), seed number per cyme (7−50.2), 1000 seed weight (10.2−30.8 g), seed color/shape/surface, amongst different common buckwheat accessions. Subsequently, however, Shershand and Ujihara (1995) reported lesser variation in days to flowering (27−38) and number of branches per plant (2.2−7.0) but more in number of flower clusters per plant (10−91). The Tartary strains from the Himalayan regions at an altitudinal range of 1000−3800 m and at latitude 26°N−36°N were collected and monitored by Yoshida et al. (1995) for determining the days from sowing to first flower. The duration in days ranged from 26 to 52 for west Nepalese strains, 28 to 39 for those from Tibet, Bhutan, east Nepal, and central Nepal, and 35 to 46 days for mid-west Nepal. Baniya et al. (1995) analyzed 150 Tartary buckwheat landraces and found differences in some qualitative traits such as stem color (red in 67% land races), leaf color at maturity (green in 62% land races), petiole color (green in 63% land races), flower stalk color (green in 43%, red in 38%, and pink in 19% land races), flower color (green in 95% and pink/red in 5% landraces), seed color (gray in 42%, brown in 30%, blackish gray in 18%, and black in 10% landraces) and seed surface and type (rough surface in 71% races and winged seeds in 27% races). A frequency distribution analysis done on the characterization and evaluation of Nepalese landraces revealed a bimodal distribution pattern in Tartary buckwheat for most of the traits analyzed (Baniya et al., 1992) which has been attributed to a subset of very early and very short landraces. In general, Tartary buckwheat has been reported to exhibit more variability in the yield and yield-related descriptors than common buckwheat. Unlike Indian landraces, collections from Nepal have not been utilized fully for crop improvement. The collections made from Bhutan have also not been well characterized or evaluated.

## Cytological Variability

Because of the buckwheats having very small chromosomes that are difficult to stain, not much is known about their cytology. Of the total 29 *Fagopyrum* species reported so far, chromosome numbers of about 17 namely, *F. esculentum, F. kashmirianum, F. cymosum, F. sagittatum, F. tataricum, Fagopyrum homotropicum, Fagopyrum gracilipes, Fagopyrum qiangcai, F. crispatofolium, Fagopyrum wenchuanense, Fagopyrum pugense, Fagopyrum rotundatum, Fagopyrum densovillosum, Fagopyrum urophyllum, Fagopyrum leptopodum, Fagopyrum statice,* and *Fagopyrum capillatum* have been documented (Stevens, 1912; Jaretzky, 1928; Sharma and Chatterjee, 1960; Gohil and Rathar, 1981; Gohil et al., 1983; Chen, 1999; Ohnishi and Asano, 1999; Nagano et al., 2000; Yang et al., 2010; Zhou et al., 2012).

Most of the species of this monobasic genus ($x = 8$) are diploid. Three species, namely *Fagopyrum zuogogense*, *F. gracilipes*, and *Fagopyrum crispatifolium* are tetraploid, while two, *F. cymosum* and *F. homotropicum*, exist in both diploid and tetraploid forms. The tetraploids of *F. cymosum* are distributed more widely in southwest China, Thailand, the Himalayan hills and Tibet, and diploid populations are restricted to Sichuan and Yunnan provinces and eastern Tibet in China. On the contrary, the tetraploid *F. homotropicum* has limited distribution in the northwestern part of Yunnan province and the southwestern part of Sichuan province (Yamane and Ohnishi, 2003), whereas diploid populations are found throughout Yunnan and Sichuan provinces. The rare occurrence of phenomenon of polyploidy in this genus is in consonance with the annual habit of most of its species.

The somatic complements of five species from Kashmir Himalayas have been worked out in detail (Table 15.1) by Gohil et al. (1983) and Gupta (2001).

As is evident from Table 15.1, chromosome complements of *F. esculentum* and *F. cymosum* comprise of longest and smallest chromosomes, respectively. Interestingly, despite being a tetraploid, the total chromosome length of the latter species is only a little more than the diploid *F. esculentum*. It appears that the advantage gained by the addition of genomes in *F. cymosum* is nullified by the decrease in their size. At the same time, an increment in ploidy level is of little significance since *F. cymosum* is sterile.

Somatic complements of five *Fagopyrum* species from northwestern Himalayas have largely median and/or submedian chromosomes with the exception of *F. cymosum* which has two subterminal chromosomes. Intraspecific karyotypic differences are also on record for common buckwheat, with the Kashmir population karyotyped by Gohil et al. (1983) having 10 metacentric and 6 submetacentric chromosomes, and that by Gupta (2001) having 16 submetacentric chromosomes. Amongst the four diploids, markers (satellites/secondary constrictions) have been reported in *F. esculentum* (Gupta, 2001) and *F. sagittatum* (Gohil et al., 1983). Secondary constrictions were observed in six submedian chromosomes of *F. sagittatum* (Gohil ct al., 1983) and on two in *F. esculentum*. The karyotypes of tetraploid *F. cymosum* from Darjeeling and Kashmir also have marker chromosomes (Sharma and Chatterjee, 1960; Gohil et al., 1983). The number, form, and size of the chromosomes, however, vary in the two reports. From the Darjeeling population, secondary constrictions have been reported on eight large chromosomes (Sharma and Chatterjee, 1960) against only two subterminal chromosomes having satellites in the Kashmir collections (Gohil et al., 1983).

The 16 chromosomes in all the diploid collections from the northwestern Himalayas could be grouped into eight homomorphic pairs, indicating their normal diploid nature. Tetraploid *F. cymosum* is an exception; its 32 chromosomes could be arranged into seven homomorphic and only one

**TABLE 15.1** Salient Features of Somatic Chromosomes of Five Species of *Fagopyrum*

| Species | ChrNo. | Chromatin Length (μm) | | Chromosome Size (μm) | | L/S Ratio | Karyotypic Formula | Authority |
| | | Total | Mean | Longest | Smallest | | | |
|---|---|---|---|---|---|---|---|---|
| *Fagopyrum esculentum* | 16 | 43.72 | 2.73 | 3.60 | 2.16 | 1.67 | 10M + 6SM | Gohil et al. (1983) |
| F. esculentum | 16 | N.M. | N.M. | 3.9 | 2.6 | 1.5 | 14SM + 2SM[SC] | Gupta (2001) |
| *Fagopyrum kashmirianum* | 16 | 37.5 | 2.34 | 2.75 | 1.89 | 1.45 | 16SM | Gohil et al. (1983) |
| *Fagopyrum sagittatum* | 16 | 38.54 | 2.41 | 2.75 | 2.06 | 1.33 | 10SM + 6SM[SC] | Gohil et al. (1983) |
| *Fagopyrum tataricum* | 16 | 35.78 | 2.23 | 2.58 | 1.89 | 1.36 | 16SM | Gohil et al. (1983) |
| *Fagopyrum cymosum* | 32 | 47.2 | 1.47 | 1.72 | 1.2 | 1.43 | 8M + 22SM + 2ST[Sat] | Gohil et al. (1983) |

heteromorphic quadruple. The heteromorphic one comprised of two median and two submedian chromosomes. On the basis of these observations, this cytotype was assigned autoploid status (Gohil et al., 1983). Marked chromosomal heteromorphism in the smallest group was attributed either to the differences in the constituent genomes of this species at the time of its origin or post ploidy modifications in its karyotype. Contrary to this, measurement of chromosomes by Gohil et al. (1983) of Fig. 8a of Sharma and Chatterjee (1960) of the northeastern collection of *F. cymosum* revealed a lot of chromosomal heteromorphicity, indicating its origin through amphiploidy. Although these observations point towards separate evolution of the two cytotypes, it would be interesting to study their similarities and dissimilarities in detail using modern techniques.

Chromosomes of *F. esculentum*, *F. kashmirianum*, *F. sagittatum*, and *F. tataricum* from Kashmir behave normally during reduction division with eight bivalents observed at diplotene and metaphase I (Gohil et al., 1983; Gupta, 2001). Chiasmata frequency per cell varied from 12.5 for *F. kashmirianum* to 18 for *F. esculentum* and *F. sagittatum*. Anaphasic I segregations were normal in all except *F. tataricum* wherein laggards were occasionally formed. Anomalies at stages subsequent to the first meiotic division were observed in a single plant of common buckwheat by Gupta (2001). In some pollen mother cells of this plant, chromosomes were present in five groups at anaphase II. While some had 2−6 microspores of varying size at tetrad stage, pollen stainability was quite low (56.5%) as compared to other plants.

All the 15 lines of 6 varieties of *F. esculentum*, cytologically analyzed by Narain (1987), were diploid. With the intent to increase seed size and protein content, Narain (1987) induced colchiploidy in three lines. The pollen mother cells of the autotetraploid lines had multivalents along with bivalents and univalents at metaphase I leading to irregular segregations at anaphase I. These anomalies, as expected, resulted in high pollen and ovule abortion and very low seed set.

In case of tetarploid *F. cymosum*, Gohil et al. (1983) noticed a peculiar type of meiosis. Though at diplotene, 16 clear bivalents were observed, stickiness was prevalent at metaphase I with some bivalents lying towards the periphery. It could be because of the coupled effect of precocious separation and non-disjunction. This anomaly upset the future route of meiosis, characterized by the formation of chromatin bridges and laggards at anaphase I and anaphase II, leading to sterility. This stickiness is either caused by environmental factors operating in the microenvironment of the pollen sac or controlled by some late acting genes. Removal of this stickiness could possibly make this species amphimictic thereby conferring the advantages of both vegetative and sexual reproduction on this taxon.

## Physiochemical and Molecular Variability

Indian buckwheat species, exhibiting little cytological differences, show wide variations in physiochemical and molecular traits. Farooq and Tahir (1987) compared four cultivated species from Kashmir, viz. *F. esculentum, F. sagittatum, F. kashmirianum,* and *F. tataricum,* for leaf size, plant height, and dry weight of roots, leaves, and branches. Of the four, *F. tataricum* with a longer growth span (25−28 weeks) acquired maximum leaf area, height, and dry weight as compared to the other three species having a shorter growth span (10−12 weeks). In all the four species, a larger proportion of dry weight was partitioned into top parts than into roots, making them susceptible to lodging (Tahir and Farooq, 1991). Further analysis of the four taxa for growth and yield characteristics under pot culture conditions revealed a high percentage (78%−94%) of seed emergence in *F. esculentum, F. sagittatum,* and *F. kashmirianum,* and lowest (38%−48%) in *F. tataricum.* Amongst these, *F. esculentum,* which bloomed first and flowered profusely, as well as formed grains first of all, gave minimum yield, owing to high incidence of flower abortion (Tahir and Farooq, 1983). On the other hand, late flowering and subsequently late maturing, both shattering and non-shattering *F. tataricum* produced more grains. Under field conditions, however, the latter species gave the lowest grain yield owing to shattering and heavy lodging (Tahir and Farooq, 1983). *F. cymosum,* the wild species from Kashmir, produces only a few small sized seeds that fail to germinate (Tahir and Farooq, 1989a). This observation is in conformity with its cytological status and abnormal meiosis (Gohil et al., 1983).

Other traits like leaf thickness, stomatal size and index, and photosynthetic pigments, vary between four buckwheat taxa. *F. esculentum* possessing thicker leaves, larger stomata, higher stomatal index, and greater content of chlorophyll and carotenoids has a greater potential for growth. In fact, maximum dry mass, leaf area, and plant height was attained much earlier in this species than the remaining three (Tahir and Farooq, 1989b). The growing potential of grains of cultivated buckwheats and their efficiency to accumulate assimilates were determined by Tahir and Farooq (1989c) as weight per grain at three different stages of development (shriveled, partially-filled, and filled) along with their nitrogen and sugar contents. The common buckwheat, producing the heaviest grains and with higher nitrogen and sugar content, was found to be more sink efficient.

Investigations on biochemical characteristics of buckwheat have confirmed its nutritive value. The four cultivated species also differ in nature and composition of grains produced by them. Comparison of the same for the macromolecules revealed lowest content of phenolics in *F. esculentum,* which accounts for its better palatability (Tahir and Farooq, 1985, 1987b). The grains of *F. kashmirianum* and *F. esculentum* contain low fat, low sugar but higher starch which ranges from 37.23% to 46.99%, respectively. Of the four, *F. tataricum* has least protein content while *F. sagittatum* and *F. kashmirianum* have nearly the same amount (Farooq and Tahir, 1982). Amongst different protein fractions, the amount of prolamins is low and nearly the same in all the four taxa, whereas that of albumin-globulin is lower in the grains of *F. esculentum* than the remaining three species (Tahir and Farooq,

1985). The seeds of common buckwheat are also rich in iron (60−100 ppm) and zinc (20−30 ppm), and leaves are rich in rutin, an important antioxidant (Ahmad and Raj, 2012). While comparing polygraphs based on grain and leaf zymograms, *F. esculentum* turned out to be distinct from *F. sagittatum*, *F. tataricum*, and *F. kashmirianum*, with the latter exhibiting great deal of resemblance (Farooq and Tahir, 1987).

Since quantitative morphological traits are susceptible to environmental influences, it is important to supplement phenotypic traits with protein and DNA based markers for characterizing genetic diversity. Rout and Chrungoo (2007) have carried out cluster analysis based on SDS PAGE protein profiles of 11 different accessions each of *F. esculentum*, *F. tataricum*, and one of *F. cymosum*. Similarity estimates obtained thereupon made them (Rout and Chrungoo, 2007) segregate Indian Himalayan buckwheats into three broad clusters; cluster 1 included *F. esculentum* accessions, cluster 2 *F. tataricum* accessions, and *F. cymosum* remained as a separate group. In cluster 1, three subgroups were identified. Subgroup 1 included Local, Kamroo local 2, OC-2 and VL7 varieties which belong to central Himalayas, with VL7 standing apart from the remaining three. Subgroups II and III clubbed the accessions with winged seeds and those having stripes on the seed coat, respectively. The *F. tataricum* accessions in cluster 2 showed 100% similarity, most probably because of their practicing self-fertilization. Endosperm protein profiles of *F. esculentum* analyzed by Rout and Chrungoo (2007) revealed the presence of 42 and 31 kDa bands in accessions having winged grains and those having prominent striations, respectively. Another 41/39 kDa duplex band was found in the protein profile of VL7, a high yielding, early-maturing cultivar suitable for cultivation in mid hills. Such protein band/s can be used for the identification of accessions with specific morphological markers and also for tagging specific elite cultivars. So far as the extent of similarity between different species is concerned, *F. cymosum* showed the least similarity (39.4%) with *F. esculentum* and the maximum (57%) with *F. tataricum*. Further, principal component analysis on the SDS PAGE profiles place VL7 (released by VPKAS, Almora) and EC323729 (from Eastern Europe) accessions of common buckwheat are distinct from the others of the same species (Rout and Chrungoo, 2007).

Rout and Chrungoo (2007) also carried out random amplified polymorphic DNA (RAPD) analysis in 23 accessions of *F. esculentum*, *F. tataricum*, and *F. cymosum*, using 20 primers. While 1490 and 300 bp RAPD were detected only in *F. tataricum*, a 250 bp RAPD in one of its cultivars (KBB-3) and 1150 bp RAPD was observed in all Tartary buckwheat accessions except Shimla B-1 (early maturing and high yielding). Such diagnostic markers can be used for cultivar identification. RAPD profiles used further for generating dendrograms placed different accessions in three groups: I, II, and III. Group one included all the common buckwheat accessions which, however, were placed in different subgroups on the basis of similarity values that ranged from 47% to 100%. *F. tataricum* accessions placed in the second group were further segregated into two subgroups, with one including KBB-3, Sangla 1, 2, 3, 5, 6, 7, DRLT-1274 and another having Himpriya, Kuppa local, and Shimla B-1. Between different Tartary buckwheat accessions, the extent of similarity varied from 69% to 100%. The similarity matrix generated on RAPD using OPC-13 primer revealed closer similarity of *F. cymosum* to *F. tataricum* than to *F. esculentum*.

Senthilkumaran et al. (2008) have also characterized 29 populations of *F. esculentum*, 20 of *F. tataricum*, and 2 of *F. cymosum*, using RAPD analysis; more variability was found in accessions from Uttaranchal than those from Himachal Pradesh. Compared to *F. tataricum* accessions, those of *F. esculentum* were more diverse and as expected exhibited more heterozygosity. Further, differentiation among the populations of different zones (J&K, H. P., and Uttaranchal) of individual species is low; whereas differentiation between the species is relatively strong.

## Susceptibility to Diseases

In many parts of the world, buckwheat has not been willingly cultivated as it is considered a troublesome crop. The undersurface of leaves of Kashmir buckwheats as reported by Tahir et al. (1985) are infested by *Aphis gossypii* which lead to their downward curling, wilting, and finally drying up. The aphid infestation and consequent damage is reported to be more severe in *F. tataricum* than the other three species. Aphids through their secretions create an environment favorable for fungal growth and lead to the occurrence of smooth mold. The aphid population in nature is interestingly controlled by 14 predator species belonging to orders Diptera, Neuroptera, and Coleoptera (Bhat et al., 1986) that either consume this pest or its larvae.

## GERMPLASM CONSERVATION AND VARIETAL IMPROVEMENT

In order to carry out systematic research work for improvement of under-utilized crops and management of their genetic resources, an All India Coordinated Project was initiated in 1982 and buckwheat was included in the same in 1984. A large number of accessions of *F. esculentum*, *F. tataricum*, *F. giganteum*, and *F. cymosum*, collected from different agro-ecological regions, are maintained at various institutes under varying conditions (Joshi and Rana, 1994).

Ex situ accessions, as such, provide important backups to the in situ conservation at farm level, where species are expected to continuously evolve and generate more variability. In areas where buckwheat is grown as one of the major crops with rich diversity like Bharmour, Pangi, Sangla and Nesang of Himachal Pradesh, Kargil and Drass of Jammu and Kashmir and Kapkote, Darma valley and Garbyang in Uttar Pradesh, some efforts have been made at grassroots level to promote on-farm conservation by making locals aware of its culinary, medicinal, and other uses (Rana, 2004).

For analyzing and tapping the variability in buckwheat, workers (Joshi and Rana, 1994; Joshi, 1995, 1999) from NBPGR (Shimla), a rich repository of Indian buckwheat races, evaluated 408 accessions collected from different agro-ecological regions of the Himalayas. Taking into consideration the plant/leaf/flower/seed characters, they placed them in seven species, viz. *F. esculentum*, *F. emarginatum*, *F. tataricum*, *F. himalianum*, *F. cymosum*, and *F. giganteum*. These workers (Joshi and Rana, 1995) carried out morphological analysis and characterized 40 elite and ecologically diverse common buckwheat lines. Of these, the 13 most promising ones, with greater yield, were multiplied and grown under 15 environmental conditions at three locations. Four of these, viz. ICI3374, ICI3411, Kulu Gangri, and VL7, were high yielding over the years and showed consistent performance. Other varieties released include "Himgiri" of *F. tataricum* by NBPGR, Shimla, VL7 of *F. esculentum* by VPKAS, Almora, "PRB1" of *F. esculentum* by Govinda Vallab Pant Agriculture University and Technology, Ranichauri (Hore and Rathi, 2002), and Sangla 1 of *F. tataricum* by Himachal Pradesh Agriculture University, Palampur. For Tartary buckwheat known to have a tightly adhering hull which gives bitter taste to the flour, two lines (IC329457 and IC341679) with loose hulls have been documented. The hull of these genotypes, commonly called "rice buckwheat," can be removed by rubbing with the hands. Similarly, one Tartary (IC341671) and two common buckwheat lines (EC323729 and EC323731) have been selected for early maturity (60−65 days) as compared to the other lines (90−105 days). Recently, early maturing varieties have been in great demand in buckwheat growing areas because of the introduction of other cash crops. Since in the high Himalayas a gap of nearly 2−3 months exists between harvesting of the pea and the onset of snowfall, farmers have the option to use this period for cultivating early buckwheat varieties (Rana et al., 2012). Despite the release of elite varieties, people of the northeastern states are continuing with the time-tested material that has been available to them for so long (Hore and Rathi, 2002).

In Nepal, 318 working collections of *F. esculentum* and 193 of *F. tataricum* are maintained at the National Hill Crops Research Program at Kabre, 80 and 92 respective accessions at Central Plant Breeding and Biotechnology Division, Kathmandu, 200 accessions at National Agriculture Research Council at Khumaltar, some at Godawari, and 288 at Kirtinagar (Arora et al., 1995). A glance at the buckwheat accessions maintained in Nepal and India indicate that those of *F. tataricum* are less than that of *F. esculentum*. This is largely because of *F. esculentum* being the economically most important species and partly owing to difficulties associated with collecting *F. tataricum*, which grows in mountainous areas (Campbell, 1997).

Some steps were also taken to evolve technology for buckwheat cultivation in the plains (Narain, 1983). These initiatives, with some positive results, raised the scope of extending cultivation of *Fagopyrum* to the plains in India.

## FUTURE PROSPECTS

In the Indian subcontinent, on account of low yield coupled with low acceptability, this pseudo-cereal has not attracted the attention it deserves. While many as yet inaccessible areas remain to be scanned for variants, even the collections available have not been studied and tapped for its improvement. There is an urgent need to complete this work. It is also important to improve its acceptability by making people aware of its nutritional qualities and by providing them with the end product after making some value additions.

## ACKNOWLEDGMENTS

Authors wish to record gratefully the help rendered by Dr. Veenu Kaul in giving final shape to the manuscript.

## REFERENCES

Ahmad, F., Raj, A., 2012. Buckwheat: a legacy on the verge of extinction in Ladakh. Curr. Sci. 1, 13.

Arora, R.K., Engels, J.M.M., 1992. Buckwheat genetic resources in the Himalayan region: present status and future thrust. Buckwheat Genetic Resources in East Asia. Proc. of IBPGR Workshop, Ibarski, Japan, 18−20 September 1991. International Crop Networt Series. No. 6, IBPGR, Rome, pp. 87−91.

Arora, R.K., Baniya, B.K., Joshi, B.D., 1995. Buckwheat genetic resources in the Himalaya: their diversity, conservation and use. In: Matano, T., Ujihara, A. (Eds.), Current Advances in Buckwheat Research. Vol. I−III. Proc. 6th Intl. Symp. on Buckwheat in Shinshu. Shinshu University Press, Japan, pp. 39−45.

Baniya, B.K., Riley, K.W., Dongol, D.M.S., Shershand, K.K., 1992. Characterisation and evaluation of Nepalese buckwheat (*Fagopyrum* spp.) landraces. In: Rufa, L., Mingde, Z., Yongru, T., Jianying, L., Zongwen, Z. (Eds.), Proc. 5th Intl. Symp. on Buckwheat in Taiyun, China. Agricultural Publishing House, pp. 64−74.

Baniya, B.K., Dongol, D.M.S., Dhungel, N.R., 1995. Further characterisation and evaluation of Nepalese buckwheat (*Fagopyrum* spp.) landraces. In: Matano, T., Ujihara, A. (Eds.), Current Advances in Buckwheat Research. Vol. I−III. Proc. 6th Intl. Symp. on Buckwheat in Shinshu. Shinshu University Press, Japan, Taiyun, China, pp. 295−304.

Bhat, M.R., Bali, R.K., Tahir, I., 1986. Predator complex of melon aphid (*Aphis gosypii* Glov.), a serious pest of buckwheat in Kashmir (India). Fagopyrum 6, 12.

Campbell, C.G., 1997. Buckwheat *Fagopyrum esculentum* Moench, Promoting the Conservation and Use of Underutilised and Neglected Crop, vol. 19. International Plant Genetic Resources Institute, Rome, Itlay.

Chen, Q.F., 1999. A study of resources of *Fagopyrum* (Polygonaceae) native to China. Bot. J. Linn. Soc. 130, 53−64.

Farooq, S., Tahir, I., 1982. Grain characteristics and composition of some buckwheats (*Fagopyrum* Garetn.) cultivated in Kashmir. J. Econ. Tax. Bot. 3, 871−881.

Farooq, S., Tahir, I., 1987. Species relationship in buckwheat (*Fagopyrum* spp.) through electrophoresis of proteins and enzymes. Genetika 19 (I), 61−65.

Gohil, R.N., 1984. Buckwheat in India-past, present and future. Fagopyrum 4, 3−6.

Gohil, R.N., Misri, B., 1986. Comparative performance of eleven local buckwheat germplasms under uniform environmental conditions. In: Proc. 3rd International Symposium on Buckwheat, Poland.

Gohil, R.N., Rathar, G.M., 1981. Cytogenetic studies on some members of Polygonaceae of Kashmir II. Buckwheats. J. Cytol. Genet. 16, 59−63.

Gohil, R.N., Rathar, G.M., Tahir, I., Farooq, S., 1983. Comparative cytology, growth and grain composition of West Himalayan buckwheats. In: Nagatomo, T., Adachi, T.

(Eds.), Buckwheat Research, Proc. 2nd Intl. Symp. on Buckwheat in Miyazaki, pp. 87−102.

Gupta, D., 2001. Studies on Inter- and Intraspecific Variation in the Cultivated Species of Fagopyrum Mill. M. Phil. Dissertation, University of Jammu Jammu (J&K).

Hore, D., Rathi, R.S., 2002. Collection, cultivation and characterisation of buckwheat in North-eastern region of India. Fagopyrum 19, 11−15.

Jaretzky, R., 1928. Histologische and Karyologische studies an Polygonaceae. Jahrb. F. Wiss. Bot. 69, 357−490.

Joshi, B.D., Rana, J.C., 1995. Stability analysis in buckwheat. Ind. Jour. Agri. Sci. 65, 32−34.

Joshi, B.D., 1999. Status of buckwheat in India. Fagopyrum 16, 7−11.

Joshi, B.D., Paroda, R.S., 1991. Buckwheat in India. Sciences Monograph. National Bureau of Plant Genetic Resources, New Delhi, p. 117.

Joshi, B.D., Rana, R.S., 1994. Buckwheat Genetic Resources in South Asia: A Status Report. National Bureau of Plant Genetic Resources, New Delhi.

Munshi, A.H., 1982. A new species of *Fagopyrum* from Kashmir Himalaya. J. Econ. Tax. Bot. 3, 627−630.

Munshi, A.H., 1983. Studies on Himalayan polygonaceae-V *Fagopyrum sagittatum* Gilib.—a new record for India. J. Econ. Tax. Bot. 4, 959−960.

Nagano, M., All, J., Campbell, C.G., Kawasaki, S., 2000. Genome size analysis in the genus *Fagopyrum*. Fagopyrum 17, 35−39.

Narain, P., 1983. Buckwheat in India. Fagopyrum 3, 7−12.

Narain, P., 1987. Studies on buckwheat improvement. Fagopyrum 7, 19−20.

Ohnishi, O., 1989. Cultivated buckwheat and their wild relatives in the Himalayan and Southern China. Proc. 4th International Symposium on Buckwheat. USSR, Orel, pp. 562−571.

Ohnishi, O., 1991. Discovery of wild ancestor of common buckwheat. Fagopyrum 11, 5−10.

Ohnishi, O., 1992. Buckwheat in Bhutan. Fagopyrum 12, 5−13.

Ohnishi, O., Asano, N., 1999. Genetic diversity of *Fagopyrum homotropicum*, a wild species related to common buckwheat. Genet. Resour. Crop Evol. 46, 389−398.

Rathar, G.M., 1982. Cytogenetic Studies of Some Members of Polygonaceae of Kashmir. Ph.D. Thesis, University of Kashmir, India.

Rana, J.C., 2004. Buckwheat genetic resources management in India. In: Proceedings of the 9th International Symposium on Buckwheat, Prague, pp. 271−282.

Rana, J.C., Chauhan, R.C., Sharma, T.R., Gupta, N., 2012. Analysing problems and prospects of buckwheat cultivation in India. Eur. J. Plant Sci. Biotechnol. 6, 50−56.

Rout, A., Chrungoo, N., 2007. Genetic variation and species relationships in Himalayan buckwheat as revealed by SDS-PAGE of endosperm proteins extracted from single seeds and RAPD based fingerprints. Genet. Resour. Crop Evol. 54, 767−777.

Senthilkumaran, R., Bisht, L.S., Bhat, K.V., Rana, J.C., 2008. Diversity in buckwheat (*Fagopyrum* spp.) landrace populations from north-western Indian Himalaya. Genet. Resour. Crop Evol. 55, 287 302.

Sharma, A.K., Chatterjee, T., 1960. Chromosomes studies of some members of Polygonaceae. Caryologia 13, 486−506.

Shershand, K., Ujihara, A., 1995. Studies on some morphological and agronomical traits of buckwheat in Nepal. In: Matano, T., Ujihara, A. (Eds.), Current Advances in Buckwheat Research. Vol. I−III. Proc. 6th Intl. Symp. on Buckwheat in Shinshu. Shinshu University Press, Japan, pp. 285−393.

Stevens, N.F., 1912. Observations on the heterostylous plants. Bot. Gaz. 53, 277−308.

Tahir, I., Farooq, S., 1983. Growth and yield characteristics of some buckwheats (*Fagopyrum* Gaertn.) from Kashmir. Geobios 10, 193−197.

Tahir, I., Farooq, S., 1985. Grain composition in some buckwheat cultivars (*Fagopyrum* spp.) with particular reference to protein fractions. Qual. Plant. Plant Foods Hum. Nutr. 35 (2), 153−158.

Tahir, I., Farooq, S., 1987a. Breeding systems in buckwheat (*Fagopyrum* spp.) cultivated in Kashmir. Genetika 19 (3), 205−208.

Tahir, I., Farooq, S., 1987b. Occurrence of phenolic compounds in four cultivated buckwheats with particular reference to grain quality. Fagopyrum 7, 7−8.

Tahir, I., Farooq, S., 1989a. Growth and yield in four buckwheats (*Fagopyrum* spp.) grown in Kashmir. Fagopyrum 9, 44−46.

Tahir, I., Farooq, S., 1989b. Some morpho-physiological characteristics in four buckwheats (*Fagopyrum* spp.) grown in Kashmir. J. Econ. Tax. Bot. 13, 433−436.

Tahir, I., Farooq, S., 1989c. Grain analysis and leaf characteristics of perennial buckwheat (*Fagopyrum cymosum* Meissn.). Fagopyrum 9, 41−43.

Tahir, I., Farooq, S., 1991. Growth patterns in buckwheats (*Fagopyrum* spp.) grown in Kashmir. Fagopyrum 11, 63−67.

Tahir, I., Farooq, S., Bhat, M.R., 1985. Insect pollinators and pests associated with cultivated buckwheat in Kashmir (India). Fagopyrum 5, 3−5.

Yamane, K., Ohnishi, O., 2003. Morphological variation and differentiation between diploid and tetraploid cytotypes of *Fagopyrum cymosum*. Fagopyrum 20, 17−25.

Yang, X., Wu, Z.F., Chen, H., Shao, J.R., Wu, Q., 2010. Karyotype and genetic relationship based on RAPD markers of six wild buckwheat species (*Fagopyrum* spp.) from southwest of China. Genet. Resour. Crop Evol. 57, 649−656.

Yoshida, M.S., Matsudo, M.S., Hagiwara, A., Ujihara, A., Matano, T., 1995. Variation of days to flowering from various areas in the world. In: Matano, T., Ujihara, A. (Eds.), Current Advances in Buckwheat Research. Vol. I−III. Proc. 6th Intl. Symp. on Buckwheat in Shinshu. Shinshu University Press, Japan, pp. 405−410.

Zhou, M., Bai, D., Tang, Y., Zhu, X., Shao, J., 2012. Genetic diversity of four new species related to southwestern Sichuan buckwheats as revealed by karyotype, ISSR and allozyme characterization. Plant Syst. Evol. 298, 751−759.

## FURTHER READING

Tahir, I., Farooq, S., 1988. Review article on buckwheat. Fagopyrum 8, 33−53.

Tasuji, K., Yasui, Y., Ohnishi, O., 1999. Search for *Fagopyrum* species in eastern Tibet. Fagopyrum 16, 1−6.

# Chapter Sixteen

# Screening of Common Buckwheat Genetic Resources for Recessive Genes

**Vida Škrabanja[1], Ivan Kreft[2], and Mateja Germ[1]**
[1]*University of Ljubljana, Ljubljana, Slovenia,* [2]*Nutrition Institute, Ljubljana, Slovenia*

## INTRODUCTION

Common buckwheat (*Fagopyrum esculentum* Moench) is an obligatory cross-fertilized plant. The fertilization ability is regulated by a pin/thrum system, which determines not just the length of stylus and anthers, but also other characters (Kreft et al., 1979). The system is regulated by a specific locus, which is subjected to intense research—a comprehensive review and new data are reported by Ueno et al. (2016). One allele is dominant, and the other is recessive—heterozygotes express the dominant character. In common buckwheat, there is usually no self-fertilization. Because of regular outcrossing, buckwheat plants in populations, and as well cultivars, are heterozygous in many loci.

## RECESSIVE GENES IN THE POPULATION

Cross-fertilized populations are genetically and phenotypically very heterogenous. In such a way, recessive genes are hidden in the population and appear only when the gene is accidentally in a homozygous situation. The frequency of such homozygotes could be calculated by using the Hardy–Weinberg formulas, and in the case of two alleles it is as follows: $p + q = 1$; $p^2 + 2pq + q^2 = 1$; where $p$ is the relative frequency of one allele, $q$ is the relative frequency of another allele, $p^2$ is the relative frequency of one type of homozygotes, $q^2$ is the relative frequency of other type of homozygotes, and $2pq$ is the relative frequency of heterozygotes. However, if one type of heterozygotes or homozygotes as

**161**

*Buckwheat Germplasm in the World.* DOI: https://doi.org/10.1016/B978-0-12-811006-5.00016-1
© 2018 Elsevier Inc. All rights reserved.

compared to the other has a different survival rate and the different rate of effective fertilization (viable and effective pollen), the prediction of the rate of producing seeds with different genotypes by the Hardy—Weinberg formulas is not feasible. So, the relative frequency of alleles could change from generation to generation. If some type of pollen grain (haploid phase with expressed recessive genes) is less viable and with a lower expectancy of fertilization of the ovule, the genes of such pollen have a lower probability of contributing to the gene pool of the next generation. However, if a recessive gene is physiologically connected with pin/thrum phenotypic regulation, it is saved from extinction as the pin and thrum phenotypes have a tendency to be equally present in the population. So, the relative frequency of the recessive gene should be around 0.25, to give the equal relative presence (each around 0.5) of the respective phenotypes. We studied 14 domestic populations from Slovenia, 2 from Austria, and 1 from Italy (Z. Luthar and I. Kreft, unpublished results; Kreft, 2001) and no significant deviation was found from 0.5:0.5 ratio of frequencies of pin and thrum plants.

## RECESSIVE GENES AFFECTING STARCH SYNTHESIS, A CASE STUDY

Recently, buckwheat grain has been introduced to many countries because the flour and groats are nutritionally rich, and the foods prepared from buckwheat have beneficial effects on human health. Among the important health effects is low glycemic and insulin response (Skrabanja et al., 2001) after consumption of buckwheat.

Buckwheat milling products contain about 80% (w/w) starch depending on the production procedure (Skrabanja et al., 2004). The amylose content of starch is the basis for the appearance of retrograded starch during the hydrothermal processing of food materials (Skrabanja et al., 2001). Buckwheat is known as having relatively small starch granules and an amylose content of starch that is higher as compared to cereals (Skrabanja and Kreft, 1998).

In common buckwheat, the apparent amylose concentration ranges from 22% to 26% (Li et al., 1997). As the varieties have different starch characteristics, it is necessary to take this into account in processing buckwheat (Li et al., 1997, 2009). In addition to the genetic layout of buckwheat varieties, some contribution to the amylose content may have also growing conditions and availability of assimilates during the phase of buckwheat grain filling.

The extent and rate of starch digestibility, as influenced by the technological procedure, was investigated in buckwheat samples and it was established that the highest amount of resistant starch was determined in the cooked buckwheat groats, and about 6% of the total starch was resistant (Skrabanja et al., 1998, 2001). So, buckwheat groats made by the traditional method—popular in Slovenia—of boiling buckwheat grain prior to husking, followed by slow drying, contain less than 48% (in dry matter) rapidly available starch. It is much less in comparison to white wheat bread, with a respective value of around 59%.

Possible interactions of flavonoids with alpha-amylase receptors were studied by Kim et al. (2010).

Waxy (amylose-free) mutants are known to lack an effective starch granule-bound protein, known as waxy (Wx) protein, connected with amylose biosynthesis (Chrungoo et al., 2013). There is a need in buckwheat breeding to find and use waxy mutants for having alternative buckwheat material which would after hydrothermal treatment form as little as possible retrograded starch. There could be a need for waxy buckwheat grain for specific kneading and chewing properties due to the high percentage of amylopectin in waxy grain. As waxy mutants are expected to be recessive in comparison to usual alleles, a search for waxy forms by investigating the starch in pollen grains was performed. The haploid phase is expected to show full expression of recessive genes (Gregori and Kreft, 2012). Indeed, in the domestic Slovenian buckwheat variety of gray-grain buckwheat, several plants with about half of their pollen grains remaining unstained by iodine solution were found, and so the absence of amylose in affected pollen grains was confirmed. Such plants are obviously heterozygotes, with about half of their pollen grains being able to be stained with iodine solution, and other mutant half not stainable with iodine. Plants producing a part of pollen grains without amylose were crossed among themselves. In their progeny, several plants were found having only pollen grains not stained with iodine, and so being confirmed as homozygotes. However, the plants had defective pollen grains (some slightly smaller than normal) and defective endosperm, so there were problems in propagation and maintenance of the material.

Gregori and Kreft (2012) and Chrungoo et al. (2013) reported that in the low amylose homozygote recessive mutants, under the scanning electrone microscope (SEM), pinpricks on the surface of starch granules appeared visible, and within starch granules there were empty spaces between the amylopectin layers. Growth rings of alternate layers of buckwheat starch grain were reported after SEM research was combined with confocal laser microscopy (Chrungoo et al., 2013). However, even if the homozygotes for low amylose gene had poor viability, it is suggested that the method of screening the pollen grains could potentially result in finding some genes for better viable low amylose pollen, plants, and grain.

## SELF-FERTILIZATION IN BUCKWHEAT POPULATION, A CASE STUDY

At the Biotechnical Faculty, University of Ljubljana, we isolated the plants of a domestic population of gray-grained buckwheat from Vrhtrebnje, Slovenia to prevent cross-fertilization. After isolation at the time of flowering there were no seeds at all on a huge majority of the plants. The control plants, covered by the same type of water-resistant translucent paper bags, but cross pollinated, gave seeds, so the effect was not just the physiological impact of isolation condition and/or material. Only on a few plants some seeds were formed, which were the next year sown to study the subsequent generation. All progeny plants after such self-fertilization were very small, just a few centimeters in height, of unusual appearance and, if flowering, with highly deformed flowers. We were not able to find seeds on the progenies of selfed plants (Pundrić and Kreft, 1982; Kreft, 2001). The conclusion was that at least in the domestic population of common buckwheat on Vrhtrebnje there are frequent lethal or sublethal recessive genes preventing the growth and reproduction of the progeny after selfing.

## A RECESSIVE GENE AFFECTING DETERMINATE BUCKWHEAT GROWTH HABIT, A CASE STUDY

Common buckwheat plants are usually non-determinate, their and apex is able to grow further, as much as ecological conditions and physiological ability of plants allow. Buckwheat plants with a recessive gene "d" for the determinate growth habit were described by Fesenko (1968). Independently of this Russian finding, we observed few determinate plants in a domestic population of gray-grained buckwheat on Grmada mountain, Vrhtrebnje, Slovenia. The seeds of determinate plants were collected and multiplied in isolation. Most of the progeny plants were non-determinate. We understood this by condition that in the determinate plants of previous generations, their flowers were fertilized by surrounding plants without "d" genes, or being heterozygotes for this gene. We allowed the progeny of collected plants to cross-fertilize among them. After sowing the seeds for subsequent generation, we obtained many determinate plants. Non-determinate individuals were removed during flowering time, as soon as we established that they were non-determinate. In such a way, we minimized the possibility of plants with a dominant "D" allele to be spread in the material. After few generations of strict selection and removal of non-determinate plants, we obtained the population of only determinate plants. Besides the determinate growth habit, we observed more intense side branching in comparison to the previous domestic population. Side branching appeared because of missing apical dominance, otherwise the effect of active apex prevents the appearance of too many side branches.

## CONCLUSION

In screening buckwheat genetic resources for valuable recessive genes, it is important to take into account that recessive genes could be hidden in the population as a part of heterozygotes and that, according to the Hardy–Weinberg formulas (in the case of two alleles it is: $p + q = 1$; $p^2 + 2pq + q^2 = 1$), it is expected that they rarely appear expressed—so great number of individual plants should be screened in order to find those with expressed recessives. An alternative method is to self-fertilize plants, which, at least in the domestic population of Siva dolenjska buckwheat, gave a few progenies which expressed only sublethal and other deleterious properties. After the regeneration of haploid plants (tissue culture of pollen grains) doubled to dihaploids, the same types of problems are expected as after self-fertilization. One method is to screen pollen grains—a haploid phase—but only a few recessive genes for characters (for example amylose/amylopectin ratio) could be scored in this haploid phase.

## ACKNOWLEDGMENTS

This study was financed by the Slovenian Research Agency, through programs P3-0395 "Nutrition and Public Health" and "Biology of plants" P1-0212, and projects L4-7552 and J4-5524, supported by EUFORINNO 7th FP EU Infrastructure Programme (RegPot No. 315982) and co-financed by the European Community under project No. 26220220180: Building Research Centre "AgroBioTech." Ms. Ksenja Pundrić is thanked for performing isolation of the plants and for scoring the phenotypes of progeny plants. Prof. Zlata Luthar, University of Ljubljana, Biotechnical Faculty, is thanked for the nice cooperation in field experiments and in the evaluation of buckwheat phenotypes.

## REFERENCES

Chrungoo, N.K., Devadasan, N., Kreft, I., Gregori, M., 2013. Identification and characterization of granule bound starch synthase (GBSS-I) from common buckwheat (*Fagopyrum esculentum* Moench). J. Plant Biochem. Biotechnol. 22, 269–276.

Fesenko, N.V., 1968. Geneticheskij faktor, obuslovlivajushij determinantnij tip rastennija u grechihi. Genetika 4, 165–166.

Gregori, M., Kreft, I., 2012. Breakable starch granules in a low-amylose buckwheat (*Fagopyrum esculentum* Moench) mutant. Int. J. Food Agric. Environ. 10, 258–262.

Kim, H., Kim, J.K., Kang, L., Jeong, K., Jung, S., 2010. Docking and scoring of quercetin and quercetin glycosides against alpha-amylase receptor. Bull. Korean Chem. Soc. 31, 461–463.

Kreft, I., 2001. Morfološki znaki heterostilije in končne rasti pri navadni ajdi (*Fagopyrum esculentum* Moench) v Sloveniji. Razprave IV. Razreda SAZU 42, 143–151.

Kreft, I., Pundrić, K., Predalič, B., 1979. Prispevek k proučevanju dimorfizma peloda pri ajdi. Genetika 11, 9–14.

Li, W.D., Lin, R.F., Corke, H., 1997. Physicochemical properties of common and Tartary buckwheat starch. Cereal Chem. 74, 79–82.

Li, Y., Gao, F., Gao, F., Shan, F., Bian, J., Zhao, C., 2009. Study on the interaction between 3 flavonoid compounds and alpha-amylase by fluorescence spectroscopy and enzyme kinetics. J. Food Sci. 74, C199–C203.

Pundrić, K., Kreft, I., 1982. Expression of flower types and other characters in the progeny of self-pollinated buckwheat plants. Fagopyrum 2, 20.

Skrabanja, V., Kreft, I., 1998. Resistant starch formation following autoclaving of buckwheat (*Fagopyrum esculentum* Moench) groats. An in vitro study. J. Agric. Food Chem. 46, 2020–2023.

Skrabanja, V., Laerke, H.N., Kreft, I., 1998. Effect of hydrothermal processing of buckwheat (*Fagopyrum esculentum* Moench) groats on starch enzymatic availability in vitro and in vivo in rats. J. Cereal Sci. 28, 209–221.

Skrabanja, V., Elmstahl, H.G.M.L., Kreft, I., Bjorck, I.M.E., 2001. Nutritional properties of starch in buckwheat products: studies in vitro and in vivo. J. Agric. Food Chem. 49, 490–496.

Skrabanja, V., Kreft, I., Golob, T., Modic, M., Ikeda, S., Ikeda, K., et al., 2004. Nutrient content in buckwheat milling fractions. Cereal Chem. 81, 172–176.

Ueno, M., Yasui, Y., Aii, J., Matsui, K., Sato, S., Ota, T., 2016. Genetic analyses of the heteromorphic self-incompatibility (S) locus in buckwheat. In: Zhou, M., Kreft, I., Woo, S.H., Chrungoo, N., Wieslander, G. (Eds.), Molecular Breeding and Nutritional Aspects of Buckwheat. Academic Press, London, pp. 323–353.

# Chapter Seventeen

# Genetic Polymorphism of Common Buckwheat (*Fagopyrum esculentum* Moench.), Formed from Evolutionary Mutation Reserves of the Cultivar

**Lyubov K. Taranenko[1], Oleh L. Yatsyshen[2], Pavlo P. Taranenko[1], and Taras P. Taranenko[1]**

[1]*Scientific-Production enterprise Antaria, SEO, Ukraine, [2]National Scientific Center "Institute of Agriculture" of the National Academy of Agricultural Sciences, Ukraine*

## INTRODUCTION

The buckwheat species (*Fagopyrum esculentum* Moench.) has low competitiveness in phytocenosis in comparison with other grain crops due to specific architectonics of plants in connection with a high level of partitation.

Taking into consideration this fact, breeders of buckwheat are faced with the necessity of the improvement of buckwheat's genome potential. There are three main directions: (1) using evolutionarily obtained mutations, (2) receiving valuable recombination by means of intra- and interspecific hybridization (recombigenesis), and (3) the selection of prospective forms arising from the inbreeding of populations.

*Buckwheat Germplasm in the World.* DOI: https://doi.org/10.1016/B978-0-12-811006-5.00017-3
© 2018 Elsevier Inc. All rights reserved.

For the identification of mutant forms it is necessary to study hidden genetic reserves on the basis of inbreeding, to select valuable genotypes from ecological-geographical groups, to select different forms of genetic recombigenesis, and to study genetic nature with the goal of determining of a strategy of use in the breeding practice.

For the future prospects in the development of breeding, important traits are those that go beyond the limitations of the morphological and physiological characteristics of species (self-compatibility, determined growth, rational homeostasis of bearing). The only source of the traits is the evolutionarily-formed reserve mutations of buckwheat that are sufficiently different from intraspecific polymorphism.

Initial attempts to study gene mutations in buckwheat are described by Altgauzen (1910), Egiz (1924−1925), and Shubina and Tikhonova (1937).

Mutant forms of buckwheat with the following traits were selected: neoteny (Samoylovich, 1951); homostyly (Marshall, 1969; Zamiatkin et al., 1971; Fesenko and Antonov, 1973; Fesenko, 1983; Zakharov, 1980; Kovalenko and Shymova, 1986) limited branching (Fesenko and Antonov, 1973); short stem (Sabitov, 1984; Fesenko and Koblev, 1981; Zakharov, 1982; Taranenko, 1989); self-fertility (Bober and Taranenko, 1976; Anokhina, 1977); green flowers (Alekseeva, 1975); narrow leaves (Fesenko, 1986; Fesenko et al., 1991); "brachytic main stalk with high quantity of grains" form, form without leaves and with high quantity of grains (Taranenko, 1989), and others.

That is why one goal of the studies was the detection of spontaneous mutations—which are major source of genetic material—in addition to studying their genetic nature with the goal of determining a strategy of using them in breeding practice.

## MATERIALS AND METHODS

Taking place in the National Scientific Centre "Institute of Agriculture of UAAS", the following methods were used for the generation and selection of the gene diversity of buckwheat:
- The detection and identification of genotypes from different ecological-geographical groups
- Genetic recombigenesis with the help of intra- and interspecific hybridization with subsequent identification of a genotype's diversity by analysis of progeny.
- Inbreeding as a formative factor for differentiation of buckwheat's population by diversity of genotypes.

## RESULTS

Due to the use of inbreeding as a formatting process for the differentiation of buckwheat's population by diversity of genotype the following evolutionary mutant forms were selected: determinant forms with three types of inflorescences, three types of "green flower" traits (Figs. 17.1—17.3), "red flower" trait (Fig. 17.4) from the collection of VIR, shorter central stem (Fig. 17.5), brachysm (Fig. 17.6), fasciations (Fig. 17.7), and others (Table 17.1).

Gene diversity was identified in the following traits: determinate growth, green flowers, red flowers, self-compatibility, dwarf, fasciations, and "brachytic central stalk with inflorescences with high percentage of seeds". These were studied according to the "System of genetic studying of initial material" (Merezhko, 1994), and the methods of Serebrovsky modified by Rokitsky (1966).

**FIGURE 17.1** The inflorescences of self-compatible form with green flowers.

**FIGURE 17.2** The inflorescences of self-incompatible form with pink flowers.

**FIGURE 17.3** The inflorescences of self-compatible form with green flowers with "hat-like" inflorescences.

**FIGURE 17.4** The inflorescences of buckwheat with red flowers.

**FIGURE 17.5** Morphotype of a buckwheat (right) with brachytic main stalk.

**FIGURE 17.6** Dwarfish form of buckwheat.

**FIGURE 17.7** Form of buckwheat with stalk with fasciations.

**TABLE 17.1** Diversity and Spectrum of Buckwheat's Traits Detected by Different Methods

| Trait | Expression of Morphological or Biochemical Trait | Method of Receiving or Origin |
|---|---|---|
| Green flowers | • self-compatibility<br>• with pink ends of flowers<br>• self-incompatibility<br>• hat-like inflorescences | Inbreeding |
| Determinate growth | • single fringe-like inflorescence<br>• double fringe-like inflorescence<br>• shield-like inflorescence | Inbreeding |
| Brachysm | Different height of plants | Inbreeding |
| Red flowers | Flowers with different intensity of coloring | VIR collection, Ustimivska Experimental Station |
| Brachytic central stalk with inflorescences with high percentage of seeds | Morphological trait | Inbreeding |
| Fasciation | Morphological trait | VIR collection |

For determination of the character of inheritance of different types inflorescence and determinate forms, diallelic crosses between respective determinants and indeterminants (1, 2, and 4−5 inflorescences) were carried out.

The following genotypes were crossed:

1. Form with long single fringe-like inflorescence (Fig. 17.8)
2. Form with double fringe-like inflorescence which grow from the same axil of stalk (Fig. 17.9)
3. Bundle-like form with 4−5 inflorescences (Fig. 17.10)
4. Wild-type indeterminant form

By the character of inheritance of a type of inflorescences, which are different for determinant forms, in F1 an incomplete dominance of single fringe-like inflorescences over double fringe-like appeared. Complete dominance of double fringe-like inflorescences was observed over bundle-like inflorescences. Complementary interaction of the genes in F1 was observed in crosses between determinant plants with single fringe-like inflorescences and plants with bundle-like inflorescences.

According to the index of direct traits of productivity, determinate forms with different types of inflorescences had advantages over indeterminate wild-type forms. The greatest advantages were had by plants with double fringe-like inflorescences.

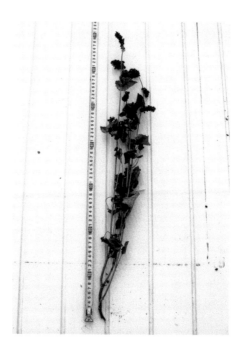

**FIGURE 17.8** Determinant plant of buckwheat with single fringe-like inflorescence.

**FIGURE 17.9** Determinant plant of buckwheat with double fringe-like inflorescences.

**FIGURE 17.10** Determinant plant of buckwheat with bundle-like inflorescences.

The analysis F1 of crosses between determinate and indeterminate forms showed that determinism is the recessive monogenic trait.

The value in forms with green flowers is in the fact that they have 4−6 fibro-vascular bundles in the pedicle while wild-type plants have only 2−4 bundles. It makes these forms a valuable source for breeding against the fall of grain. Also detected were differences in the intensity of coloring and the shape of flowers.

It was revealed that in the crosses between wild-type forms with white flowers and green flowers the development of green flowers is controlled by a recessive allele.

Studies are being conducted into the nature of the genetic control of green flower development from different sources, with the goal of finding a strategy for using the trait of green flowers in breeding programs for buckwheat resistance to the fall of grain.

The identified forms with red flowers are valuable due to their resistance to drought. Flower color varies from light pink to dark red and substantially depends on environmental conditions. An inheritance of the trait is intermediate (Fesenko and Antonov, 1973; Zheleznov, 1966) and according to the data of Zheleznov (1966) is controlled by two polymeric genes.

An expressivity of green and red flower coloring is, from our point of view, under strong influence of physiological and biochemical processes of adaptation to environmental conditions.

A form was detected which was called "Brachytic central stalk with inflorescences with high percentage of seeds". This form is characterized by shield-like inflorescences with a high quantity of grains. The grains are drop-shaped and well filled. The leaves are intensive green up to the end of ripening. This form is a very valuable "physiological recombination" with a specific type of regulation of photosynthesis, that ensures intensive fruitification that lasts up to the final ripening.

In studies of penetration of the trait in F1, received in crosses of wild-type indeterminate forms with the new selected form, the complete dominance of wild-type was observed.

In F2, for the trait "brachytic central stalk", from 212 descendants 187 were wild-type and 65 had brachytic central stalk. It is evidence of monogenic recessive control of the trait.

The other brachytic forms, differentiated by length of stalk, were selected and characterized. A brachytism can be controlled by recessive as well as by dominant genes.

Self-fertility of *Fagopyrum homotropicum* and its hybrids is defined by homostyly, which crosses with *F. esculentum* and is inherited as a monogenic trait. The hybrids F1 have Gg-type homostylic flowers because in crosses *F. esculentum* × (D-form-1, *F. homotropicum*) as *F. homotropicum* as *F. esculentum* carry the locus of heterostyly (recessive) that is repressed by the epistatic gene of homostyly (Fig. 17.11).

**FIGURE 17.11** Wild-type self-compatible plant with hat-like inflorescences.

That is why monogenic segregation between homostyly and absence of homostyly is observed in hybrids formed with long stamens × *F. homotropicum*. In this case, the absence of homostyly is expressed as long stamens.

After pollination of D-plants the transmission of the flower's characteristics was not disturbed (monogenic segregation 1:1). The segregation in F2 (*F. homotropicum/F. esculentum* D-form) had a ratio of 32:36 with homostyly with long stamens.

In the results of our studies were established particularities of inheritance of the following traits: determinant form, green flowers, red flowers, brachytism, wild-type homostyly (Table 17.2).

## CONCLUSIONS

Results identified mutations including three types of inflorescences, three types of green flowers, red flowers, "Brachytic central stalk with inflorescences with high percentage of seeds", and fasciations that were determined and classified as the natural diversity of buckwheat traits that is a widely occurring phenomenon in the plant's phylogenesis.

Patterns of gene control of inheritance of such traits as determinant form, green flowers, red flowers, brachytism, and wild-type homostyly were established.

**TABLE 17.2** Genetical Nature of Traits of Detected Gene Diversity of Buckwheat, 2014

| Trait | Alternative Manifestation of Trait | Method | Manifestation in F1 | Segregation in F2 | Gene Control |
|---|---|---|---|---|---|
| Determinant form | Types of inflorescences<br>Single<br>Double<br>Multy (4) | Inbreeding | Recessive | 187: 61 = 3.06:1<br>82:27 = 3.04:1<br>59:19 = 3.10:1<br>46:15 = 3.06:1 | Monogenic<br>Monogenic<br>Monogenic<br>Monogenic |
| Green flowers<br>13/2011<br>Dec-11<br>Mar-11 | Types of inflorescences<br>Self-compatible<br>Self-incompatible with red ends<br>Hat-like type | Inbreeding<br>Inbreeding<br>Inbreeding | Recessive | 208:68=3.06:1<br>73:24=3.04:1<br>91:30=3.03:1<br>44:15=3.11:1 | Monogenic<br>Monogenic<br>Monogenic<br>Monogenic |
| Red flowers | Color of flowers with different intensity | VIR collection<br>Ustimivska Experimental Station | Recessive | 2 genes (according [24]) | – |
| "Brachytic central stalk" | Morphological | Inbreeding | Recessive | 187:65=2.85:1 | Monogenic |
| Homostyly, self-fertility, uniformity of wild-type *Fagopyrum homotropicum* | Morphological | University of Kyoto (Japan) | Recessive | – | – |
| Dwarf form | Morphological | Inbreeding | Recessive | 74:19=3,8:1 | Monogenic |
| Brachytic form | Morphological | Inbreeding | Recessive | 74:19=3,8:1 | Monogenic (in testing) |
| Leaf-free form | Morphological | Inbreeding | – | – | – |

# REFERENCES

Alekseeva, E.S., 1975. Some features of green-flower buckwheat. Problems of Plant Oncology and Teratology. Nauka, Leningrad, pp. 87−89.

Altgauzen, L.F., 1910. From variety investigations of buckwheat. In: Report from the Office for Agriculture and Soil Science of the Scientific Committee Directorate-General agriculture and land management. Sankt-Peterburg, pp. 1−8.

Anokhina, T.A., 1977. Manifestation of self-incompatibility in monomorphic and dimorphic populations of buckwheat. Genetics 16 (1), 136−142.

Bober, A.F., Taranenko, L.K., 1976. The polycross method in buckwheat breeding. Genetics, Breeding, Seed Production and Cultivation of Buckwheat. Moscow, Kolos, pp. 93−95.

Egiz, S., 1924−1925. Experiments on the rationale of buckwheat breeding techniques. Trudy po prikladnoy botanike, genetike i selektsii 14, 1−17.

Fesenko, N.V., 1983. Buckwheat Breeding and Seed Production. Moscow, Kolos.

Fesenko, N.V., 1986. Narrow-leaved form of buckwheat "Goreths". Bulletin nauchno-tekhnicheskoy informatsii SRI zernobobovykh kultur 35, 40−41.

Fesenko, N.V., Antonov, V.V., 1973. A new homostylous form of buckwheat. Bulletin nauchnotekhnicheskoy informatsii SRI zernobobovykh kultur 5, 12−14.

Fesenko, N.V., Koblev, S.,Yu., 1981. Short-stem mutant "Orlovskiy Karlik" − donor of non-lodging. Selektsya i semenovodstvo 12, 20−22.

Fesenko, N.V., Suvorova., G.N., Фесенко, Н.В., 1991. Narrow-leaved form of buckwheat "Treugolnaya". Genetics, Breeding, Seed Production and Cultivation of Cereal Crops. Kishinev, pp. 25−29.

Kovalenko, V.I., Shymova, S.V., 1986. Destruction of heterostylism and main stages of S-locus evolution in buckwheat. Doklady of Moscow Society of Naturalists. General Biology, Moscow, pp. 66−68.

Marshall, H.G., 1969. Isolation of self-fertile homomorphic forms in buckwheat *Fagopyrum sagittatum* Gilib. Crop Sci. 9 (5), 651−653.

Merezhko, A.F., 1994. Program of phenogenetic screening of intraspecific diversity of cultivated plants. Genetics 30, 99.

Sabitov, A.M., 1984. Results and prospects of buckwheat breeding in the steppe zone of the Bashkir ASSR. In: Breeding, seed production and varietal agrotechnics in Bashkiria. pp. 98−102.

Samoylovich, I.F., 1951. One-stem buckwheat. Trudy Molotovskogo selskokhoziaystvennogo instituta 8, 18−24.

Shubina, A.F., Tikhonova, T.O., 1937. Biology of buckwheat flowering. Selektsya I semenovodstvo 10, 14−16.

Taranenko, L.K., 1989. Genetic rationale to improve buckwheat breeding methods *Fagopyrum esculentum* Moench. [dissertation], Kharkiv.

Zakharov, N.V., 1980. A new homostylous form of buckwheat and evaluation of it as a donor selfcompatibility. Bulletin nauchno-tekhnicheskoy informatsii SRI zernobobovykh kultur 26, 38−42.

Zakharov, N.V., 1982. Correlative variability of structural elements of buckwheat productivity. In. Breeding, seed production and cultivation technique of buckwheat. Orel, pp. 32−37.

Zamiatkin, V.F., Belilovskaya, A.S., Zamiatkin, F.E., 1971. Self-pollinating buckwheat. In: Breeding, genetics and biology of buckwheat. Bulletin nauchno-tekhnicheskoy informatsii SRI zernobobovykh kultur. Orel, pp. 103−111.

Zheleznov, A.V. Experimental generation of mutations in buckwheat. In: Experimental Mutagenesis in Agricultural Plants and Its Use in Breeding. Moscow, 1966. pp. 164−170.

# FURTHER READING

Fesenko, N.V., Antonov, V.V., 1975. Breeding of heterosis buckwheat hybrids based on periodic selection. Selskokhoziaystvennaya Biol. 10 (4), 605−609.

Rokitskiy, P.F., 1974. Introduction to Statistical Genetics. Vysheyshaya shkola, Minsk.

# Chapter Eighteen

# Cultivation of *Fagopyrum tataricum* and *Fagopyrum esculentum* in Order to Obtain Raw Material with High Rutin Content in the Far East of Russia

**Alexey Grigoryevich Klykov**

*Federal State Budget Scientific Institution "Far Eastern Regional Agricultural Scientific Centre" and Federal State Budget Scientific Institution "Primorsky Scientific Research Institute of Agriculture", Russia*

## INTRODUCTION

Species of *Fagopyrum* Mill. genus have valuable edible and medicinal traits. Medicinally, the raw leaves and tops of the shoots in bloom are used (Hinneburg and Neubert, 2005; Kreft et al., 2006). Representatives of the genus *Fagopyrum* are promising sources of flavonoids, particularly 3-O-rutinoside quercetin (rutin and vitamin P) (Odetti et al., 1990; Kreft, Fabjan, and Yasumoto, 2006). *Fagopyrum esculentum* Moench and *Fagopyrum tataricum* (L.) Gaertn. are two major sources of flavonoids. These species are rich in phenolic compounds, among which is rutin (Kreft et al., 1999), and therefore they can be used for the

**179**

Buckwheat Germplasm in the World. DOI: https://doi.org/10.1016/B978-0-12-811006-5.00018-5
© 2018 Elsevier Inc. All rights reserved.

production of functional foods (Brunori et al., 2009). Rutin, as a medical product, is included in many preparations: "Rutin", "Venoruton", "Ascorutin". Antidiabetic effects of rutin were also noticed (Srinivasan et al., 2005), along with protective effects, preventing hemoglobin oxidation (Grinberg et al., 1994). Rutin decreases the capillary fragility associated with hemorrhagic changes and has antiinflammatory effects (Guardia et al., 2001). It also increases the organism's resistance to cardiovascular disease and cancer (He et al., 1995; Liu, 2002; Zhou and Song, 2009). *Fagopyrum tataricum* and *Fagopyrum cymosum* Meissn. are used as an edible and medicinal crop in South-Eastern Asia (China, India). *Fagopyrum esculentum* is a cereal and melliferous crop which is widely cultivated in many countries around the world. In Russia, common buckwheat is mostly used as a food crop for the production of cereals (groats).

Plants of species of *F. esculentum* are widely used in popular medicine. Furthermore, the content of the flavonoids in buckwheat is influenced by various environmental factors as well. Fabjan et al. (2003) suggested that the content of rutin varied from 1.22% to 1.66%, which implied the complexity of the crop development and the integrated conditions under specific circumstances that may influence the chemical ingredients simultaneously, such as moisture of the soil, etc.

It was also important to explore the possibility of the use of the secondary resources of buckwheat for practical purposes. In production, a significant amount of waste (secondary resources) is generated in the form of straw and fruit (buckwheat hull), which so far has not been effectively implemented. The straw proportion in the total aboveground mass of the plant depends on the variety and amounts to 40%−60% whilst the hull is 20%−30% (from the weight of grain). The main advantage of the secondary raw materials is their annual reproducibility and low cost. Therefore, one of the most promising directions of the complex usage of secondary raw material resources in the agro-industrial complex and for the introduction of ecologically safe methods for their utilization could be usage of secondary resources as an additional source of raw material for the pharmaceutical industry to obtain rutin and micro-fertilizers, as well as to solve problems of environmental pollution. This necessitates the development of cultivation technology of buckwheat to produce the plant raw material on an industrial scale with maximum rutin content.

## Materials and Methods

*Plant materials.* Field experiments were carried out from 1997−2016 in the Primorsky Scientific Research Institute of Agriculture (44.34°N, 131.58°E), Ussuriysk district, Primorsky krai, Russia and in the Pacific Institute of Bioorganic Chemistry named after G.B. Elyakov, the Far Eastern Branch of Russian Academy of Sciences. Members of the family Polygonaceae Juss were used as objects of the study: cultivated species of the genus *Fagopyrum* Mill. (*F. esculentum* Moench)—Izumrud variety, Russia (*F. tataricum* (L.) Gaertn.)-k-17, China.

The soil of the experimental field was meadow-brown, bleached. Power of the arable layer is 22 cm, humus content is 3.9%, $P_2O_5$—19.1 mg/100 g of soil, $K_2O$—15.2 mg/100 g of soil, pH—5.8. The field experiment for the study of the technological influence upon rutin content in buckwheat and crop productivity took place on an experimental plot with an area of 10 $m^2$ with 6 repetitions. Study of planting dates effecting rutin content in the aboveground mass of *F. esculentum* included variations, with sowing on May 30, 15, June 30, 15, and July 30. Study of sowing norm effect and effect of doses of nitrogen fertilizers on the background of phosphate and potash upon rutin content in aboveground mass of *F. esculentum* and *F. tataricum* has two factors: Factor A— Sowing norm: 1.0, 1.5, and 2.0 million viable seeds per hectare; Factor B— Dose of fertilizers: without fertilizers (control), $N_{30}P_{60}K_{60}$, $N_{60}P_{60}K_{60}$, and $N_{90}P_{60}K_{60}$. Mineral fertilizers were applied in the form of ammonium nitrate, potassium chloride, and ammophos. Calculation of applied fertilizers was done on the active substance, taking into account the composition of each fertilizer. Experiments on micro-elemental effects upon rutin content in aboveground mass of *F. esculentum* and *F. tataricum* included the following variants: the plants treated with distilled water (control); with 0.05% solution of $CuCl_2$; with 0.05% solution of Co $(CH_3COO)_2$ $4H_2O$, with 0.05% solution $H_3BO_3$, with 0.05%-s solution of $MnSO_4$.

*Definition of rutin.* Rutin was defined by the method of M.M. Anisimova (2011). Max. 30 plants of each genotype were used for the analysis. Dry matter content and therefore the so-called total rutin were determined in whole plants. Absorption of the extracted solution was measured at 360 nm spectrophotometer Shimadzu UVmini on-1240 (Japan) and compared to that of a standard curve rutin. For rutin identification NMP [1]H spectrums were used that were registered on spectrometer AC-250 in $CDCl_3$ and $d_6$. Then they compared the spectrums with the pure rutin ("Chemopol", Czech Republic). Mass spectrums were mass-produced on the equipment LKB-9000S (Sweden) with straight input under the energy of ionizing electrons equal to 18 and 70 eV (electron volts). Absorption spectrum of the buckwheat alcoholic extract was recorded on a spectrophotometer Shimadzu UVmini-1240 (Japan). In all the variants, the rutin content in the stems was determined both in the mature phase and in buckwheat seeds. Reliability of the results between control and test samples was evaluated using Student *t*-test ($P < 0.05$).

*Statistical analysis.* The means and standard deviations were calculated using Microsoft Office Excel 2003. Significant differences in these data were calculated using analysis of variance (ANOVA-Duncan's multiple test, SIGMASTAT 9.0). The reliability of the results between the control and experimental samples was evaluated using Student's *t*-test.

## Results and Discussion

*Influence of planting terms, seeding rate, fertilizers, and microelements upon rutin content.* In the conditions of Primorski Krai (Russian Far East) seeding of *F. esculentum* was conducted over five dates: May 30, June 15 and 30, and July 15 and 30. It was determined that rutin content increases from May 30 until July 15. The maximum amount of rutin was observed in the planting term of July 15 (37.5 mg/g). In the later sowing term (July 30) rutin content in the plant did not exceed 34.1 mg/g. High yield of green mass (2.2 kg/m$^2$) and dry matter (300 g/m$^2$) were formed when sown on June 30. In the conditions of Primorski Krai (Russia), sowing of *F. esculentum* for usage as raw materials for the pharmaceutical industry is advisable to perform from June 30 until July 15. The study showed that one of the most effective factors effecting rutin accumulation is the seed norm. The highest rutin content of *F. tataricum* (42.5 mg/g) was observed when the sowing norm was equal to 1 million viable seeds per 1 ha. Increase of the sowing norm to 2 million viable seeds per 1 ha reduces rutin content by 29%. This proves that the sowing norm has a significant influence upon rutin accumulation in *F. tataricum*. While examining ways of planting and the sowing norms of *F. esculentum* it was noted that the largest number of plants with a predominance of anthocyanin coloration were formed in the wider spaced rows of crops, with 45 cm of inter row space and with a sowing norm of 1.0 and 1.5 million of seeds per hectare. It was revealed that only plants in the well-lit parts of the crop (especially at the edge of the field) have anthocyanin coloring. Towards the center of the field, where they are thick, there is only soft sun lighting, and the plants are green-red either with little visible sign of anthocyanin or with no sign (green).

There was a defined relation between color of plants and rutin content in the aboveground mass. Plants with red coloring contain more rutin (Klykov and Moiseenko, 2005). Ways of planting and sowing norms can create conditions for a higher accumulation of rutin in aboveground mass of *F. esculentum* in order to obtain raw materials for the pharmaceutical industry. Knowledge of the variable nature of the trait provides the opportunity to use directly it for plant evaluation in selection and seed breeding. When you use the best backgrounds of nutrition and sowing norms, seed production of *F. esculentum* increases by 1.3 times.

According to Zaprometov (1993), nitrogen nutrition has a significant effect upon the formation of phenolic compounds. Deficiency of nitrogen in soil stimulates accumulation of phenolic compounds in various organs of plants and excess of nitrogen nutrition is accompanied by intensive development of overground mass and decrease of phenolic compounds in it.

Among researchers, points of view on effectiveness of application of mineral fertilizers for *F. esculentum* are different. In more detail, the effect of mineral nutrition elements upon the formation of phenolic compounds in whole plants or in some organs of plants was examined in the review of McClure (1979). Our research shows that mineral fertilizers effect rutin content in aboveground mass at *F. esculentum* and *F. tataricum*. In increasing doses of nitrogen fertilizers on the background of $P_{60}K_{60}$, rutin content is reduced (Table 18.1). Maximum rutin content in *F. tataricum* (45.7 mg/g) and

**TABLE 18.1** Rutin Content in Aboveground Mass and the Dry Matter Content Depending on the Use of Fertilizers and Microelements in the Phase of Mass Flowering of *Fagopyrum esculentum* and *Fagopyrum tataricum*

| Variant | *F. esculentum* Rutin (mg/g) | *F. esculentum* Dry Matter (g/m$^2$) | *F. tataricum* Rutin (mg/g) | *F. tataricum* Dry Matter (g/m$^2$) |
|---|---|---|---|---|
| Control (no treatment) | $36.0 \pm 0.1$ | $262.1 \pm 6.7$ | $42.1 \pm 0.2$ | $314.7 \pm 8.3$ |
| $N_{30}P_{60}K_{60}$[a] | $39.3 \pm 0.2$ | $289.7 \pm 5.3$ | $45.7 \pm 0.2$ | $335.3 \pm 9.3$ |
| $N_{60}P_{60}K_{60}$[a] | $35.4 \pm 0.1$ | $295.9 \pm 7.8$ | $39.8 \pm 0.1$ | $352.6 \pm 9.9$ |
| $N_{90}P_{60}K_{60}$[a] | $33.8 \pm 0.1$ | $300.8 \pm 9.0$ | $34.6 \pm 0.1$ | $359.5 \pm 9.2$ |
| $CuCl_2$[b] | $38.9 \pm 0.2$ | $270.5 \pm 8.1$ | $42.9 \pm 0.2$ | $360.9 \pm 9.7$ |
| $Co\,(CH_3COO)_2 \cdot 4H_2O$[b] | $39.5 \pm 0.2$ | $272.3 \pm 7.7$ | $43.4 \pm 0.2$ | $320.1 \pm 8.1$ |
| $H_3BO_3$[b] | $36.7 \pm 0.1$ | $279.7 \pm 5.8$ | $46.7 \pm 0.2$ | $340.7 \pm 8.5$ |
| $MnSO_4$[b] | $38.6 \pm 0.2$ | $275.6 \pm 6.8$ | $47.9 \pm 0.2$ | $360.8 \pm 9.0$ |

[a] Applying fertilizer into the soil before sowing.
[b] Plants treatment by the solution in concentration of 0.05% at the budding stage.

*F. esculentum* (39.3 mg/g) was observed in the plants on the background of $N_{30}P_{60}K_{60}$, minimal rutin content was observed on the background of $N_{90}P_{60}K_{60}$ (34.6 and 33.8 mg/g, respectively). While examining rutin accumulation in the green mass of the crop, we paid attention to the fact that mineral nutrition in a dose of $N_{30}$ on the background of $P_{60}K_{60}$ has a positive effect on the concentration of the substance throughout the growing season. Biosynthesis of flavonoids is weakened by the influence of nitrogen, because two systems of the plants are competing for the amino acid phenylalanine, i.e., protein and phenol synthesis. In our experiments at the period of green mass formation, the mineral nutrition in the dose of $N_{30}P_{60}K_{60}$ enhanced rutin accumulation. This is due to the effective usage of the main nutrition elements by the plants on meadow-brown soil. Along with general activation in growth and other vital processes (including protein synthesis), there are simultaneously activated enzyme systems participating in the formation of molecules of flavonoids. These systems have extremely high efficiency even in the most adverse conditions (enhanced capture of phenylalanine for construction of protein molecules). Specifically, at the flowering stage, when polyphenol exchange is especially amplified, under the influence of mineral nutrition in aboveground mass of *F. esculentum* and *F. tataricum*, increased content of the most oxidized flavonoids-Flavonols (rutin) is more evident. Increase in doses of nitrogen fertilizers on the background of phosphate and potash increases productivity of green mass and dry matter.

The highest yield of green mass and dry matter of *F. esculentum* and *F. tataricum* was formed due to applying fertilizers in dose $N_{90}P_{60}K_{60}$. In order to clarify the relation between rutin content and nitrogen, the steam method of correlation was used. A negative correlation ($r = -0.63$) was defined in above-ground mass between rutin content and nitrogen applied with mineral fertilizers in high doses $N_{60}$ and $N_{90}$. This was due to the decrease of rutin biosynthesis under the effect of nitrogen in high doses. In addition, it was observed that in the conditions of abundant nitrogen nutrition of plant, natural anthocyanin coloration weakens, i.e., plants change the usage of nitrogens for protein molecule formation. Maximum rutin output of *F. esculentum* and *F. tataricum* per $m^2$ was obtained at the stage of mass flowering under applying fertilizers in dose $N_{30}P_{60}K_{60}$. Application of balanced mineral nutrition gives the possibility to smooth the competitive relationship between the processes of protein biosynthesis and the accumulation of flavonoid substances to ensure higher levels of rutin accumulation in the green mass of *F. esculentum*.

For high yield formation of aboveground mass applying only nitrogen, phosphorus, and potassium is not enough. All elements of mineral nutrition including microelements are necessary. Study of the influence of microelements (boron, copper, manganese, cobalt) on rutin content in aboveground mass of buckwheat defined that the greatest amount (39.5 mg/g) was noted in *F. esculentum* under the treatment of plants by cobalt solution at a concentration of 0.05%. When *F. tataricum* was treated by the solution of manganese at a concentration of 0.05% its rutin content reached 47.9 mg/g. Foliar fertilizing with cobalt and boron ensures more intensive growth and increases biomass and dry matter of *F. esculentum*. Applying copper and manganese ensures more intensive growth and increases biomass and dry matter of *F. tataricum*. The study revealed that the most rutin output per $m^2$ was obtained from *F. esculentum* under treatment with cobalt (10.8 g) and manganese (10.7 g). As for *F. tataricum*, the dosage of the treatment consisted of manganese (15.5 g) and boron (14.3 g). When *F. esculentum* plants were treated by cobalt and manganese, and *F. tataricum* plants were treated by boron and manganese, they increased both rutin and productivity of aboveground mass. The study results can be practically applied for cultivation of *F. esculentum* and *F. tataricum* to obtain rutin.

*Influence of biologically active substances upon rutin content in Fagopyrum esculentum.* Application of natural, ecologically pure biological preparations in agriculture allows you not only to increase productivity and protect plants from diseases, but also to get clean, safe products. In recent years researchers have intensively studied the biological activity of chemical compounds derived from algae which belong to different classes: the low molecular metabolites of different classes of lipids, pigments, polysaccharides (Mori et al., 2004; Khan et al., 2007).

Research has identified the varietal specificity of *F. esculentum* towards the growth regulator. There was studied effect of aqueous extracts of red algae, *Grateloupia divaricata, Chondrus pinnulatus, Ahnteltiopsis flabelliformis, Neorhodomela larix, Tichocarpus crinitus*, brown algae *Stephanocystis crassipes, Coccophora langsdorfii, Sphaerotrichia divaricata, Saccharina japonica, Sargassum pallidum, Chorda filum*, and green algae *Ulva fenestrata* and

*Codium fragile,* collected in October 2011, red algae *N. larix, T. crinitus,* brown algae *S. japonica, S. pallidum,* and green algae *U. fenestrata* and *C. fragile,* collected in November 2011, in January, May, and August 2012, upon the growth of seedling roots of *F. esculentum* (Izumrud variety). The article shows that the most expressed stimulating effect was observed in extracts of red algae *A. flabelliformis, N. larix,* collected in October 2011. These algae extracts increased root growth of sprouts of *F. esculentum* at a maximum of 16%−20% compared with the control. Extracts from the other studied algae collected during this period showed weak activity (Anisimov and Klykov, 2014). In field experiments, aqueous extract of *N. larix* stimulated productivity of *F. esculentum* by 46% and rutin content in grainby by 20%. We studied the influence of ethanolic extract, fractions of polyphenolic compounds, total lipids; different classes of lipids monogalactosyldiacylglycerol, digalactosyldiacylglycerol, sulfoquinovosyldiacylglycerol; fatty acids; and pigments: chlorophyll and fucoxanthin derived from *Laminaria cichorioides* Miyabe, which is brown algae, widespread in the Far Eastern seas, with effects upon the growth of seedling and productivity of *F. esculentum.* Maximum stimulating effect upon root growth of seedlings of *F. esculentum* provided: ethanolic extract, extracts of chlorophyll, fucoxanthin, digalactosyldiacylglycerol and sulfoquinovosyldiacylglycerol at concentration of 1 μg/mL. Fractions of polyphenolic compounds, monogalactosyldiacylglycerols, sulfoquinovosyldiacylglycerols, fatty acids and fucoxanthin at concentration of 100 μg/mL. The field experiments proved that seed treatment of *F. esculentum* by ethanolic extract of *L. cichorioides* at a concentration of 2 mg/mL had a stimulating effect upon the morphological traits of *F. esculentum*: plant height (9.2%), the number of inflorescences per plant with fruit (48.6%), productivity of one plant (30.6%), and rutin content in aboveground mass (22.2%).

Brown algae polysaccharides are characterized by great diversity. Fucoidans differ in monosaccharides composition, apart from in that fructose in their structure may be present—(Galactose, Xylose, mannose, arabinose, rhamnose, glucuronic acid), the degrees of sulfating and molecular weight differ too. The influence of polysaccharides derived from algae was studied: laminaran from *Laminaria* (*L. cichorioides*), (*L. gurjanovae* A. Zin.); AntiVir and 1.3; 1.6-β-ᴅ glucooligosaccharides-products of the enzymatic transformation of laminaran from *Laminaria*; fucoidan from *Laminaria*; *L. japanese* (*L. japonica* Aresch), *Fucus* (*F. evanescens* C. Ag.); as well as mixtures of laminaran and fucoidan from *Laminaria* (4:1); *Laminaria* (*L. gurjanovae* A. Zin.) (5:1) and mixtures of fucoidan and polimannuronic acid (2:1) of *Undaria pinnatifida* (Harv.) Sur. on seed germination and seedling growth of *F. esculentum.* It was defined that fucoidan and fucoidan mixture and polimannuronic acid (2:1) stimulated germination of seeds and seedling root growth of *F. esculentum.* Sprouts root was 15.4% and 22.9% longer than controls when the substance concentration was 1 and 10 μg/mL, respectively. Under fucoidan effect roots were 14.5% shorter than the control when the substance concentration was $100\,\mu g/mL$. 1.3; 1.6-$\beta$-D-glucooligosaccharide accelerate the germination of seeds of *F. esculentum*, increasing germination energy, size, and total weight of the roots of seedlings at the early stages of development (1-2-e day). Samples of branched 1.3; 1.6-$\beta$-D-glucooligosac-charids cause the formation of strong roots, as well as short and persistent hypocotyl, preventing excessive growth of sprouts of *F. esculentum*.

Also studied was the effect of algae extract *Grateloupia divaricata*-$10^{-9}$ g/mL and $10^{-10}$ g/mL; *N. larix*—$10^{-7}$ g/mL and $10^{-8}$ g/mL; extract derived from the population of red-flowered and red-stemmed *F. esculentum*-$10^{-8}$ g/mL and a population of pink-flowered-$10^{-6}$ g/mL; alkaloid verruculo-gen (3) out for *Aspergillus fumigatus* $10^{-7}$ m and $10^{-8}$ m for economically valuable traits of *F. esculentum* (Klykov et al., 2014). The study showed that the alkaloid verruculogen and aqueous extracts of the seaweed and *F. esculentum* have a stimulating effect upon 1000 grains mass, on the number of inflorescences per plant, and the productivity of *F. esculentum* (Table 18.2).

**TABLE 18.2** Effect of Extracts Derived From Marine Algae, Buckwheat, and Alkoloid From Fungus *Aspergillus fumigatus* Upon Productivity, Morphological Characteristics, and Rutin Content in *Fagopyrum esculentum*

| Variant | Mass 1000 Seeds (g) | The number of Inflorescences Per Plant, PCs | Productivity of Plant (g) | Rutin (mg/g) |
|---|---|---|---|---|
| Control (no treatment) | $30.0 \pm 0.1$ | $19.1 \pm 0.2$ | $0.81 \pm 0.03$ | $14.0 \pm 0.1$ |
| **Extract derived from *Grateloupia divaricata*** | | | | |
| $10^{-9}$ g/mL | $31.8 \pm 0.2$ | $20.9 \pm 0.2$ | $0.96 \pm 0.02$ | $16.7 \pm 0.1$ |
| $10^{-10}$ g/mL | $31.6 \pm 0.1$ | $20.0 \pm 0.2$ | $0.91 \pm 0.03$ | $15.8 \pm 0.1$ |
| **Extract derived from *Neorhodomela larix*** | | | | |
| $10^{-7}$ g/mL | $32.0 \pm 0.2$ | $22.0 \pm 0.3$ | $1.15 \pm 0.05$ | $14.9 \pm 0.1$ |
| $10^{-8}$ g/mL | $31.6 \pm 0.1$ | $22.6 \pm 0.3$ | $1.18 \pm 0.04$ | $17.6 \pm 0.1$ |
| **Extract derived from *F. esculentum*** | | | | |
| (Population red-stemmed, red-flowered)-$10^{-8}$ g/mL | $31.7 \pm 0.1$ | $25.0 \pm 0.4$ | $1.35 \pm 0.06$ | $17.8 \pm 0.1$ |
| (Population pink-flowered)-$10^{-6}$ g/mL | $35.3 \pm 0.3$ | $21.4 \pm 0.2$ | $1.17 \pm 0.03$ | $14.9 \pm 0.1$ |
| **Alcaloid, verruculogen (3) from *A. fumigatus*** | | | | |
| $10^{-7}$ M | $34.5 \pm 0.2$ | $22.7 \pm 0.2$ | $1.19 \pm 0.04$ | $16.5 \pm 0.1$ |
| $10^{-8}$ M | $35.2 \pm 0.2$ | $23.5 \pm 0.3$ | $1.29 \pm 0.06$ | $17.9 \pm 0.1$ |

Alkaloid verruculogen derived from *A. fumigatus* in concentration of $10^{-8}$ g/mL and water extract of red-stemmed population of *F. esculentum* in concentration of $10^{-8}$ g/mL stimulated productivity of one plant by 59% and 67%, respectively. The largest number of plants with red and red-green color with high rutin content was defined under treatment by the extract in red-stemmed and red-flowered populations of *F. esculentum* $10^{-8}$ g/mL under treatment by sea algae extract *N. larix*-$10^{-8}$ g/mL and alkaloid, verruculogen (3) in the dose of $10^{-7}$ M. The control plants were with low rutin content, consisting of plants with green and red-green color. Therefore the color of the plants can be used as a visual diagnostic sign for the selection of forms with a high content of rutin. The maximum rutin content in plants of *F. esculentum* (21.4% more than in the control) was obtained when spraying their extracts of *N. larix* ($10^{-8}$ g/mL), on red-stemmed and red-flowered populations of *F. esculentum* ($10^{-8}$ g/mL) and alkaloid verruculogen (3)-$10^{-8}$ M. The study revealed that the most promising, i.e., extract for *F. esculentum* (red-stemmed and red-flowered populations)-$10^{-8}$ g/mL; extract of sea algae *N. larix*-$10^{-8}$ g/mL; alcaloid, verruculogen (3)-$10^{-8}$ M, which help to increase rutin in aboveground mass of *F. esculentum*, positively affect the number of inflorescences per plant, weight of 1000 grains, and productivity. The study proved that the biologically active substances increase rutin in seeds and overground mass of *F. esculentum* (up to 20%), positively effect upon morphological traits of plants and technological indicators, and increase productivity.

*Prospects for usage of production wastes of Fagopyrum esculentum.* Buckwheat nowadays is widely used in various industries. It seems to us that in the near future the most promising will be the involvement of fruit shells (the hull that accumulates in large numbers on cereal plants) into processing. The Institute of Chemistry of the Far Eastern Branch of the Russian Academy of Sciences together with FSBSI "Primorsky Scientific Research Institute of Agriculture" studied content and composition of polysaccharides of fruit shells and straw of Izumrud and Pri7 varieties of buckwheat derived using sequential extraction with water, and solutions of ammonium oxalate and sodium hydroxide (Zemnuhova et al., 2004).

Our research proved that the content of polysaccharides depends on the type of raw material (fruit shell or straw) and the variety. The largest output of polysaccharides from the wastes was achieved in all cases with the first (aquatic) extraction. The color of the dry product depends on the initial raw material and the method of obtaining polysaccharides and has a white, light-brown, or nearly black color. Polysaccharides of water and oxalic extracting of all the samples are characterized by a high content of glucose. Polysaccharides of alkaline extraction have complex and monosaharide content. They also contain residues of rhamnose, arabinose, xylose, mannose, glucose, and galactose. Inositol is found in trace amounts. Uronic acids are presented by glucuronic acid galacturonic acids. While extracting raw materials into the solution, along with organic substances, there are also extracted metals from the plant content. The data of the study prove that polysaccharides derived from the extracts in solid state can absorb metals in the solution. The rest of the raw material, nondissolved under extraction, practically has no ash.

Studies showed that the ash content received after burning fruit shells of *F. esculentum* (at 600°C) was an average of 2%. In the ash there were found the following elements: potassium, natrium, copper, silver, calcium, magnesium, zinc, aluminum, manganese, iron, nickel, chromium, and phosphorus, and their concentration depends on the variety and type of raw material of *F. esculentum*. It is significant from the point of view of usage of the fruit coat of *F. esculentum* as a secondary raw material. To increase *F. esculentum* productivity the pre-sowing preparation of seeds is important. We found that the seeds' treatment by the ash of fruit shells of *F. esculentum* in the norm of 1 kg/ton has provided the greatest harvest increase (2.2 kg/ha). The analysis of the literature and experience on the possibility of use of the wastes *F. esculentum* as a raw material for obtaining valuable chemical products shows that although the list of proposed usage methods of waste is great, and that there should be no problem of recycling, waste has so far not been fully exploited.

## CONCLUSION

Results of the study of *F. esculentum* and *F. tataricum* as raw materials to obtain rutin, as well as data on the maintenance of production by the raw materials, can be used in practical application in the pharmaceutical industry. The research found that rutin content in aboveground mass of *F. esculentum* depends on ecological conditions. Plants of *F. esculentum* (with a high rutin content) have red coloration in crops with good lighting (inter row space is 45 cm), while in the thick sowings (inter row space is 15 cm) with less lighting plants have a green-red and green color, indicating decrease in rutin content and pointing to the importance of the effect of environmental factors upon the accumulation of flavonoids in the tissues of plants. In aboveground mass of *F. esculentum* and *F. tataricum* the greatest amount of rutin was observed in crops under the sowing norm of 1.5 million viable seeds per hectare on a background of mineral fertilizers $N_{30}P_{60}K_{60}$. Application of micro-elements (Cu, B, Co) increases rutin content in aboveground mass of *F. esculentum* and *F. tataricum*. Maximum rutin content in aboveground mass of *F. esculentum* was observed when spraying a solution of cobalt at a concentration of 0.05%. Based on the analysis a complex of traits was developed to guide in the evaluation of the plants of *F. esculentum*. They include morphological (stem color) and chemical (rutin content) indices as diagnostic signs of the forms of selection of buckwheat with high rutin content, and resistance to adverse abiotic and biotic factors of the environment.

# REFERENCES

Anisimov, M.M., Klykov, A.G., 2014. Metabolites of terrestrial plants and marine organisms as potential regulators of growth of agricultural plants in the Russian Far East. J. Agric. Sci 6 (11), 88−102.

Anisimova, M.M., 2011. Pharmacological research of *Fagopyrum sagittatum* Gilib. Synopsis of a thesis on competition of a scientific degree of the candidate of pharmaceutical sciences. Samara, 25p. (in Russian).

Brunori, A., Baviello, G., Zannettino, C., Corsini, G., Sándor, G., Végvári, G., 2009. The use of *Fagopyrum tataricum* Gaertn. Whole flour to confer preventive contents of rutin to some traditional Tuscany biscuits. Food Technol. 34 (1), 38−41.

Fabjan, N., Rode, J., Kosir, I.J., Wang, Z., Zhang, Z., Kreft, I., 2003. Tartary buckwheat (*Fagopyrum tataricum* Gaertn.) as a source of dietary rutin and quercitrin. J. Agric. Food Chem. 51, 6452−6455.

Grinberg, L.N., Rachmilewitz, E.A., Newmark, H., 1994. Protective effects of rutin against hemoglobin oxidation. Biochem. Pharmacol. 48, 64−649.

Guardia, T., Rotelli, A.E., Juarez, A.O., Pelzer, L.E., 2001. Antiinflammatory properties of rutin, quercetin and hesperidin on adjuvant arthritis in rat. Farmaco 56, 683−387.

He, J., Klag, M.J., Whelton, P.K., Mo, J.P., Chen, J.Y., Qian, M.C., et al., 1995. Oats and Buckwheat intake and cardiovascular disease risk factors in an ethnic minority of China. Am J. Clin. Nutr. 6, 366−372.

Hinneburg, I., Neubert, Reinhard H.H., 2005. Influence of extraction parameters on the phytochemical characteristics of extracts from buckwheat (*Fagopyrum esculentum*) herb. J. Agric Food Chem. 53, 3−7.

Khan, M.N.A., Cho, J.-Y., Lee, M.-C., Kang, J.-Y., Park, N.G., Fujii, H., et al., 2007. Isolation of two antiinflammatory and one proinflammatory polyunsaturated fatty acids from the brown seaweed *Undaria pinnatifida*. J. Agric. Food Chem. 55, 6984−6988.

Klykov, A.G., Moiseyenko, L.M., 2005. Patent 2255466 RU: IPC[7] A 01 H 1/04. Method of choice of *F. esculentum* Plants with high Rutin Content in the Overground Mass. (in Russian).

Klykov, A.G., Anisimov, M.M., Moiseenko, L.M., Chaikina, E.L., Parskaya, N.S., 2014. Effect of biologically active substances on morphological characteristics, rutin content and productivity of *Fagopyrum esculentum* Moench. Agr. Sci. Dev. 3 (1), 139−142.

Kreft, I., Fabjan, N., Yasumoto, K., 2006. Rutin content in buckwheat (*Fagopyrum esculentum* Moench) food materials and products. Food Chem. 98 (3), 508−512.

Kreft, S., Knapp, M., Kreft, I., 1999. Extraction of rutin from buckwheat (*Fagopyrum esculentum* Moench) seeds and determination by capillary electrophoresis. J. Agric. Food Chem. 47 (11), 4649−4652.

Liu, L., 2002. Buckwheat deep process in china. J. Northwest A & F University 30, 83−85.

McClure, J.W., 1979. Biochemistry of Plant Phenolics. Plenum Press, New York.

Mori, K., Ooi, T., Hiraoka, M., Oka, N., Hamada, H., Tamura, M., et al., 2004. Fucoxanthin and its metabolites in edible brown algae cultivated in deep seawater. Mar. Drugs 2, 63−72.

Odetti, P.R., Borgoglio, A., Pascale, A.D., Rolandi, R., Adezad, L., 1990. Prevention of diabetes increased aging effect on rat collagen-linked fluorescence by aminoguanidine and rutin. Diabetes 39, 796−801.

Srinivasan, K., Kaul, C.L., Ramarao, P., 2005. Partial protective effect of rutin on multiple low dose streptozotocin−induced diabetes in mice. Indian J. Pharmacol. 37, 327−328.

Zaprometov, M.N., 1993. Phenolic compounds: distribution, metabolism, and functions in plants. Moscow.

Zemnukhova, L.A., Tomshich, S.V., Shkorina, E.D., Klykov, A.G., 2004. Polysaccharides from buckwheat production waste. Russian J. Appl. Chem. 77 (7), 1178−1181.

Zhou, X., Song, X.A., 2009. Study of the effect of bourgeon on the nutrients in the sprouts of plant seeds and food of germinated buckwheat. J. Shanghai Inst. Technol. 9 (3), 171−174.

# Chapter Nineteen

# Bioactive Compounds and Their Biofunctional Properties of Different Buckwheat Germplasms for Food Processing

**Oksana Sytar[1,2], Wioletta Biel[3], Iryna Smetanska[4], and Marian Brestic[1]**

[1]*Slovak University of Agriculture in Nitra, Nitra, Slovak Republic,* [2]*Taras Shevchenko National University of Kyiv, Kyiv, Ukraine,* [3]*West Pomeranian University of Technology in Szczecin, Szczecin, Poland,* [4]*Hochschule Weihenstephan-Triesdorf, Weidenbach, Germany*

The most widely-grown buckwheat species include common buckwheat (*Fagopyrum esculentum*) and Tartary buckwheat (*Fagopyrum tataricum*). Buckwheat grain is a source of valuable protein and starch with low glycemic index or high amount of unsaturated fatty acids. It contains compounds with prophylactic value, too. The highest content of dietary fiber is in bran fraction, where it counts for 40%. Present phytosterols are useful in lowering blood cholesterol. Buckwheat is a better source of magnesium, potassium, phosphorus, zinc, manganese, and copper than other cereals. Among vitamins the most abundant is pyridoxin. Buckwheat is effective in the management of many diseases, mainly cardiovascular and digestion disorders, cancer, diabetes, and obesity. In the last decades buckwheat is an interesting material not only for the development of new functional foods, but for the preparation of

**191**

*Buckwheat Germplasm in the World.* DOI: https://doi.org/10.1016/B978-0-12-811006-5.00019-7
© 2018 Elsevier Inc. All rights reserved.

concentrates with healing buckwheat components, too. Buckwheat is ubiquitous but grows mainly in the northern hemisphere. For many years, the cultivation of buckwheat was in decline, yet recently it has been observed to be on the increase because of the health-promoting properties of its grains. Production of buckwheat is highly beneficial for both food security and ecological health, because buckwheat (1) has modest demands on soil quality (Li and Zhang, 2001); (2) naturally eliminates weeds (Bjorkman et al., 2008; Pilipavicius et al., 2009); (3) does not require fertilizers—neutral effect of nitrogen fertilization, high or low (Christensen et al., 2010) and has allelopathic potential (Choi et al., 1991); (4) does not require pesticides, because it suppresses pests by attracting an increased population of pest parasites (Lee and Heimpel, 2005); (5) is a crop providing extensive forage resources to both honeybees and wild pollinators (Cawoy et al., 2009); (6) and is gluten-free (Wronkowska et al., 2013; Giménez-Bastida et al., 2015; Kaur et al., 2015), with studies showing health benefits in reducing risk of cancer (Cade et al., 2007), heart disease (Anderson, 2004), diabetes, and immune disorders (Lee et al., 2012; Stringer et al., 2013).

## NUTRITIONAL COMPONENTS

Buckwheat grains contain a variety of nutrients, the main compounds being proteins, polysaccharides, dietary fiber, lipids, rutin, polyphenols, and elements (Kim et al., 2006; Qin et al., 2010). It has been reported that buckwheat protein has many unique physiological functions, such as curing chronic human diseases, decreasing blood cholesterin, constipation, mammary carcinogenesis and colon carcinogenesis, and restraining gallstone. In rat feeding experiments, studies have proven that buckwheat protein extract has hypocholesterolemic, anticonstipation, and antiobesity activities (Zhang et al., 2012). In the literature, the protein content of buckwheat grains has been reported to range from 8.5% to 18.9% dry weight (d.w.) (Krkošková and Mrázová, 2005; Eggum et al., 1981; Wei et al., 2003). Biel and Maciorowski (2013) evaluated the chemical composition of buckwheat products (groat, bran, and hull) and the greatest buckwheat protein content was found in bran (183.4 g/kg d.w.). The protein in groat was 148.3 g/kg d.w., which confirms the results of Wei et al. (2003). Protein content in buckwheat is particularly high when compared to other grains such as wheat, rice, maize, and sorghum (Shewry et al., 2009; Fabian and Ju, 2011) (Table 19.1).

Contrary to most common grains, the proteins in buckwheat are composed mainly of globulins and albumins, and contain very little or no storage prolamin proteins, which are the main storage proteins in cereals, and also the toxic proteins in celiac disease. The amino acid composition of globulins and albumins differs significantly from that of prolamins, which has implications in relation to their nutritional quality. Globulins and albumins contain less glutamic acid and proline than prolamins, and more essential amino acids such as lysine. In addition, protein bioavailability in buckwheat is high, and

**TABLE 19.1** Comparison of Buckwheat Seed Composition (g/kg Dry Weight) With Other Commonly Used Cereals and Pseudo-Cereals

| Item | Protein (N × 6.25) | Crude Ash | Crude Fibre | Crude Fat | Total Carbohydrates |
|---|---|---|---|---|---|
| Tartary buckwheat | 150 | 24 | 26 | 28 | 772 |
| Common buckwheat | 124 | 22 | 23 | 28 | 803 |
| Wheat (*Triticum aestivum* L.) | 142 | 18 | 23 | 19 | 798 |
| Husked oat (*Avena sativa* L.) | 115 | 23 | 136 | 48 | 678 |
| Amaranth (*Amaranthus* spp.) | 146 | 32 | 39 | 88 | 695 |
| Quinoa (*Chenopodium quinoa*) | 138 | 33 | 23 | 51 | 755 |

*Alvares-Jubete, L., Wijngaard, H., Arend, E.K., Gallagher, E., 2010. Polyphenol composition and in vitro antioxidant activity of amaranth, quinoa, buckwheat and wheat as affected by sprouting and baking. Food Chem. 119, 770–778; Valcárcel-Yamani, B., and Lannes, S.C.S., 2012. Applications of quinoa (Chenopodium quinoa Willd.) and amaranth (Amaranthus spp.) and their influence in the nutritional value of cereal based foods. Food Public Health 2(6), 265–275.*

it has been shown in several animal studies to be superior to that of common cereals, and close to the quality of animal proteins (Kayashita et al., 1999; Tomotake et al., 2006).

Buckwheat is a rich source of dietary fiber. Bonafaccia et al. (2003) analyzed the composition of dietary fiber of Tartary buckwheat. Total dietary fiber (TDF) content of the seeds was 26%, with that of soluble (SDF) and insoluble (IDF) fibers being 0.54% and 24%, respectively. The composition of dietary fiber of Tartary buckwheat appeared to be similar to that of common buckwheat. Yang et al. (2014) found that, in one Tartary buckwheat variety, the total dietary fiber content was 8.4%, of which 8.2% was insoluble fiber and 0.2% was soluble fiber. Bran with hull fragments contains 400 mg/g TDF, of which 250 mg/g is SDF, while bran without hull fragments contains 160 mg/g TDF, of which 750 mg/g is soluble. The content and composition of crude fiber is crucial for the quality of buckwheat products. However, the content and proportions of various carbohydrates in buckwheat may change during hydrothermal treatment. Cell wall carbohydrates can be quantified by determination of neutral detergent fiber (NDF), which includes cellulose, hemicellulose, and lignin as the major components (Van Soest et al., 1991). The highest content of a neutral detergent fiber (NDF) is in the hull (834 g/kg d.w.), while the smallest is in whole groats (56 g/kg d.w.). The quality of fiber is greatly determined by lignin content. Acid detergent lignin (ADL) content is the highest in hulls and lowest in groats (respectively 353 and 5.6 g/kg d.w., Biel and Maciorowski, 2013). Hemicelluloses and pectins are components that are potent absorbents of heavy metals (Andersen et al., 2004). Although cellulose is only slightly digested in the small intestine, it supports peristalsis. The highest content of the cellulose and hemicellulose fractions, similar to other fiber fractions, is in the hull, while the smallest is in the groats. Small lignin and cellulose content is due to the removal of lignin-rich outer parts of the grain during groat production (Nyman et al., 1984). Depending on the type of technology used in groats production, the content and fractional composition of fiber, and thus the functional properties, may vary. Resistant starch is considered another type of dietary fiber with diverse health benefits such as reducing the risk of colon cancer and obesity. Resistant starch content in the flour of Tartary buckwheat ranged from 13.1% to 22.5%, which was lower than that of common buckwheat flour (Qin et al., 2010). Dietary fiber content is significantly higher in buckwheat seeds in comparison with amaranth and quinoa, which have fiber levels comparable to those found in common cereals (Jancurová et al., 2009; Aguilar et al., 2015; Alvarez-Jubete et al., 2010). It is important to note that buckwheat fiber is free of phytic acid, a major anti-nutritional factor in common wheat (Steadman et al., 2001).

The amino acid composition of buckwheat proteins is well balanced and of a high biological value (Kato et al., 2001). The buckwheat seeds have a balanced amino acid spectrum with high lysine, isoleucine, tryptophan, valine, histidine, and phenyloalanine. The content of essential amino acids (EAA) and total amino acids (AA) in buckwheat is higher than in common cereals. The concentration of several essential amino acids in buckwheat is much higher than in wheat (Biel and Jacyno, 2014). This is especially the case with some of the most important amino acids such as lysine, threonine, tryptophan, and the sulfur-containing amino acids. The content of methionine and cysteine in the protein of the buckwheat seed is high, the average is 4 g per 16 g N, which makes this seed a good complementary protein source to legumes (Sujak et al., 2006). The main outstanding characteristic of the buckwheat proteins is the high lysine content (average of 5,14 g/100 g protein), which is higher than in any of the cereal grains and pseudo-cereals. The content of lysine in buckwheat protein is more than double in wheat protein (Wei et al., 1995). Buckwheat is an excellent source of tryptophan—Biel and Maciorowski (2013) showed that the most abundance was in groats and bran, respectively 1.45 and 1.41 g per 16 g N, but the least was in the hull (0.62 g per 16 g N). The contents of Val, Leu, and Ile in the protein of buckwheat seeds are respectively 29%, 4%, and 30% higher than in the protein of the wheat grains. The buckwheat has a high protein content which is higher in comparison with other commonly-used cereals. Low lys/arg and met/arg ratios may suggest that buckwheat products have blood cholesterol-lowering properties. As shown by Kayashita et al. (1995) the use of buckwheat protein preparations in the hamster diet significantly contributed to the reduction in cholesterol in the blood, liver, and gall bladder, and also inhibited the formation of gall stones as a result of changes in cholesterol metabolism. It is believed that protein buckwheat extracts reduce the level of LDL and VLDL, and prevent the development of colorectal cancer by reducing the proliferation of tumor cells. The buckwheat seed has a high-protein content with an abundance of essential amino acids (almost 27% more in comparison to the wheat grain). Since Arg and Gly are involved in the synthesis of other nitrogenous compounds that are important to physiological viability, these non-essential amino acids are important as part of the diet when optimum growth conditions are targeted. Compared to cereal grains, buckwheat proteins contain relatively low concentrations of glutamic acid and proline, but are rich in arginine and aspartic acid. Buckwheat is considered as one of the best seed-protein concentrate sources and so has the potential as a protein substitute for the food and pharmaceutical industry.

## ELEMENTS

Buckwheat is a source of minerals, valuable both for nutritional and technological reasons. Mineral content, expressed as total ash, averaged 23 g/kg d.w. (Table 19.1). Buckwheat is rich in potassium, phosphorus, magnesium, calcium, and sodium (Khan et al., 2013). Potassium, phosphorus, and magnesium are most concentrated in the bran, particularly in the bran from which the hulls were removed before milling the grains (Steadman et al., 2001). A high concentration of potassium, at almost seven times that of sodium, was recorded (5.65 and 0.788 g/kg d.w., respectively), indicating a good source for patients with hypertension. Buckwheat is high in potassium content (average of 5.65 g/kg d. w.).Minerals like molybdenum, zinc, iron, copper, manganese, etc., are detected in appreciable amounts in buckwheat. Steadman et al. (2001) founded 16.4 g/kg d.w. of Mn in buckwheat whole groats. Buckwheat seeds are also an important dietary source of selenium (0.029 mg/kg dry weight). It is noteworthy that Ikeda et al. (2006) compared the composition of Fe, Zn, Cu, Mn, Ca, Mg, K, and P of buckwheat flour to cereal flours by using an in vitro enzymatic digestion technique. The results showed a higher content of essential minerals in buckwheat flour in comparison with other cereal flours. Buckwheat is a richer source of nutritionally important minerals than many common cereals, with the exception of calcium.

## PHENOLIC COMPOUNDS OF BUCKWHEAT EXTRACTS

Polyphenols contribute to about 20% of the total antioxidant activity (AA) in common buckwheat, whereas in Tartary buckwheat they are responsible for 85%−90% of the total AA (Morishita et al., 2007). Information on flavonoid and phenolic composition of several buckwheat varieties can be found, information on phenolic acids is lacking. Phenolics such as rutin, quercetin, chlorogenic acid, and such C-glycosylflavones as orientin, isoorientin, vitexin, and isovitexin, were identified. The total phenolics and rutin contents in the plants' anatomical parts were higher, in the order of: inflorescences > leaves > stems and flowers. The orientin, isoorientin, vitexin, and isovitexin contents in common buckwheat were higher than those in Tartary buckwheat. At the same time, the rutin content, which assumed for >90% of the total phenolics, was higher in Tartary buckwheat. (Seo et al., 2013). The phenolic compounds (rutin, isoorientin, and orientin) in buckwheat sprouts have superoxide anion radical-scavenging activities (Watanabe, 2007). Rutin, quercetin, orientin, vitexin, isovitexin, and isoorientin have been isolated and identified in the grains of buckwheat (Table 19.2).

In the buckwheat seeds the flavonoids were presented only by rutin and isovitexin but the buckwheat hulls contains also quercetin, orientin, and vitexin. De-hulling the grain by using various temperature regimes resulted in the significant decreasing of the total flavonoid concentration in the grain by 75% compared to the control. In the hulls of buckwheat under the same treatments conditions found a smaller but significant (15%−20%) reduction (Dietrych-Szostak and Oleszek, 1999). The flavonoids orientin, isoorientin rutin, quercetin, vitexin, and isovitexin were found in the hulls of five buckwheat varieties of Polish buckwheat.

The antioxidant activity (AA) of analyzed flavonoids and synthetic antioxidant BHT (butyl-hydroxytoluene) could be sequenced in the following way: vitexin < isovitexin < orientin < isoorientin < rutin < quercetin < BHT (Dietrych-Szóstak, 2004). The content of phenolic antioxidants in the hulls of buckwheat were from highest to lowest in order: protocatechuic acid > 3,4-dihydroxybenzaldehyde (> hyperin > rutin > quercetin). The content of rutin

**TABLE 19.2** Comparable Characteristics Identified Phenolic Compounds in the Different Organs of Common (CB) and Tartary Buckwheat (TB)

| Item Flavonoids[a] | Stems | | Leaves | | Flowers | | Grain | | Hulls | |
|---|---|---|---|---|---|---|---|---|---|---|
| | CB[b] | TB[c] | CB[d] | TB[c] | CB[e] | TB[c] | CB[f] | TB[g] | CB[h] | TB[i] |
| Orientin | − | − | + | − | − | − | + | a | + | a |
| Rutin | + | + | + | + | + | + | + | + | a | + |
| Quercetin | + | + | + | + | + | + | a | + | + | + |
| Quercitrin | a[b] | a | a | a | a | a | a | + | a | a |
| Isovitexin | + | − | + | − | + | − | + | a | a | a |
| Isoorientin | − | − | + | − | − | − | + | a | a | a |
| Vitexin | − | − | + | − | − | − | + | a | a | a |
| Hyperin | a | a | a | a | a | a | a | a | + | a |
| Protocatechuic acid | a | a | a | a | a | a | a | a | + | a |
| 3,4-dihydroxy-benzaldehyde | a | a | a | a | a | a | a | a | + | a |

[a]a-not determined.
[b]Seo et al. (2013).
[c]Seo et al. (2013), Kim et al. (2006).
[d]Seo et al. (2013), Koyama et al. (2011).
[e]Seo et al. (2013), Koyama et al. (2011).
[f]Dietrych-Szostak and Oleszek (1999).
[g]Fabjan et al. (2003).
[h]Watanabe et al. (1997).
[i]Steadman et al. (2001).

and quercetin was 4.3 mg/100 g and 2.5 mg/100 g, respectively. Among the identified compounds vitexin and isovitexin in the extract did not show peroxyl radical-scavenging activity (Watanabe et al., 1997).

## BUCKWHEAT FOR FOOD PROCESSING

Nowadays buckwheat is receiving increased attention as a potential functional food. Several research studies have reported the use of buckwheat products such as flour, flakes, and extrudates as substitutes for wheat flour in bakery, confectionery, and pastry manufacturing technology (Gavrilova, 2008; Yildiz and Bilgicli, 2012; Baljeet et al., 2010; Lin et al., 2009; Wronkowska et al., 2013). It is well known that processing can induce chemical changes in food products; therefore, it is important to consider the effects of processing on the bioactive compounds of buckwheat (Sytar et al., 2016). Thermal techniques such as cooking and baking as well as application of high pressure and microwaves are used for the processing of buckwheat in the food industry (Giménez-Bastida et al., 2015).

Microwave heating has gained popularity in food processing due to its ability to achieve high heating rates, significant reduction in cooking time, more uniform heating, and safe handling. This technique might change flavor and nutritional qualities of food to a less significant extent as opposed to conventional heating during the cooking process (Chandrssekaran et al., 2013). However, the data on the effect of thermal treatments on the antioxidant capacity of buckwheat and its products are still limited. Buckwheat belongs to pseudo-cereals and therefore is gluten-free. The most common form of gluten intolerance is celiac disease (Fasano, 2001). Gliadin is the generally accepted factor causing celiac disease (Virta et al., 2009). Globally coeliac disease affects between 1 in 100 and 1 in 170 people—rates do, however, vary between different regions of the world from as few as 1 in 300 to as many as 1 in 40 (Szajewska et al., 2016). The alternative to consuming wheat products can be the substitution with some other cereals, like buckwheat. Cereals like rye, oat, and barley and pseudo-cereals like buckwheat cannot be used for baking without improvers (Payne et al., 1981).

## BUCKWHEAT FLOUR

The raw material for buckwheat flakes and buckwheat flour is buckwheat grain. The technology for making such buckwheat products provides hydrothermal treatments (HTT). HTT is interpreted as grain handling by water and heat for direct change (improvement) of the whole technological complex (flour milling, bakery, and groat properties) (Kazakov and Karpilenko, 2005). HTT, which is commonly used during the processing of groats, strongly affects the rheological properties of dough made from groats flour (Zheng et al., 1997). This operation leads to partial denaturing of protein, starch gelatinization, and an increase in the amount of dextrin formation. In the production of flakes more stringent HTT processing procedures are used than for flour (Babich et al., 2001). Buckwheat flour is rich in thiamine, riboflavin, and pyridoxine (Fabjan et al., 2003).

The rising amount of buckwheat flour in gluten-free bread formulation caused a decrease in crumb hardness during storage. At the same time this was in agreement with the decrease in starch gelatinization enthalpy with the increasing amount of buckwheat flour in gluten-free formula in comparison with the control sample. Buckwheat flour could be incorporated into gluten-free formula and have a positive influence on bread texture and delaying its staling (Wronkowska et al., 2013). It was established that buckwheat hull hemicelluloses can be used to improve the quality of bread to prepare from medium-quality wheat flours. 0.5% of content of buckwheat hull hemicelluloses had a considerable effect on bread flour quality in relation to the resistance to extension and fermentation of dough. The buckwheat hull hemicelluloses in content 0.3%−0.5% advanced improvement of sensory properties of fresh bread and higher scores for overall acceptability. In long-term storage, all breads containing buckwheat hull hemicelluloses exhibited higher softness and elasticity than the control (Hromádková et al., 2007).

## DOUGHS AND PASTRIES

Buckwheat, added to food as a supplement, can prevent food from oxidation during processing. Buckwheat flour could be incorporated into bread recipes, providing buckwheat-enriched wheat bread with more sugars, a stronger umami taste and a more characteristic aroma (Bastida et al., 2015). It has been developed in recipes for biscuits with partial and complete replacement of wheat through buckwheat flour (Kilian et al., 2016). This product can provide a healthy alternative to traditional pastries for consumers with food intolerances. The high quality for gluten-free products means the taste and texture seem for consumers to be very similar to products with wheat. The dough has a pulpy, lubricating consistency in muffins consisted of buckwheat flour. This can be explained by the strong water-binding properties of buckwheat flour. The smallest pore formation was observed in the final buckwheat product compared with wheat products, since buckwheat bakery firstly contained no gluten.

Investigation of the structural and mechanical properties of dough showed an increase in water-absorbing capacity in all samples when adding buckwheat products (Lin et al., 2013). Moreover, dough made with buckwheat flakes has a lower value of mixing tolerance index (by 47%) than dough made from buckwheat flour, and a higher valorimetric value (by 20%). Determination of dough properties by amylogram has shown that a sample containing buckwheat flakes has a higher maximum viscosity than a sample containing buckwheat flour. Determination of the gas-production and gas-retention capacity of dough is also presented, along with an analysis of the quality of finished products based on the results of laboratory baking tests.

The samples of bread supplemented with buckwheat flakes have better shape stability (by 21%), specific volume (by 12%), and porosity (by 11%) than bread made from buckwheat flour.

## NONBAKERY PRODUCTS

The known buckwheat derived non-bakery products are buckwheat honey, roasted groats (kasha), and from raw groats—pasta, noodles, and tea infusion. Buckwheat tea can be made of common or Tartary buckwheat (Zielińska et al., 2009; Qin et al., 2013). Park et al. (2008) investigated the effect of boiling on rutin content in tea made from flowers and dried leaves of different species of buckwheat. Although it has been described that buckwheat tea shows a lower antioxidant capacity and a lower content of total phenolic compounds than green tea, this product may be offered to consumers as a new type of tea enriched in flavonoids, especially rutin and other compounds, such as quercetin and flavone C-glucosides (Qin et al., 2013). The production of steamed and roasted whole Tartary buckwheat tea samples was tracked to explore the effects of steaming, roasting, and reconstructed granulation on the aroma and nutritional components of Tartary buckwheat tea. Roasted whole Tartary buckwheat tea as a mixture of roasted Tartary buckwheat grains and tea-containing reconstructed granules had considerably better quality and flavor than its steamed counterpart. The major aroma compounds of steamed Tartary buckwheat tea were alkanes and alkenes, while roasted Tartary buckwheat tea contained aldehydes and alkanes as major aroma compounds and was also rich in phenols, alcohols, ethers, ketones, and esters, which caused better flavor in roast Tartary buckwheat tea and than the steamed one (Sui et al., 2012). It is important to consider what the effects on the bioactive compounds from buckwheat will be during the processing of buckwheat into tea. To use inflorescences is a new step in developing organic tea production with buckwheat plants in European countries (Thwe et al., 2013). In this process many steps are involved in creating tea from raw buckwheat seeds. Raw whole seeds are first soaked in water, then steamed, then dried, before they are removed from their hulls. The dehulled groats are then roasted so that the tea can be made (Qin et al., 2011).

## PERSPECTIVES AND CONCLUSIONS

It is generally accepted that not only the seeds but also the other parts of buckwheat plants are rich in compounds with a high biological value. The seeds used for food contain proteins which have high lysine levels and specific amino acid content, giving them a high nutritional value. The absence of gluten in buckwheat flour determines its potential use in gluten-free diets, but the quality of protein and the antioxidants composition of different varieties of buckwheat can vary—which is important to know for flour production. Other valuable bioactive compounds in seeds include resistant starch, dietary fiber, rutin, and anotherpolyphenols. The vegetative mass of buckwheat

plants has a higher content of biologically active compounds such as phytosterols, flavonoids, fagopyrins, phenolic acids, lignans, and vitamins. The high
content of these compounds should become the major goal in the future
selection of genetic resources and breeding aimed at nutritional effects.
Based on recent knowledge, the most valuable sources of fagopyrin, orientin,
rutin, isovitexin, and quercetin are the vegetative mass of common buckwheat
(*F. esculentum*). At the same time, the seeds of *F. cymosum* and *F. tataricum*
have the highest rutin content among the investigated buckwheat species.
Other important bioactive compounds are phenolic acids, which are abundant
mainly in the leaves and inflorescences of buckwheat plants—but their content varies depending on genotype. A detailed experimental analysis of different buckwheat species regarding phenolic acids composition is lacking. Apart
from the genotypic variation of the total content of bioactive compounds in
buckwheat the environment plays a crucial role. The characterization of existing buckwheat varieties and the introduction of buckwheat-based products
from elite varieties into the modern food chain are needed to fill the current
gap. In this review, we have presented the high diversity of bioactive compounds in buckwheat varieties, which could be broadly exploited for food
and medicinal use.

## REFERENCES

Aguilar, E.G., Albarracín, G.J., Uñates, M.A., Piola, H.D., Camiña, J.M., Escudero, N.L.,
2015. Evaluation of the nutritional quality of the grain protein of new amaranths varieties. Plant Foods Hum. Nutr. 70, 21–26.
Alvares-Jubete, L., Wijngaard, H., Arend, E.K., Gallagher, E., 2010. Polyphenol composition and in vitro antioxidant activity of amaranth, quinoa, buckwheat and wheat as
affected by sprouting and baking. Food Chem. 119, 770–778.
Andersen, O., Nielsen, J.B., Nordberg, G.F., 2004. Nutritional interactions in intestinal cad-
mium uptake. Possibilities for risk reduction. BioMetals 17, 543–547.
Anderson, J., 2004. Whole grains and coronary heart disease: the whole kernel of truth.
Am. J. Clin. Nutr. 80 (6), 1459–1460.
Babich, M., Bayram-Gali, V., Kalinichenko, V., 2001. Grain processing to ready-to-eat
cereal flakes and grains. Khraneniyeipererabotkazerna (in Russian) 9, 38–40.
Baljeet, S, Ritika, B, Roshan, L., 2010. Studies on functional properties and incorporation
of buckwheat flour for biscuit making. Int Food Res. J. 17 (4), 1067–1076.
Bastida, J.A.G., Piskula, M.K., Zielinski, H., 2015. Recent advances in processing and
development of buckwheat derived bakery and non-bakery products. Pol. J. Food Nutr.
Sci. 65 (1), 9–20.
Biel, W., Jacyno, E., 2014. Chemical composition and nutritive value of protein in hulled
dwarf oat lines and the effect on serum lipid profile in rats. Ital. J. Food Sci. 26 (2),
203–209.
Biel, W., Maciorowski, R., 2013. Evaluation of chemical composition and nutritional qual-
ity of buckwheat groat, bran and hull. Ital. J. Food Sci. 25 (4), 384–390.
Bjorkman, T., Bellinder, R., Hahn, R., Shail, J., 2008. Buckwheat Cover Crop Handbook.
Cornell University, p. 17.
Bonafaccia, G., Marocchini, M., Kreft, I., 2003. Composition and technological properties
of the flour and bran from common and tartary buckwheat. Food Chem. 80, 9–15.
Cade, J., Burley, V., Greenwood, D., 2007. Dietary fibre and risk of breast cancer in the
UK Women's Cohort Study. Int. J. Epidemiol. 36 (2), 431–438.

Cawoy, V, Ledent, J.F., Kinet, J.M., Jacquemart, A.L., 2009. Floral biology of common buckwheat. Eur. J. Plant Sci. Biotechnol. 3 (1), 1−9.

Chandrssekaran, S, Ramanathan, S, Basak, T., 2013. Microwave food processing-a review. Food Res. Int. 52, 234−261.

Choi, B., Kim, S., Park, K., Park, R., 1991. Acid amide, dinitroaniline, triazine, urea herbicide treatment and survival rate of coarse grain crop seedlings. Res. Rep. Rural Dev. Administration Upland Ind. Crops 33, 33−42.

Christensen, K., Kaemper, M., Loges, R., Frette, X., Christensen, L., Grevsen, K., 2010. Effects of nitrogen fertilisation, harvest time, and species on the concentration of polyphenols in aerial parts and seeds on normal and tartary buckwheat. Eur. J. Hortic. Sci. 75 (4), 153−164.

Dietrych-Szóstak, D., 2004. Flavonoids in hulls of different varieties of buckwheat and their antioxidant activity. In: Proceedings of the 9th International Symposium on Buckwheat, Prague, pp. 621−625.

Dietrych-Szostak, D, Oleszek, W., 1999. Effect of processing on the flavonoid content in buckwheat (*Fagopyrum esculentum* Möench) grain. J. Agric. Food Chem. 47 (10), 4384−4387.

Eggum, B.O., Kreft, I., Javornik, B., 1981. Chemical composition and protein quality of buckwheat (*Fagopyrum esculentum* Moench). Qual. Plant Plant Foods Hum. Nutr. 30, 175−179.

Fabian, C., Ju, Y-H., 2011. A review on rice bran protein: its properties and extraction methods. Crit. Rev. Food Sci. Nutr. 51 (9), 816−827.

Fabjan, N, Rode, J, Kosir, IJ, Zhang, Z, Kreft, I., 2003. Tartary buckwheat (*Fagopyrum tataricum* Gaertn.) as a source of dietary rutin and quercitrin. J. Agric. Food Chem. 51, 6452−6455.

Fasano, A., 2001. Current approaches to diagnosis and treatment of celiac disease. Evolving Spectrum 120, 636−651.

Gavrilova, O.M., 2008. Technology development of bread with buckwheat flour application.Synopsis Ph.D. sci. diss. Moscow State University of Food Production, Moscow, 25 p (in Russian).

Giménez-Bastida, J.A., Piskuła, M., Zieliński, H., 2015. Recent advances in development of gluten-free buckwheat products. Trends Food Sci. Technol. 44, 58−65.

Hromádková, Z, Stavová, A, Ebringerová, A, Hirsch, J., 2007. Effect of buckwheat hull hemicelluloses addition on the bread-making quality of wheat flour. J. Food Nutr. Res. 46 (4), 158−166.

Ikeda, S., Yamashita, Y., Tomura, K., Kreft, I., 2006. Nutritional comparison in mineral characteristics between buckwheat and cereals. Fagopyrum 23, 61−65.

Jancurová, M., Minarovičová, L., Dandár, A., 2009. Quinoa − a review. Czech J. Food Sci. 27 (2), 71−79.

Kato, N., Kayashita, J., Tomotake, H., 2001. Nutritional and physiological functions of buckwheat protein. Rec. Res. Dev. Nutr. 4, 113−119.

Kaur, M., Sandhu, K.S., Arora, A., Sharma, A., 2015. Gluten free biscuits prepared from buckwheat flour by incorporation of various gums: physicochemical and sensory properties. LWT-Food Sci. Technol. 62, 628−632.

Kayashita, J., Shimaoka, I., Nakajoh, M., 1995. Production of buckwheat protein extract and its hypocholesterolemic effect. Current Advances in Buckwheat Research 919−926.

Kayashita, J., Shimaoka, I., Nakajoh, M., Kondoh, M., Hayashi, K., Kato, N., 1999. Muscle hypertrophy in rats fed on a buckwheat protein extract. Biosci. Biotechnol. Biochem. 63 (7), 1242−1245.

Kazakov, E., Karpilenko, G., 2005. Biochemistry of grains and their products, third ed. St. Petersburg, GIORD Publ., 512 p. (in Russian).

Khan, F., Arif, M., Khan, T.U., Khan, M.I., Bangash, J.A., 2013. Nutritional evaluation of common buckwheat of four different villages of Gilgit-Baltistan. J. Agric. Biol. Sci. 8 (3), 264−266.

Kilian, M, Stankowski, S, Groß, E, Smetanska, I., 2016. Product design in the development of recipes of gluten-free and sugar- reduced bakery products. Mech. Technol. 1, 23–27.

Kim, S-J., Kawaharada, C., Suzuki, T., Saito, K., Hashimoto, N., Takigawa, S., et al., 2006. Effect of natural light periods on rutin, free amino acid and vitamin C contents in the sprouts of common (*Fagopyrum esculentum* Moench) and tartary (*F. tataricum* Gaertn.) buckwheats. Food Sci. Technol. Res. 12 (3), 199–205.

Koyama, M., Nakamura, C., Nakamura, K., 2011. Changes in phenols contents from buckwheat sprouts during growth stage. J. Food Sci. Technol. 50 (1), 86–93.

Krkošková, B., Mrázová, Z., 2005. Prophylactic components of buckwheat. Food Res. Int. 38, 561–568.

Lee, Ch, Hsu, W.-H., Shen, S.-R., Cheng, Y.-H., Wu, S.-Ch, 2012. *Fagopyrum tataricum* (buckwheat) improved high-glucose-induced insulin resistance in mouse hepatocytes and diabetes in fructose-rich diet-induced mice. Exp. Diabetes Res. 1–10.

Lee, J., Heimpel, G., 2005. Impact of flowering buckwheat on *Lepidopterian* cabbage pests and their parasitoris at two spatial scales. Biol. Control. 134, 290–301.

Li, S.Q, Zhang, Q.H., 2001. Advances in the development of functional foods from buckwheat. Crit. Rev. Food Sci. Nutr. 41 (6), 451–464.

Lin, L, Liu, H, Yu, Y, Lin, S, Mau, J., 2009. Quality and antioxidant property of buckwheat enhanced wheat bread. Food Chem. 112 (4), 987–991.

Lin, L-Y, Wang, H-E, Lin, S-D, Liu, H-M, Mau, J-L., 2013. Changes in buckwheat bread during storage. J. Food Process. Preserv. 37 (4), 285–290.

Morishita, T., Yamaguchi, H., Degi, K., 2007. The contribution of polyphenols to antioxidative activity in common buckwheat and tartary buckwheat grain. Plant Prod. Sci. 10 (1), 99–104.

Nyman, M., Siljerstorm, M., Pederson, B., Bach Kundsen, K.E., 1984. Dietary fibre content and composition in six cereals at different extraction rates. Cereal Chem. 61, 14–19.

Park, CH, Kim, YB, Choi, YS, Heo, K, Kim, SL, Lee, KC, et al., 2008. Rutin content in food products processed from groats, leaves, and flowers of buckwheat. Fagopyrum 17, 63–66.

Payne, P, Corfield, K, Holt, L, Blackman, J., 1981. Correlations between the inheritance of certain high molecular weight subunits of glutenin and breadmaking quality in progencies of six crosses of bread wheat. J. Sci. Food Agric. 32, 51–60.

Pilipavicius, V., Lazauskas, P., Jasinskaite, S., 2009. Weed control by two layer ploughing and post-emergence crop tillage in spring wheat and buckwheat. Agron. Res. 7, 444–450.

Qin, P., Wang, Q., Shan, F., Hou, Z., Ren, G., 2010. Nutritional composition and flavonoids content of flour from different buckwheat cultivars. Int. J. Food Sci. Technol. 45 (5), 951–958.

Qin, P., Li, W., Yang, Y., Guixing, R., 2011. Changes in phytochemical compositions, antioxidant and α-glucosidase inhibitory activities during the processing of tartary buckwheat tea. Food Res. Int. 50 (2), 562–567.

Qin, P., Wu, L., Yao, Y., Ren, G., 2013. Changes in phytochemical compositions, antioxidant and β-glucosidase inhibitory activities during the processing of tartary buckwheat tea. Food Res. Int. 50, 562–567.

Seo, J.M., Lee, D.B., Arasu, M.V., Wu, Q., Suzuki, T., Yoon, Y-H., et al., 2013. Quantitative differentiation of phenolic compounds in different varieties of buckwheat cultivars from China, Japan and Korea. J. Agric. Chem. Environ. 4, 109–116.

Shewry, P.R., D'Ovidio, R., Lafiandra, D., Jenkins, J.A., Mills, E.N.C., Békés, F., 2009. Wheat grain proteins. Wheat: chemistry and technology. pp. 223–298.

Steadman, K.J., Burgoon, M.S., Lewis, B.A., Edwardson, S.E., Obendorf, R.L., 2001. Buckwheat seed milling fraction: description, macronutrient composition and dietary fibre. J. Cereal Sci. 33, 271–278.

Stringer, D.M., Taylor, C.G., Appah, P., Blewett, H., Zahradka, P., 2013. Consumption of buckwheat modulates the post-prandial response of selected gastrointestinal satiety hormones in individuals with type 2 diabetes mellitus. Metabolism 62 (7), 1021–1031.

Sui, X-F, Li, X, Qin, L-K, Zhao, Y, Lin, M., 2012. Analysis of volatile aroma compounds in steamed and roasted whole tartary buckwheat teas and distribution of major chemical components during the production process. J. Food Sci. 33 (22), 269−273.

Sujak, A., Kotlarz, A., Strobel, W., 2006. Compositional and nutritional evaluation of several lupin seeds. Food Chem. 98, 711−719.

Sytar, O, Brestic, M, Zivcak, M, Tran, L-SP, 2016. The contribution of buckwheat genetic resources to health and dietary diversity. Curr. Genomics 17, 193−206.

Szajewska, H., Mearin, S.R., Koninckx, C.R., Catassi, C., Domellöf, M., Fewtrell, M., et al., 2016. Gluten introduction and the risk of coeliac disease. J. Pediatr. Gastroenterol. Nutr. 1, 24−36.

Thwe, AA, Kim, JK, Li, X, Bok Kim, Y, Romij Uddin, M, Kim, SJ, et al., 2013. Metabolomic analysis and phenylpropanoid biosynthesis in hairy root culture of tartary buckwheat cultivars. PLOS ONE 8 (6), e65349.

Tomotake, H., Yamamoto, N., Yanaka, N., Ohinata, H., Yamazaki, R., Kayashita, J., et al., 2006. High protein buckwheat flour suppresses hypercholesterolemia in rats and gallstone formation in mice by hypercholesterolemic diet and body fat in rats because of its low protein digestibility. Nutrition 22 (2), 166−173.

Valcárcel-Yamani, B., Lannes, S.C.S., 2012. Applications of quinoa (*Chenopodium quinoa*Willd.) and amaranth (*Amaranthus* spp.) and their influence in the nutritional value of cereal based foods. Food Public Health 2 (6), 265−275.

Van Soest, P.J., Robertson, J.B., Lewis, B.A., 1991. Methods for dietary fiber, neutral detergent fiber, and nonstarch polysaccharides in relation to animal nutrition. J. Dairy Sci. 74, 3583−3597.

Virta, LJ, Kaukinen, K, Collin, P., 2009. Incidence and prevalence of diagnosed coeliac disease in 11289 Finland: results of effective case finding in adults. Scand. J. Gastroenterol. 44, 933−938.

Watanabe, M., 2007. An anthocyanin compound in buckwheat sprouts and its contribution to antioxidant capacity. Biosci. Biotechnol. Biochem. 71 (2), 579−582.

Watanabe, M., Ohshita, Y., Tsushida, T., 1997. Antioxidant compounds from buckwheat (*Fagopyrum esculentum* Moench) hulls. Agric. Food Chem. 45 (4), 1039−1044.

Wei, Y., Zhang, G.Q., Li, Z.X., 1995. Study on nutritive and physico-chemical properties of buckwheat flour. Nahrung 39, 48−54.

Wei, Y., Hu, X., Zhang, G., Ouyang, S., 2003. Studies on the amino acid and mineral content of buckwheat protein fractions. Nahrung/Food 47, 114−116.

Wronkowska, M, Haros, M, Soral-Śmietana, M., 2013. Effect of starch substitution by buckwheat flour on gluten-free bread quality. Food Bioprocess. Technol. 6, 1820−1827.

Yang, N., Li, Y.M., Zhang, K., Jiao, R., Ma, K.Y., Zhang, R., et al., 2014. Hypocholesterolemic activity of buckwheat flour is mediated by increasing sterol excretion and down-regulation of intestinal NPC1L1 and ACAT2. J. Funct. Foods 6, 311−318.

Yildiz, G, Bilgicli, N., 2012. Effects of whole buckwheat our on physical, chemical, and sensory properties of at bread, lavas. Czech J. Food. Sci. 30, 534−540.

Zhang, Z.L., Zhou, M.L., Tang, Y., Li, F.L., Tang, Y.X., Shao, J.R., et al., 2012. Bioactive compounds in functional buckwheat food. Food Res. Int. 49 (1), 389−395.

Zheng, GH, Sosulski, FW, Tyler, RT., 1997. Wet-milling, composition and functional properties of starch and protein isolated from buckwheat groats. Food Res. Int. 30 (7), 493−502.

Zielińska, D, Szawara-Nowak, D, Zieliński, H., 2009. Antioxidative and anti-glycation activity of bitter buckwheat tea. Eur. J. Plant Sci. Biotechnol. 3, 79−83.

# FURTHER READING

Dietrych-Szostak, D, Oleszek, W., 1998. The composition and concentration of flavonoid s in buckwheat groats and hulls. Pol. J. Food Nutr. Sci. 7/4 (8), 151−153.

Seguro, K., Kumazawa, Y., Kuraishi, Ch., Sakamoto, H., Motoki, M., 1996. The epsilon-(gamma-glutamyl)lysine moiety in crosslinked casein is an available source of lysine for rats. J. Nutr. 126 (10), 2557–2562.

Vršková, M., Bencová, E., Foltys, V., Havrlentová, M., Čičová, I., 2013. Protein quality evaluation of naked oat (*Avenanuda* L.) and buckwheat (*Fagopyrum esculentum* Moench) by biological methods and PDCAAS method. JMBFS 2 (1), 2079–2086.

# Chapter Twenty

# Rutin Content Assessment of Tartary Buckwheat Germplasm Cultivated in Italy

**Andrea Brunori[1], Chiara Nobili[1], Gerardo Baviello[1], Silvia Procacci[1], and György Végvári[2]**

[1]*ENEA, Rome, Italy*, [2]*Kaposvár University, Kaposvár, Hungary*

## INTRODUCTION

Tartary buckwheat is a minor underutilized crop—traditionally cultivated in Asia but less well known in the European scenario—characterized by rusticity and low input requirements, thus matching the European Union's (EU) most recent expectations of promoting increased sustainability of agriculture and a bio-based economy.

This species is richer in dietary beneficial components compared to common buckwheat. In particular it contains high amounts of flavonoids (Fabjan et al., 2003) capable of expressing antioxidant properties, modulating detoxification enzymes, stimulating the immune system, and decreasing platelet aggregation (Daglia, 2011). Classified as phytochemicals, these biomolecules may exert a stress-protective role on human health when conspicuously taken within the diet (Scalbert and Williamson, 2000).

In addition to their employment as dietary supplements, the use of phytochemicals has been successfully introduced in several more technological areas thanks to their ability to enhance food preservation (Rodríguez Vaquero et al., 2010), favoring protective action on human skin (Nichols and Katiyar, 2010) and exerting biological pest control (Mošovská and Bírošová, 2012).

In view of taking advantage of the beneficial properties of flavonols, Tartary buckwheat grain may represent an environmentally friendly and natural source of such biomolecules, among which the most represented is rutin.

**205**

*Buckwheat Germplasm in the World.* DOI: https://doi.org/10.1016/B978-0-12-811006-5.00020-3
© 2018 Elsevier Inc. All rights reserved.

Field trials carried out in the Italian high hills and mountainous areas (Brunori et al., 2005; Brunori et al., 2006) have proved that Tartary buckwheat can be profitably grown, confirming even more the characteristic high content of rutin in the grain (Fabjan et al., 2003).

Nevertheless, grain rutin content showed appreciable differences between the accessions under evaluation. Therefore it becomes important to investigate the widest gene pool available.

In order to promote the use of Tartary buckwheat as a natural source of rutin, a large collection of Tartary buckwheat varieties and strains were grown at the experimental site of Terranova del Pollino in the summers of 2008 and 2009 at an altitude above sea level of 950 and 1300 m respectively, and were assessed for grain rutin content.

## BUCKWHEAT VARIETIES AND GROWING CONDITIONS

Tartary buckwheat varieties were kindly provided by research Institutes and germplasm Banks (Table 20.1), due to the unavailability of local genetic resources.

Genetic resources, undergoing a process of multiplication, were cultivated in single small plots of 1.5 m$^2$ at the location of Terranova del Pollino. In 2008, 49 accessions were grown on a heavy clay soil placed at 950 m above the sea level, in the following year the number of entries was doubled to a total of 98 and the experimental site was moved to 1300 m above sea level on a similar soil type. Both years a supplement of 100 Kg/ha of N as $NH_4NO_3$ and 90 kg/ha of P as $P_2O_5$ was provided at the time of seed bed preparation, sowing took place around mid-June, hand weeds control was secured during the first month from seedlings emergence, and the crop was harvested during the second half of September.

The rainfall and temperatures registered during crop growth are reported in Fig. 20.1.

## GRAIN RUTIN CONTENT

Rutin content of the grain was evaluated on wholemeal samples obtained from clean grains by the use of a FOSS TECATOR CYCLOTEC 1093 laboratory mill.

Three replicated samples of 200 mg were extracted with 4 mL of methanol (HPLC grade). Extraction was performed in the dark, for 24 hours, at room temperature.

Rutin content, shown in Table 20.2, was determined by the HPLC method according to the procedure previously described (Brunori and Végvári, 2007). Results were subjected to statistical analysis using the $t$-test ($P = 0.05$).

In 2008, grain rutin content varied from highs of 1694 mg/100 g d.w. of PI427235 and 1643 mg/100 mg d.w. of PI451723 to lows of 941 mg/100 g d. w. of Wei 93-8 and 997 mg/100 g d.w. of N8614 Lukla.

**TABLE 20.1** Buckwheat Varieties and Strains Utilized: Origin and Seed Source

| Variety or Strain | Origin | Source |
|---|---|---|
| Golden | Bosnia-Herzegovina | Parco Scientifico e Tecnologico del Molise, Campobasso, Italy |
| Q0001120 | China | Department of Biology, Honghe University, Yunnan, China |
| PI481632, PI481634, PI481636, PI481637, PI481641, PI481643, PI481644, PI481645, PI481648, PI481649, PI481650, PI481651, PI481652, PI481653, PI481654, PI481655, PI481656, PI481658, PI481659, PI481660, PI481661, PI481662, PI481663, PI481664, PI481665, PI481666, PI481667, PI481668, PI481669, PI481671, PI481672, PI481673, PI481674, PI481675 | Bhutan | Northeast Regional PI Station, USDA, Agricultural Research Service, Plant Genetic Resources Unit, Geneva, New York, USA |
| PI451723 | Mexico | |
| PI427235, PI427237, PI427238, PI427239, PI427240 | Nepal | |
| PI199769, PI 476852 Madawaska, PI503879 | USA | |
| Hei Feng, Hei Qiao-4, Wei 93-8, Xinong 9909 | China | Hodowli Róslin Palikije, Wojciechów, Poland |
| 01Z5100012 | Czech Republic | Department of Gene Bank, Division of Genetics and Plant Breeding, Research Institute of Crop Production, Prague-Ruzyne, Czech Republic |
| 01Z5100014 | USA | |
| B9121 Drukyal Dzong, B9132 Jakar, B9133 Chumey, B9138 Rukubji, B9123 Thimphu | Bhutan | Plant Germ-Plasm Institute, Graduate School of Agriculture, Kyoto University, Japan |
| C8816 Malong, C9708 Meigu, C9713 Zhagiao, C9717 Kuer, C9222 Qinlin | China | |
| N7605 Chumoa | Nepal | |
| P9324 Karimabad | Pakistan | |
| Donan, Ishisoba | Japan | Plant Genetic Resources Laboratory, Dept. of Upland Agriculture, National Agricultural Research Center for Hokkaido Region, Shinsei, Memuro-cho, Kasai-gun, Hokkaido, Japan |
| FAG 48, FAG 100 | Belarus | The Leibniz Institute of Plant Genetics and Crop Plant Research (IPK) in Gatersleben, Germany |
| FAG 50, FAG 99 | China | |
| FAG 143 | Italy | |
| FAG 98, FAG 111, FAG 112, FAG 113 | Slovakia | |
| FAG 21, FAG 26 Welsford, FAG 27, FAG 34, FAG 40 Welsford, FAG 49, FAG 149 | Unknown | |
| RCAT 039997 | Romania | The Institute for Agrobotany at Tápiószele, Hungary |
| RCAT 039994 Weswod ican, RCAT 039995, RCAT 040105, RCAT 040127, RCAT 051652, RCAT 051653, RCAT 061058, RCAT 061811, RCAT 062180, RCAT 062213, RCAT 062357, RCAT 065291, RCAT 065394, RCAT 069534 Welsford, RCAT 072708, RCAT 073948 | Unknown | |

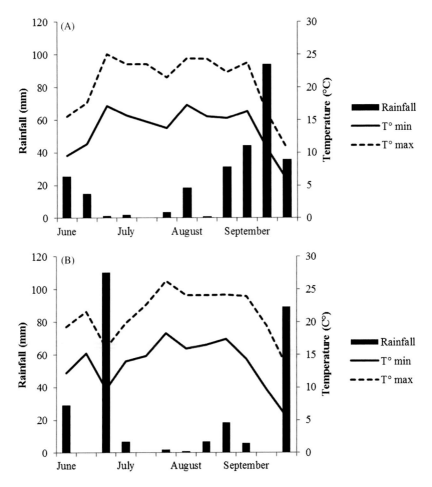

**FIGURE 20.1** Temperatures (min. and max.) and rainfall trend at Terranova del Pollino during experimentation intervals of 2008 (A) and 2009 (B).

The following year, the values obtained ranged from the highest value (1544 mg/100 g d.w.) of RCAT 061058 followed by 01Z5100012 (1516 mg/100 g d.w.) and FAG 50 (1513 mg/100 g d.w.), to the lowest value of the accessions P9324 Karimabad (577 mg/100 g d.w.) and PI481632 (705 mg/100 g d.w.).

Data obtained in 2009, although in line with a similar investigation (Kitabayashi et al., 1995), resulted in somewhat lower levels than those observed in 2008.

In the second year of investigation the experimental site was moved 350 extra meters above sea level, to observe a potential increase in grain rutin content, as it is known that solar irradiance may positively influence this quality trait (Kreft et al., 2002). This did not occur, suggesting that this quality of parameter is of a highly complex nature, possibly influenced by several other factors. In 2009, achenes ripening occurred in late September. The

**TABLE 20.2** Grain Rutin Content of Tartary Buckwheat Varieties and Strains Grown at the Location of Terranova del Pollino at 950 m Above the Sea Level in 2008 and at 1300 m Above Sea Level in 2009

| Variety or Strain | Grain Rutin Content* (mg/100 g Dry Weight) 2008 | 2009 | Variety or Strain | Grain Rutin Content* (mg/100 g Dry Weight) 2008 | 2009 |
|---|---|---|---|---|---|
| 0125100012 | | $1516 \pm 12.68^{ab}$ | PI 476852 Madawaska | | $1316 \pm 16.40^{fg}$ |
| 0125100014 | $1279 \pm 7.55^{h}$ | $1223 \pm 11.36^{hi}$ | PI 481632 | | $705 \pm 9.07^{q}$ |
| B9121 Drukyal Dzong | $1170 \pm 10.48^{kl}$ | $1307 \pm 15.30^{fg}$ | PI 481634 | $1173 \pm 5.21^{k}$ | $1018 \pm 10.39^{lmn}$ |
| B9123 Thimphu | | $773 \pm 12.61^{p}$ | PI 481636 | $1602 \pm 20.02^{b}$ | $1142 \pm 13.01^{ijk}$ |
| B9132 Jakar | $1498 \pm 8.57^{cd}$ | $1094 \pm 7.18^{kl}$ | PI 481637 | | $938 \pm 16.65^{no}$ |
| B9133 Chumey | $1286 \pm 10.48^{h}$ | $1083 \pm 8.95^{kl}$ | PI 481641 | | $1126 \pm 16.20^{jk}$ |
| B9138 Rukubji | $1364 \pm 18.84^{fg}$ | $954 \pm 18.37^{no}$ | PI 481643 | | $1084 \pm 16.53^{kl}$ |
| C8816 Malong | $1430 \pm 11.22^{e}$ | $1147 \pm 8.31^{jk}$ | PI 481644 | | $1152 \pm 14.60^{ijk}$ |
| C9222 Qinlin | | $756 \pm 3.16^{p}$ | PI 481645 | | $984 \pm 17.72^{mno}$ |
| C9708 Meigu | $1094 \pm 12.41^{l}$ | $1203 \pm 15.00^{hij}$ | PI 481648 | $1429 \pm 12.33^{e}$ | $1212 \pm 15.60^{hi}$ |
| C9713 Zhagiao | $1100 \pm 9.33^{l}$ | $1183 \pm 6.37^{ij}$ | PI 481649 | $1507 \pm 6.57^{c}$ | $1174 \pm 5.47^{ij}$ |
| C9717 Kuer | $1367 \pm 14.75^{fg}$ | $1225 \pm 16.02^{hi}$ | PI 481650 | $1426 \pm 15.63^{ef}$ | $1071 \pm 9.17^{kl}$ |
| Donan | $1314 \pm 8.08^{gh}$ | $1487 \pm 13.40^{bc}$ | PI 481651 | $1204 \pm 20.80^{ijk}$ | $1003 \pm 14.87^{mn}$ |
| FAG 21 | | $1077 \pm 6.31^{kl}$ | PI 481652 | $1416 \pm 14.89^{ef}$ | $999 \pm 18.91^{lmn}$ |
| FAG 26 Welsford | | $1448 \pm 0.74^{cd}$ | PI 481653 | $1450 \pm 17.32^{de}$ | $1053 \pm 17.27^{klm}$ |

*Continued*

**TABLE 20.2** continued

| Variety or Strain | Grain Rutin Content* (mg/100 g Dry Weight) 2008 | 2009 | Variety or Strain | Grain Rutin Content* (mg/100 g Dry Weight) 2008 | 2009 |
|---|---|---|---|---|---|
| FAG 27 | | $1305 \pm 12.38^{fg}$ | PI 481654 | $1471 \pm 18.68^{cde}$ | $1090 \pm 20.83^{klm}$ |
| FAG 34 | | $1425 \pm 12.72^{cde}$ | PI 481655 | $1146 \pm 22.36^{kl}$ | $1012 \pm 9.37^{mn}$ |
| FAG 40 Welsford | | $1400 \pm 9.52^{e}$ | PI 481656 | $1361 \pm 8.08^{fg}$ | $1066 \pm 6.95^{kl}$ |
| FAG 48 | | $1272 \pm 14.60^{gh}$ | PI 481658 | $1385 \pm 4.63^{f}$ | $1163 \pm 16.99^{ijk}$ |
| FAG 49 | | $1174 \pm 2.03^{ij}$ | PI 481659 | $1361 \pm 18.68^{fg}$ | $1142 \pm 14.02^{ijk}$ |
| FAG 50 | | $1513 \pm 11.30^{ab}$ | PI 481660 | $1406 \pm 14.22^{ef}$ | $874 \pm 21.01^{o}$ |
| FAG 98 | | $1335 \pm 15.41^{fg}$ | PI 481661 | $1441 \pm 20.66^{cdef}$ | $920 \pm 12.96^{o}$ |
| FAG 99 | | $1372 \pm 20.38^{def}$ | PI 481662 | $1265 \pm 3.67^{h}$ | $1041 \pm 7.63^{lmn}$ |
| FAG 100 | | $1132 \pm 13.09^{jk}$ | PI 481663 | $1183 \pm 8.89^{kl}$ | $1019 \pm 4.01^{mn}$ |
| FAG 111 | | $1330 \pm 8.91^{fg}$ | PI 481664 | $1232 \pm 4.36^{l}$ | $1018 \pm 11.57^{lmn}$ |
| FAG 112 | | $1267 \pm 12.82^{gh}$ | PI 481665 | $1126 \pm 16.58^{kl}$ | $998 \pm 10.86^{mn}$ |
| FAG 113 | | $1239 \pm 14.96^{gh}$ | PI 481666 | $1185 \pm 16.80^{ijkl}$ | $1003 \pm 16.23^{lm}$ |
| FAG 143 | | $1222 \pm 13.33^{hi}$ | PI 481667 | $1121 \pm 12.86^{l}$ | $1112 \pm 8.35^{jk}$ |
| FAG 149 | | $1486 \pm 7.52^{bc}$ | PI 481668 | $1098 \pm 13.17^{l}$ | $1120 \pm 17.63^{kl}$ |
| Golden | $1250 \pm 5.70^{hi}$ | $1287 \pm 5.65^{g}$ | PI 481669 | $1164 \pm 13.13^{kl}$ | $1168 \pm 1.32^{ij}$ |
| Hei Feng | $1341 \pm 8.50^{g}$ | $1438 \pm 19.67^{bcde}$ | PI 481671 | $1146 \pm 22.36^{kl}$ | $1012 \pm 9.37^{mn}$ |
| Hei Qiao-4 | $1265 \pm 4.04^{h}$ | $1228 \pm 12.11^{h}$ | PI 481672 | $1200 \pm 5.29^{j}$ | $1058 \pm 8.96^{l}$ |

| Genotype | | |
|---|---|---|
| Ishisoba | 1311 ± 10.37$^{gh}$ | 1208 ± 19.85$^{hij}$ |
| N7605 Chumoa | | 1112 ± 11.83$^{jkl}$ |
| P9324 Karimabad | | 577 ± 15.59$^{f}$ |
| PI 199769 | 1471 ± 19.14$^{cde}$ | 1318 ± 15.84$^{fg}$ |
| PI 427235 | 1694 ± 11.05$^{a}$ | 1399 ± 12.82$^{def}$ |
| PI 427237 | | 993 ± 13.03$^{mn}$ |
| PI 427238 | | 1196 ± 19.68$^{hij}$ |
| PI 427239 | | 962 ± 21.19$^{mno}$ |
| PI 427240 | | 1209 ± 10.46$^{hi}$ |
| PI 451723 | 1643 ± 13.86$^{b}$ | 1385 ± 10.15$^{ef}$ |
| RCAT 051652 | | 1432 ± 4.24$^{de}$ |
| RCAT 051653 | | 1303 ± 16.48$^{fg}$ |
| RCAT 061058 | | 1544 ± 11.83$^{a}$ |
| RCAT 061811 | | 1497 ± 13.25$^{abc}$ |
| RCAT 062180 | | 1478 ± 15.68$^{bcd}$ |
| RCAT 062213 | | 1472 ± 3.57$^{bc}$ |
| RCAT 062357 | | 1291 ± 14.11$^{g}$ |
| PI 481673 | 1142 ± 12.91$^{kl}$ | 1059 ± 18.81$^{klm}$ |
| PI 481674 | 1348 ± 8.51$^{g}$ | 975 ± 6.35$^{n}$ |
| PI 481675 | 1085 ± 9.06$^{l}$ | 1001 ± 11.74$^{mn}$ |
| PI 503879 | 1481 ± 2.96$^{d}$ | 1043 ± 13.75$^{lm}$ |
| Q0001120 | 1100 ± 18.32$^{l}$ | 1241 ± 16.27$^{gh}$ |
| RCAT 039994 Weswod ican | | 1492 ± 6.89$^{bc}$ |
| RCAT 039995 | | 1071 ± 20.61$^{klm}$ |
| RCAT 039997 | | 1367 ± 14.49$^{ef}$ |
| RCAT 040105 | | 1338 ± 10.48$^{fg}$ |
| RCAT 040127 | | 1351 ± 11.87$^{f}$ |
| RCAT 065291 | | 1326 ± 15.11$^{fg}$ |
| RCAT 065394 | | 1319 ± 13.74$^{fg}$ |
| RCAT 069534 Welsford | | 1383 ± 17.57$^{def}$ |
| RCAT 072708 | | 1132 ± 17.67$^{ijk}$ |
| RCAT 073948 | | 967 ± 6.73$^{n}$ |
| Wei 93-8 | 941 ± 16.70$^{n}$ | 1249 ± 19.68$^{gh}$ |
| Xinong 9909 | | 1248 ± 7.50$^{gh}$ |

*Within each year, grain rutin contents identified with the same letter are not statistically different according to the $t$-test with the level of significance $P = 0.05$. Values are given as the average of three replicated analyses.

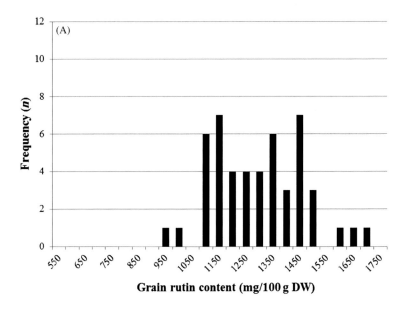

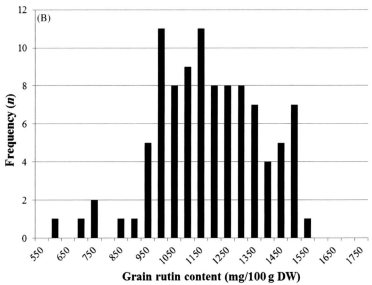

**FIGURE 20.2** Frequency distribution of grain rutin content of buckwheat varieties and genetic stocks grown in the summer of 2008 (A) and 2009 (B) in the Pollino massif at 950 and 1300 m a.s.l., respectively.

same period was characterized by adverse weather conditions, including heavy rain, which may have counteracted any possible positive effect related to altitude and limited the accumulation of rutin in the grain to below expectations.

In both years the varieties Donan, Golden, and Ishisoba, previously culti-vated at the same location with appreciable agronomic results (Brunori et al., 2006), were not among the best expressions of rutin. Furthermore, several new entries added in 2009 performed better than the stocks PI 427235 and PI 451723, which in 2008 ranked first and second respectively, thus proving that there is scope to search for the best expression of rutin content in the grain of Tartary buckwheat through the assessment of a large number of both varieties and accessions.

The large majority of entries presented grain rutin content values in the range of 1000−1500 mg/100 g d.w., as shown by the frequency distribution (Fig. 20.2A, B).

The identification of a number of varieties and genetic stocks capable of expressing satisfactory contents of rutin in the grain suggests that Tartary buckwheat may represent a valid natural source of rutin.

## REFERENCES

Brunori, A., Vegvari, G., 2007. Rutin content of the grain of buckwheat (*Fagopyrum esculentum* Moench and *Fagopyrum tataricum* Gaertn.) varieties grown in Southern Italy. Acta Agron. Hung. 53 (3), 265−272.

Brunori, A., Brunori, A., Baviello, G., Marconi, E., Colonna, M., Ricci, M., 2005. The yield of five buckwheat (*Fagopyrum esculentum* Moench) varieties grown in Central and Southern Italy. Fagopyrum 22, 98−102.

Brunori, A., Brunori, A., Baviello, G., Marconi, E., Colonna, M., Ricci, M., et al., 2006. Yield assessment of twenty buckwheat (*Fagopyrum esculentum* Moench and *Fagopyrum tataricum* Gaertn.) varieties grown in Central (Molise) and Southern Italy (Basilicata and Calabria). Fagopyrum 23, 83−90.

Christa, K., Soral-Śmietana, M., 2008. Buckwheat grains and buckwheat products − nutritional and prophylactic value of their components − a review. Czech J. Food Sci. 26, 153−162.

Daglia, M., 2012. Polyphenols as antimicrobial agents. Curr. Opin. Chem. Biol. 23, 174−181.

Fabjan, N., Rode, J., Kosir, I.J., Zhang, Z., Kreft, I., 2003. Tartary buckwheat (*Fagopyrum tartarikum* Gaertn.) as a source of dietary rutin and quercetin. J. Agric. Food Chem. 51, 6452−6455.

Kitabayashi, H., Ujihara, A., Hirose, T., Minami, M., 1995. Varietal differences and heritability for rutin content in common buckwheat, *Fagopyrum esculentum* Moench. Breed Sci. 45 (1), 75−79.

Kreft, S., Strukelj, B., Gaberscik, A., Kreft, I., 2002. Rutin in buckwheat herbs grown at different UV-B radiation levels: comparison of two UV spectrophotometric and HPLC method. J. Exp. Bot. 53 (375), 1801−1804.

Mošovská, S., Bírošová, L., 2012. Antimicotic and antifungal activities of amaranth and buckwheat extracts. Asian J. Plant Sci. 1−3.

Nichols, J.A., Katiyar, S.K., 2010. Skin photoprotection by natural polyphenols: Anti-inflammatory, anti-oxidant and DNA repair mechanisms. Arch. Dermatol. Res. 302 (2), 71.

Rodríguez Vaquero, M.J., Aredes, P., Manca de Nadra, M.C., Strasser de Saad, A.M., 2010. Phenolic compound combinations on *Escherichia coli* viability in a meat system. J. Agric. Food Chem. 58 (10), 6048−6052.

Scalbert, A., Williamson, G., 2000. Dietary intake and bioavailability of polyphenols. J. Nutr. 130 (8S Suppl), 2073S−2085SS.

# Chapter Twenty One

# Correlation Between Grain Yield and Rutin Content in Common Buckwheat Germplasm Cultivated in Southern Italy

**Andrea Brunori[1], Gerardo Baviello[1], Chiara Nobili[1], Domenico Palumbo[1], and György Végvári[2]**

*[1]ENEA, Rome, Italy, [2]Kaposvár University, Kaposvár, Hungary*

## INTRODUCTION

In Italy, buckwheat was traditionally cultivated in the Northern mountain districts from the XVI century to the beginning of the XX century. From then on, the gradual introduction of the modern high yielding wheat varieties caused a progressive loss of the cultivated area, reducing the use of this pseudo-cereal as a food source. Today the cultivation of this crop is limited to few areas of the Alpine valleys and the Apennine mountains, where buckwheat based food preparations are still consumed.

In the last decades, this species has been reconsidered and evaluated as an alternative crop to be cultivated in marginal areas, thanks to its rusticity and low input requirements (Borghi et al., 1995; Borghi, 1996).

The feasibility of buckwheat cultivation has been further explored in field trials carried out in Regions of Central and Southern Italy (Brunori et al., 2005; Brunori et al., 2006), to verify whether this crop can be profitably

**215**

*Buckwheat Germplasm in the World.* DOI: https://doi.org/10.1016/B978-0-12-811006-5.00021-5
© 2018 Elsevier Inc. All rights reserved.

grown and what kind of produce can be attained, maintaining unchanged the above-mentioned agronomic peculiarities, meanwhile preserving the several health beneficial properties of its grain (Christa and Soral-Śmietana, 2008).

Flavonoids are a class of health value added bioactive compounds found in buckwheat grain, among which is rutin, in turn credited to exert most of the health benefits. In view of this ability, the correlation between the content of this flavonoid and grain yield has been investigated from the scientific community, leading to different outcomes (Ohsawa and Tsutsumi, 1995; Morishita and Tetsuka, 2002; Brunori et al., 2012; Sobhani et al., 2014).

To broaden the knowledge on this topic, agronomic field trials were carried out at the experimental site of Terranova del Pollino (Basilicata) for two running years (2008 and 2009) to assess the grain yield and rutin content of 30 common buckwheat varieties, and the possible correlation between these parameters.

## BUCKWHEAT VARIETIES

Common buckwheat varieties were either purchased or kindly provided by research Institutes and germplasm Banks located in Countries where buckwheat is currently grown (Table 21.1), due to the unavailability of local genetic resources.

**TABLE 21.1** Buckwheat Varieties Utilized: Origin and Seed Source

| Variety | Origin | Source |
|---|---|---|
| AC Manisoba, Koban, Mancan, Springfield | Canada | Kade Research Ltd. Morden, Manitoba, Canada |
| Jana, Pyra, Špačinska | Czech Republic | University of South Bohemia, Faculty of Agriculture, České Budějovice, Czech Republic |
| Aelita | Russia | Department of Gene Bank, Division of Genetics and Plant Breeding, Research Institute of Crop Production, Prague-Ruzyne, Czech Republic |
| Emka, Hruszkowska | Poland | |
| Prego | Germany | |
| Aleksandrina, Anita Belorusskaya, Iliya, Karmen, Lena, Vlada, Zhnayarka | Belarus | RUP The Institute of Arable Farming and Plant Breeding of The National Academy of Sciences of Belarus, Zhodino, Minsk District, Belarus |
| Bamby | Austria | J. Biason, Bolzano, Italy |
| Lileja | Unknown | |
| Botan | Japan | Plant Germplasm Institute, Graduate School of Agriculture, Kyoto University, Japan |
| Darja | Bosnia and Herzegovina | Parco scientifico e Tecnologico del Molise, Campobasso, Italy |
| Kitawasesoba, Kitayuki | Japan | Plant Genetic Resources Laboratory, Dept. of Upland Agriculture, National Agricultural Research Center for Hokkaido Region, Shinsei, Memuro-cho, Kasai-gun, Okkaido, Japan |
| Kora, Luba, Panda | Poland | Stacja Hodowli Róslin Palikije, Wojciechów, Poland |
| Koto, Manor | Canada | |
| La Harpe | France | Semfor, Casaleone, Verona, Italy |

According to their geographic origin, buckwheat varieties could be divided in two groups, namely European varieties:

Aelita, Aleksandrina, Anita Belorusskaya, Bamby, Darja, Emka, Hruszkowska, Iliya, Jana, Karmen, Kora, La Harpe, Lena, Lileja, Luba, Panda, Prego, Pyra, Špačinska, Vlada, and Zhnayarka; and Pacific area (Canada and Japan) varieties:

AC Manisoba, Botan, Kitawasesoba, Kitayuki, Koban, Koto, Mancan, Manor, and Springfield.

## GRAIN YIELD

The genetic resources were cultivated in three replicated plots of 10 m$^2$ each during the summers of 2008 and 2009. Each year a first sowing took place in late May, and a second sowing was made around mid-June. The rainfall and temperatures registered during the two periods of experimentation are reported in Fig. 21.1.

The grain yield obtained after each sowing time is shown in Table 21.2.

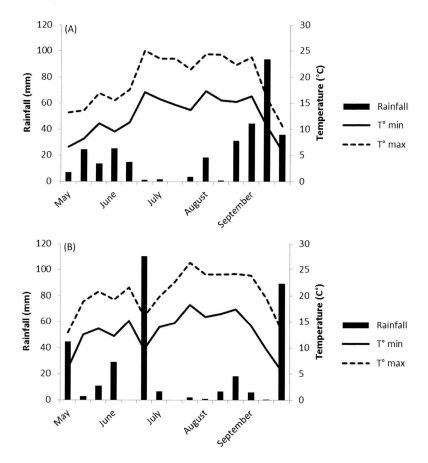

**FIGURE 21.1** Temperatures (min. and max.) and rainfall trend at Terranova del Pollino during experimentation intervals of 2008 (A) and 2009 (B).

**TABLE 21.2** Grain Yield (t/ha) of Buckwheat Varieties Grown in the Years 2008 and 2009 Observed at the Experimental Site of Terranova del Pollino (Basilicata)

| Variety | 2008 1st Sowing Time (May) | 2008 2nd Sowing Time (June) | 2009 1st Sowing Time (May) | 2009 2nd Sowing Time (June) |
|---|---|---|---|---|
| AC Manisoba | 1.66 ± 0.07 | 0.73 ± 0.06 | 1.25 ± 0.17 | 1.23 ± 0.25 |
| Aelita | 0.98 ± 0.10 | 0.95 ± 0.08 | 1.19 ± 0.15 | 0.99 ± 0.15 |
| Aleksandrina | 0.98 ± 0.07 | 0.80 ± 0.05 | 1.25 ± 0.22 | 0.96 ± 0.08 |
| Anita Belorusskaya | 1.97 ± 0.16 | 0.77 ± 0.06 | 1.15 ± 0.17 | 0.77 ± 0.12 |
| Bamby | 1.59 ± 0.06 | 0.97 ± 0.15 | 1.28 ± 0.15 | 0.95 ± 0.08 |
| Botan | 1.33 ± 0.15 | 1.05 ± 0.12 | 0.46 ± 0.14 | 1.38 ± 0.27 |
| Darja | 1.68 ± 0.10 | 0.86 ± 0.13 | 0.37 ± 0.01 | 1.02 ± 0.12 |
| Emka | 1.28 ± 0.16 | 0.99 ± 0.11 | 1.67 ± 0.02 | 1.02 ± 0.19 |
| Hruszkowska | 1.52 ± 0.13 | 1.03 ± 0.18 | 1.47 ± 0.15 | 0.84 ± 0.30 |
| Iliya | 1.03 ± 0.04 | 0.60 ± 0.20 | 1.21 ± 0.18 | 1.29 ± 0.32 |
| Jana | 1.44 ± 0.06 | 0.84 ± 0.08 | 1.52 ± 0.08 | 1.07 ± 0.29 |
| Karmen | 1.29 ± 0.13 | 0.72 ± 0.10 | 1.23 ± 0.17 | 1.10 ± 0.24 |
| Kitawasesoba | 1.48 ± 0.14 | 0.90 ± 0.14 | 1.29 ± 0.15 | 0.98 ± 0.26 |
| Kitayuki | 1.34 ± 0.14 | 1.24 ± 0.17 | 0.52 ± 0.11 | 1.41 ± 0.26 |
| Koban | 1.31 ± 0.18 | 0.75 ± 0.20 | 1.07 ± 0.17 | 1.27 ± 0.22 |
| Kora | 1.23 ± 0.05 | 0.94 ± 0.19 | 1.71 ± 0.20 | 1.22 ± 0.27 |
| Koto | 1.65 ± 0.21 | 0.81 ± 0.05 | 1.24 ± 0.17 | 1.59 ± 0.29 |
| La Harpe | 1.58 ± 0.16 | 0.94 ± 0.15 | 0.39 ± 0.12 | 1.19 ± 0.18 |
| Lena | 0.96 ± 0.17 | 0.79 ± 0.09 | 1.39 ± 0.22 | 0.69 ± 0.09 |
| Lileja | 1.55 ± 0.21 | 0.88 ± 0.16 | 1.71 ± 0.15 | 1.09 ± 0.12 |
| Luba | 1.58 ± 0.16 | 0.80 ± 0.09 | 1.58 ± 0.22 | 1.01 ± 0.21 |
| Mancan | 1.43 ± 0.18 | 1.33 ± 0.16 | 1.53 ± 0.24 | 1.48 ± 0.15 |
| Manor | 1.59 ± 0.17 | 1.23 ± 0.16 | 1.39 ± 0.26 | 1.43 ± 0.19 |
| Panda | 1.95 ± 0.16 | 0.55 ± 0.14 | 1.52 ± 0.24 | 1.11 ± 0.23 |
| Prego | 1.43 ± 0.06 | 0.84 ± 0.02 | 1.50 ± 0.30 | 1.20 ± 0.16 |
| Pyra | 1.33 ± 0.07 | 1.21 ± 0.10 | 1.19 ± 0.24 | 1.16 ± 0.11 |
| Spačinska | 1.43 ± 0.09 | 0.91 ± 0.11 | 0.56 ± 0.01 | 1.32 ± 0.11 |
| Springfield | 1.63 ± 0.12 | 1.23 ± 0.21 | 1.41 ± 0.33 | 1.59 ± 0.25 |
| Vlada | 1.69 ± 0.16 | 1.03 ± 0.08 | 1.79 ± 0.12 | 0.94 ± 0.16 |
| Zhnayarka | 1.51 ± 0.07 | 0.95 ± 0.12 | 1.47 ± 0.23 | 0.75 ± 0.11 |
| *Field average* | 1.45 ± 0.12 | 0.92 ± 0.12 | 1.24 ± 0.17 | 1.14 ± 0.19 |

In both years first sowing timing occurred in late May and second around mid-June. Values are given as the average of three replicated plots.

In 2008, the grain yield (Table 21.2) of early-sown material varied from highs of 2.0 t/ha in Anita Belorusskaya and Panda, to lows of 1.0 t/ha in Aelita, Aleksandrina, Iliya, and Lena. In general, delayed sowing brought about a reduction of grain yield—in fact the produce attained following the second harvest of this year resulted in decidedly lower levels, always below 1.3 t/ha for all varieties.

A somewhat similar trend was observed in 2009, even though the field average differences were narrower, shifting between 1.24 t/ha and 1.14 t/ha (Table 21.2).

Geographical origin appeared not to significantly influence grain yield (Fig. 21.2). Nevertheless, it can be observed how in both years the top yielders resulted among European varieties at first harvest and among Pacific Area varieties at second harvest (Fig. 21.2).

The observations made during the two campaigns invite the following considerations.

Sowing date appeared of paramount relevance. Early sowing is to be preferred to avoid drought conditions, which may intervene early in the spring season to last throughout the summer so as to prejudice seed germination and initial vegetative growth. Early sowing, meaning early ripening, is also expected to anticipate the harvest already by the beginning of August before

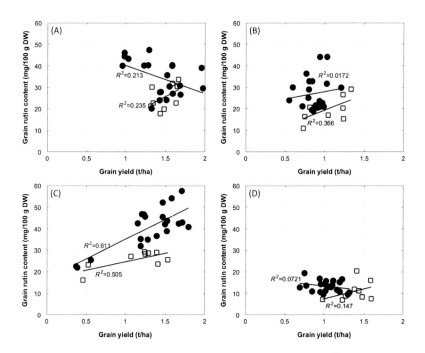

**FIGURE 21.2** Correlation (linear regression with the r-squared ($R^2$)) between grain yield potential and grain rutin content of the common buckwheat varieties grown at two sowing timings in 2008 (A and B) and 2009 (C and D). Symbols: □ Varieties originating from the Pacific area; ● Varieties originating from Europe.

the rain season starts. It is in fact a further constraint, coming along with delayed sowing, that harvesting in the autumnal season, where heavy and long rainy periods can occur, may spoil the ripening crop.

Because ploughing at the time of sowing is likely to cause a quick soil drying, it appears advisable to anticipate the ploughing in the autumn of the previous year, and to limit sowing bed preparation to a light tilling, useful also to cover the fertilizers.

Altogether, satisfactory grain yields of common buckwheat, up to 2 t/ha, were achieved, competitive when compared to world and European average values declared by FAO (FAOSTAT world 2008: 0.92 t/ha; Europe 2008: 0.96 t/ha; world 2009: 0.89 t/ha; Europe 2009: 0.95 t/ha), speaking in favor of a substantial feasibility of buckwheat cultivation in this district new to the cultivation of this crop.

Furthermore, among top yielders, varieties with appreciable rutin content could be identified (see rutin data, Table 21.3).

## GRAIN RUTIN CONTENT

For rutin (quercetin-3-rutinoside) analysis, the grain harvest of the three replicated plots was pooled and 200 mg of wholemeal samples, obtained from clean grains by the use of a FOSS TECATOR CYCLOTEC 1093 sample mill, were extracted in triplicate with 2 mL of methanol (HPLC grade). Extraction was performed in the dark, for 24 hours, at room temperature.

Rutin content was determined by the HPLC method according to the procedure previously described (Brunori and Végvári, 2007). The data obtained are shown in Table 21.3.

In 2008, the rutin content observed varied from 47 mg/100 g d.w. of Karmen to 18 mg/100 g d.w. of Mancan in the occasion of early sowing and between 44 mg/100 g d.w. of Kora and Hruszkowska, and 11 mg/100 g d.w. of AC Manisoba in the case of delayed sowing (Table 21.3).

The following year the amount of rutin detected was slightly higher at first harvest, with top yielder Kora 58 mg/100 g d.w. and bottom scorer Botan 16 mg/100 g d.w., and decidedly lower at second harvest with values ranging between 19 mg/100 g d.w. of Zhnayarka and 6 mg/100 g d.w. of Mancan.

Altogether, the data obtained confirmed the rather modest content of this compound which, however, was within the range already reported for this species (Kitabayashi et al., 1995; Ohsawa and Tsutsumi, 1995; Brunori and Végvári, 2007; Brunori et al., 2008).

In a comparison between sowing timings, in 2008 first harvest material showed mainly higher values of grain rutin content apart from a few cases with narrow differences; the following year, early sown material presented, for each variety, values decidedly higher (36 mg/100 g d.w. on average), than those obtained with late sowing (12 mg/100 g d.w. on average) (Table 21.3).

**TABLE 21.3** Grain Rutin Content (mg/100 g dry weight) of Buckwheat Varieties Grown in the Summers of 2008 and 2009 at the Experimental Site of Terranova del Pollino (Basilicata)

| Variety | 2008 | | 2009 | |
|---|---|---|---|---|
| | 1st Sowing Time (May) | 2nd Sowing Time (June) | 1st Sowing Time (May) | 2nd Sowing Time (June) |
| AC Manisoba | 33.71 ± 3.69 | 10.88 ± 1.08 | 28.31 ± 0.40 | 6.95 ± 0.63 |
| Aelita | 46.14 ± 4.89 | 21.48 ± 0.46 | 31.96 ± 0.19 | 9.65 ± 0.18 |
| Aleksandrina | 44.39 ± 7.25 | 28.99 ± 1.99 | 45.65 ± 0.48 | 14.30 ± 1.38 |
| Anita Belorusskaya | 29.63 ± 4.30 | 36.33 ± 0.91 | 42.56 ± 1.12 | 13.69 ± 0.46 |
| Bamby | 27.09 ± 3.33 | 22.63 ± 1.21 | 34.90 ± 1.20 | 10.62 ± 0.45 |
| Botan | 30.17 ± 1.17 | 16.98 ± 1.62 | 16.15 ± 0.76 | 11.97 ± 0.70 |
| Darja | 30.90 ± 4.56 | 18.80 ± 1.32 | 22.54 ± 3.47 | 11.25 ± 0.53 |
| Emka | 40.35 ± 1.53 | 21.00 ± 3.11 | 42.46 ± 0.19 | 15.77 ± 1.13 |
| Hruszkowska | 35.71 ± 2.76 | 44.09 ± 2.46 | 45.58 ± 1.67 | 10.90 ± 1.18 |
| Iliya | 43.27 ± 1.55 | 30.07 ± 1.19 | 46.83 ± 0.73 | 9.54 ± 0.40 |
| Jana | 27.83 ± 3.42 | 32.82 ± 5.44 | 38.89 ± 1.37 | 13.06 ± 0.51 |
| Karmen | 47.42 ± 3.95 | 21.06 ± 1.10 | 46.73 ± 2.79 | 13.97 ± 0.68 |
| Kitawasesoba | 19.88 ± 1.13 | 21.76 ± 1.79 | 28.71 ± 0.62 | 7.23 ± 0.27 |
| Kitayuki | 22.65 ± 0.75 | 15.38 ± 2.02 | 23.27 ± 0.60 | 9.09 ± 0.60 |
| Koban | 21.20 ± 1.39 | 16.37 ± 1.35 | 27.13 ± 0.63 | 8.79 ± 0.39 |
| Kora | 40.14 ± 2.55 | 44.12 ± 6.06 | 57.54 ± 1.52 | 16.56 ± 0.59 |
| Koto | 30.46 ± 2.09 | 20.67 ± 2.38 | 29.36 ± 1.73 | 16.06 ± 0.59 |
| La Harpe | 40.08 ± 2.15 | 23.66 ± 0.62 | 22.02 ± 3.45 | 15.19 ± 0.87 |
| Lena | 40.02 ± 4.20 | 24.99 ± 0.59 | 36.65 ± 1.05 | 12.73 ± 0.94 |
| Lileja | 30.40 ± 4.62 | 21.44 ± 2.73 | 42.92 ± 1.57 | 12.95 ± 0.85 |
| Luba | 40.24 ± 3.28 | 32.82 ± 2.50 | 54.21 ± 2.85 | 14.16 ± 0.39 |
| Mancan | 17.77 ± 1.23 | 29.19 ± 2.23 | 25.65 ± 1.62 | 8.44 ± 0.31 |
| Manor | 31.84 ± 2.11 | 26.55 ± 3.11 | 29.01 ± 1.73 | 11.20 ± 1.49 |
| Panda | 39.06 ± 2.01 | 23.91 ± 2.83 | 43.68 ± 1.58 | 15.91 ± 0.50 |
| Prego | 23.90 ± 0.77 | 19.36 ± 2.83 | 42.13 ± 2.41 | 10.82 ± 1.23 |
| Pyra | 20.05 ± 2.59 | 29.93 ± 0.27 | 35.21 ± 0.85 | 11.68 ± 0.21 |
| Spačinska | 27.70 ± 0.60 | 20.55 ± 0.55 | 25.47 ± 1.09 | 10.44 ± 0.47 |
| Springfield | 22.59 ± 1.40 | 20.29 ± 0.53 | 23.63 ± 0.82 | 7.52 ± 0.58 |
| Vlada | 26.67 ± 0.76 | 31.67 ± 1.75 | 40.81 ± 1.52 | 16.93 ± 0.34 |
| Zhnayarka | 24.11 ± 1.02 | 20.96 ± 2.12 | 52.16 ± 3.25 | 19.39 ± 1.73 |
| *Average* | 31.71 ± 2.57 | 24.79 ± 1.94 | 36.07 ± 1.44 | 12.16 ± 0.69 |

Two different sowing times were adopted: late May (first sowing) and mid-June (second sowing). Values are given as the average of three replicated analyses.

Similarly to a previous experience (Brunori et al., 2008), in 2008 the effect of sowing date on grain rutin content appeared limited and not unequivocal. A different attitude was shown in 2009, with a strong decrement—threefold on average—between early and delayed sowing. This might be explained as a possible consequence of a particularly rainy second half of September, around the time of ripening of achenes, as it has been reported that rutin may accumulate in response to solar irradiance (Kreft et al., 2002) (Fig. 21.1).

The present data on the rutin content of buckwheat grain confirm previous reports on the existence of a large variability of this characteristic among varieties of this species (Kitabayashi et al., 1995; Ohsawa and Tsutsumi, 1995; Oomah et al., 1996; Brunori and Végvári, 2007; Brunori et al., 2008).

However, despite the quite relevant extent of variation of grain rutin contents, it can be noticed how the group of European varieties is characterized on average by higher values compared to those of the Pacific area, regardless of grain yield (Fig, 21.2), as previously reported (Brunori et al., 2012).

## RELATIONSHIP BETWEEN GRAIN YIELD POTENTIAL AND RUTIN CONTENT OF THE GRAIN

As previously mentioned, the relationship between grain yield potential and rutin content observed during the two years of study is shown in Fig. 21.2. For each sowing timing, the linear regressions were calculated separately for the two different geographic origin groups.

Judging from the very low $R^2$ values, there is no apparent correlation between the two traits under investigation. An exception is notable when observing data of the 2009 first sowing, even though the correlation shown is weak, with $R^2$ values of around 0.6 for European and 0.5 for Pacific area varieties (Fig. 21.2C).

Data collected during the two years of study are evenly spread in a wide interval (Fig. 21.3) and, when taken into account as a whole, do not allow inferring any significant correlation, as previously observed by other authors (Ohsawa and Tsutsumi, 1995, Brunori et al., 2012).

Relationships of a different nature, between the two parameters, have been also observed. Studies have shown a significant direct correlation between grain yield and rutin content in one case (Sobhani et al., 2014), and the exact opposite trend in another (Morishita and Tetsuka, 2002).

Nevertheless, it appears reasonable to say that when a large number of varieties originating from several different geographical areas are compared, the large variability of grain yield potential and rutin content observed does not provide evidence for any possible correlation between these parameters.

The overall information available on the topic suggests that the relationship between grain yield potential and rutin content in common buckwheat is of a very complex nature, and difficult to dig out.

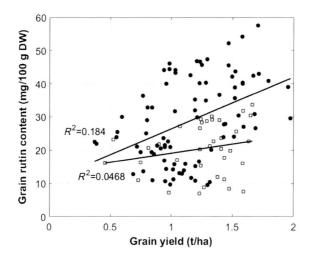

**FIGURE 21.3** Correlation between grain yield potential and grain rutin content of all the common buckwheat varieties under investigation during 2008 and 2009. Symbols: □ Varieties originating from the Pacific area; ● Varieties originating from Europe.

Possibly, further studies are required for the better understanding of the relationship between grain yield and rutin content in common buckwheat.

## REFERENCES

Borghi, B., 1996. Il grano saraceno, coltura che ha un futuro. Vita in Campagna 10, 46–48.

Borghi, B., Minoia, C., Cattaneo, M., Baiocchi, A., Kaiserman, F., Mair, V., et al., 1995. Primi risultati di prove agronomiche e varietali. L' Informatore Agrario 25, 59–63.

Brunori, A., Brunori, A., Baviello, G., Marconi, E., Colonna, M., Ricci, M., et al., 2006. Yield assessment of twenty buckwheat (*Fagopyrum esculentum* Moench and *Fagopyrum tataricum* Gaertn.) varieties grown in Central (Molise) and Southern Italy (Basilicata and Calabria). Fagopyrum 23, 83–90.

Brunori, A., Brunori, A., Baviello, G., Marconi, E., Colonna, M., Ricci, M., 2005. The yield of five buckwheat (*Fagopyrum esculentum* Moench) varieties grown in Central and Southern Italy. Fagopyrum 22, 98–102.

Brunori, A., Végvári, G., 2007. Rutin content of the grain of buckwheat (*Fagopyrum esculentum* Moench and *Fagopyrum tataricum* Gaertn.) varieties grown in Southern Italy. Acta Agron. Hung. 53 (3), 265–272.

Brunori, A., Végvári, G., Sándor, G., Xie, H., Baviello, G., Kadyrov, R., 2008. Rutin content of buckwheat grain (*Fagopyrum esculentum* Moench and *F. tataricum* Gaertn.): Influence of variety, location and sowing time. Fagopyrum 25, 21–27.

Brunori, A., Baviello, G., Kajdi, F., Teixeira da Silva, J., Györi, T., Végvári, G., 2012. Grain yield and rutin content of common and tartary buckwheat varieties grown in North-western Hungary. EJPSB 6 (Special Issue 2), 70–74.

Christa, K., Soral-Śmietana, M., 2008. Buckwheat grains and buckwheat products – nutritional and prophylactic value of their components – a review. Czech. J. Food Sci. 26, 153–162.

FAOSTAT. Available online: <http://www.fao.org/faostat/en/#data/QC>.

Kitabayashi, H., Ujihara, A., Hirose, T., Minami, M., 1995. Varietal differences and heritability for rutin content in common buckwheat, *Fagopyrum esculentum* Moench. Breed Sci. 45 (1), 75−79.

Kreft, S., Strukelj, B., Gaberscik, A., Kreft, I., 2002. Rutin in buckwheat herbs grown at different UV-B radiation levels: comparison of two UV spectrophotometric and HPLC method. J. Exp. Bot. 53 (375), 1801−1804.

Morishita, T., Tetsuka, T., 2002. Varietal differences of rutin, protein and oil content of common buckwheat (*Fagopyrum esculentum*) grains in Kyushu area. Jpn. J. Crop Sci. 71 (2), 192−197.

Ohsawa, R., Tsutsumi, 1995. Inter-varietal variation of rutin content in common buckwheat flour (*Fagopyrum esculentum* Moench.). Euphtyica 86 (3), 183−189.

Oomah, B.D., Mazza, G., Przybylski, R., 1996. Comparison of flaxseed meal lipids extracted with different solvents. Lebensm Wiss Technol. 29, 654−658.

Sobhani, M.R., Rahmikhdoev, G., Mazaheri, D., Majidian, M., 2014. Influence of different sowing date and planting pattern and N rate on buckwheat yield and its quality. AJCS 8 (10), 1402−1414.

# Chapter Twenty Two

# Buckwheat Resources in the VIR (Russia) Collection: The Photoperiod Response

**Olga I. Romanova[1], Aleksey N. Fesenko[2],**
**Nikolay N. Fesenko[2], Ivan N. Fesenko[2],**
**and Vladimir A. Koshkin[1]**

[1]*N. I. Vavilov Institute of Plant Genetic Resources (VIR), St. Petersburg, Russia*
[2]*All-Russia Research Institute of Legumes and Groat Crops, Orel, Russia*

## INTRODUCTION

The buckwheat collection in the Russian gene bank at VIR has a long history. Its first accessions were registered back in 1924. The base collection of buckwheat numbers 2,230 accessions. In terms of species, the major part of accessions belongs to common buckwheat (*Fagopyrum esculentum* Moenh). In addition, the collection includes 102 accessions of Tartary buckwheat (*Fagopyrum tataricum* (L) Gaertn.), 1 of *Fagopyrum giganteum* Krotov, another 1 of *Fagopyrum cymosum* Meissn., 2 of *Fagopyrum homotropicum* Ohnishi, as well as 1 of *F. esculentum* subsp. *ancestrale* Ohnishi. The collection contains diploid accessions and some tetraploids or tetraploid counterparts of diploids. Many important accessions of common buckwheat and some of Tartary buckwheat have been transferred to the tetraploid level by VIR researchers. One of them (VIR-17, 4n) was used to produce *F. giganteum* Krotov, a constant amphidiploid of *F. tataricum* × *F. cymosum*. In the geographic sense, all countries of the world, as well as the regions and provinces of the former USSR where buckwheat is cultivated or grows as wild/weedy species, are represented in the collection. Local buckwheats from different countries amount to 85% of the total collection. The overwhelming majority of accessions represent local populations or landraces from Russia, Ukraine,

**225**

Buckwheat Germplasm in the World. DOI: https://doi.org/10.1016/B978-0-12-811006-5.00022-7
© 2018 Elsevier Inc. All rights reserved.

Belarus, China, etc., which were included into the collection before 1941 and during the period of 1960–1970.

Accessions from the VIR collection differ by morphological and bio-chemical characters of plant and achene, by plant productivity, 1000 seeds weight, growing season duration, and by their adaptivity.

Common buckwheat is grown as a groats crop at different latitudes—from the tropics almost to the Polar Circle. As it remains sensitive to frost, it can be grown in temperate latitudes in the summer frost-free period only. So, the flowering time is the critical trait for northern buckwheat cultivars. Skripchinsky (1971) shows that the strongly pronounced short day reaction is quite regular in plants of near-tropical origin as this trait permits them to catch (and react on) the small seasonal inequality between summer and winter day length. In the case of subtropical buckwheat, its spring seedlings grow to become more late-flowering and vigorous and can produce more seeds during a long warm period; while its fall seedlings show a faster development and can set seed in a short period before the colder winter season. Such a mode of reaction is common for wild buckwheat in South China and for cultivars in low latitudes. However, such a sensitive response leads to undesirable results in high latitudes, where daylight lasts longer and the warm period is much shorter.

For instance, the populations from the Maritime Territory, China, India, and Japan fail to ripen in the Orel Region in Central Russia (Fesenko et al., 2006). Buckwheat originates from the subtropics of China, the circumstance of which determines not only the thermophilic and hydrophilic nature of the crop, but also its day-length sensitivity (Ohnishi, 1991). Most strongly it is displayed by the buckwheat genotypes adapted to cultivation at low latitudes (in tropics and subtropics) when moved to higher latitudes, where they demonstrate from delayed ripening up to a complete inability to set seed under long day (LD) conditions (Fesenko et al., 2006). The day length response is a stable character which can be easily predicted from geographic location. However, according to Skripchinskiy (1971), any taxon can contain biotypes that would differ in the degree and sign of their photoperiodic response.

## SOME DETAILS OF THE EXPERIMENT

The response to day length variation was analyzed in buckwheat populations originating from various soil and climatic regions of Russia, from nearby and distant foreign countries, as well as in buckwheat cultivars certified for commercial use in Russia. It was done by means of a vegetation test that was performed at Pushkin town (Leningrad Province) from 1999 through 2015. Over 300 common buckwheat accessions from the VIR collection have been evaluated under natural LD (18 hours) conditions and under "artificial" short day (SD) conditions from 9 PM to 9 AM for 30 days after seedlings emergence. The difference in duration of the "sprouting to early flowering" period of an accession that depended on day length conditions and was expressed as a photoperiodic sensitivity index ($I_{PhS}$) was used as a measure of plant photoperiodic sensitivity. $I_{PhS} = T_1/T_2$, where $T_1$ and $T_2$ are the duration of the "sprouting to early flowering" period (days) for buckwheat plants grown under natural LD and artificial SD, respectively. The buckwheat accessions that had an $I_{PhS}$ value of $1.00-1.10$ were classified as less photoperiod sensitive. The use of the index makes it possible to compare results obtained during different years of study (Koshkin et al., 1994).

## POLYMORPHISM FOR PHOTOPERIODIC RESPONSE AMONG THE COLLECTION ACCESSIONS

It has been previously established that local common buckwheat populations significantly vary in terms of PS (Romanova and Koshkin, 2010).

The highest PS is characteristic of the common buckwheat accessions from the subtropics (Southern China, Nepal, Bhutan, India, etc.) (Matano and Ujihara, 1981; Baniya et al., 1992). According to our results, the $I_{PhS}$ values for the Chinese local populations varied in the $1.06-1.65$ range, for the Korean ones it was within the $1.10-1.56$ range, for the Japanese it was between 1.31 and 1.70, and for the Indian it fell within the range of 1.18 to 1.70 (Table 22.1). On the average, $I_{PhS}$ values for the populations from China, Korea, Japan, and India were between 1.35 and 1.51 on average (Table 22.1).

In Japan, common buckwheat cultivars were initially differentiated into autumn and summer types on the basis of PS. The more sensitive cultivars of the autumn type, which are used as the backup crop, ripen only when sown in autumn. When sown in spring, under LD, the buckwheat plants show retarded development, produce excessive biomass instead of normal grain yield, and turn out to be low in productivity. The summer type cultivars are less photoperiod sensitive and produce sufficient grain yield when sown in spring. After a more detailed study, a part of cultivars has been allocated to the intermediate type group (Matano and Ujihara, 1981). Therefore, Japanese common buckwheat cultivars can be differentiated at least in three PS groups. However, even the least sensitive summer type cultivars from Northern Japan (Hokkaido, 44°N) were found to be unable to form normal seed yield in Russia (Orel, 53°N) because of the sharply retarded development (Fesenko et al., 1998). The Russian cultivars used in the same test showed weak reaction to day length and ripened much earlier in both conditions.

Accessions belonging to the East European agroecotype, including many Russian cultivars, were the least sensitive to day length. It should be noted that the lowest $I_{PhS}$ values among the studied accessions were recorded for cultivars Skorospelaya 86 and Molva with limited secondary branching (Romanova and Koshkin, 2010).

The performed study showed that the expansion of common buckwheat northwards was accompanied by a decrease of PS in plants: in the high-latitude regions located above 50°N, the shares of populations with a very low ($I_{PhS} = 1.00-1.10$) and low ($I_{PhS} = 1.11-1.20$) PS were equal and amounted to

**TABLE 22.1** Photoperiodic Sensitivity of Common Buckwheat Accessions

| Cultivar Origin | Accessions Number | Average | $I_{PhS}$[a] Min | Max |
|---|---|---|---|---|
| **Europe** | | | | |
| Sweden | 1 | 1.12 | | |
| Estonia | 2 | 1.1 | 1.09 | 1.1 |
| Latvia | 3 | 1.15 | 1.1 | 1.24 |
| Lithuania | 4 | 1.08 | 1.04 | 1.13 |
| Belarus | 16 | 1.13 | 1.05 | 1.24 |
| Poland | 2 | 1.14 | 1.1 | 1.18 |
| Ukraine | 66 | 1.14 | 1.03 | 1.38 |
| France | 4 | 1.22 | 1.1 | 1.29 |
| Croatia | 3 | 1.24 | 1.22 | 1.26 |
| Italy | 4 | 1.24 | 1.05 | 1.33 |
| **North America** | | | | |
| USA | 3 | 1.27 | 1.16 | 1.36 |
| Canada | 3 | 1.44 | 1.13 | 1.63 |
| **Asia** | | | | |
| Kazakhstan | 2 | 1.19 | 1.16 | 1.21 |
| China | 9 | 1.4 | 1.06 | 1.65 |
| Korea | 3 | 1.35 | 1.1 | 1.56 |
| Japan | 3 | 1.51 | 1.31 | 1.7 |
| Nepal | 1 | 1.27 | | |
| India | 5 | 1.37 | 1.18 | 1.7 |

[a]Photoperiodic Sensitivity Index.

44.3% (Fig. 22.1A). In the relatively low-latitude regions, below 50°N, the share of such populations amounted to only 29.7% and 25.7%, respectively (Fig. 22.1B).

Though the major part of local buckwheat populations from Russia, Ukraine, Belarus, and the Baltic states are those with low PS (see Table 22.1), populations with a significantly higher PS may be found among the accessions from the North Caucasus, East Siberian, and especially the Far East regions of Russia, as well as from of Ukraine (Fig. 22.2). The average $I_{PhS}$ for the populations from different regions of Russia varied from 1.08 to 1.38. At the same time, no complete PS decrease was discovered in the populations from the more northern regions: those from the central region

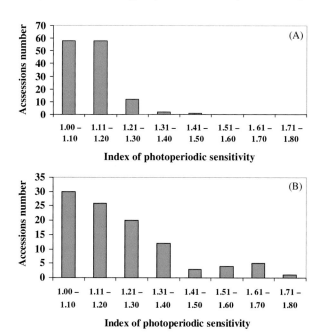

**FIGURE 22.1** (A) Photoperiodic sensitivity of local populations of *Fagopyrum esculentum* from the regions located above 50°N. (B) Photoperiodic sensitivity of local populations of *F. esculentum* from the regions located below 50°N

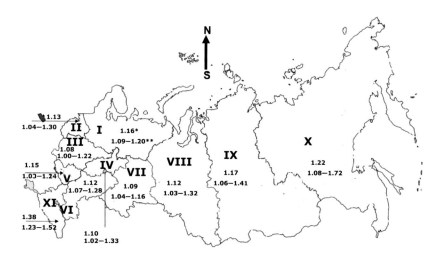

**FIGURE 22.2** Photoperiodic sensitivity of common buckwheat accessions from the Russian Federation.

I. Northern Reg.; II. Northwestern Reg.; III. Central Reg.; IV. Volga-Vyatka Reg.; V. Central Black Soil Reg.; VI. Middle Volga Reg.; VII. Urals Reg.; VIII. West Siberian Reg.; IX. East Siberian Reg.; X. Far East Reg.; XI. North Caucasus Reg.

* average $I_{PhS}$ values, ** limits $I_{PhS}$ values.

were found to have the minimum average $I_{PhS}$ of 1.08, while the values for populations from the northwestern and northern regions were somewhat higher (1.16 and 1.13, respectively).

## ON THE INHERITANCE OF PHOTOPERIODIC SENSITIVITY

When studying the differences between diploid cultivars of the summer and autumn types, T. Nagatomo came to a conclusion about the dominance of the summer type when the control is monogenic (according to: Lachmann and Adachi, 1992). This conclusion has not been disproved by crosses of tetraploid accessions (Lachmann and Adachi, 1992).

We have discovered the overwhelming contribution of one locus to the differences between a late, moderately sensitive line, and an early, with a very low PS line, however with a remark about the complex structure of the locus and the existence of the intralocus recombination. Since the $F_1$ hybrid was fertile under LD, the conclusion about the summer type dominance is quite appropriate (Fesenko et al., 2006).

In contrast, interspecific hybridization has revealed the dominance of high PS. An analysis of the day length response variation in the synthetic amphidiploid *F. giganteum* (VIR-109) and in its parental forms *F. tataricum* (VIR-108) and *F. cymosum* (VIR-4231) has shown the intermediate type of inheritance of the vegetative period duration under SD (Table 22.2), while under LD the plants of *F. giganteum* started flowering simultaneously with those of the most late parent, that is, the dominance of high PS was observed.

Under SD, the plants of *F. tataricum* displayed a compact habitus and were noted for the clearly expressed determinate stem growth and racemes on very long pedicels. The fairly wide diversity in PS even among populations from the most northern regions of Russia allows us to suppose a complex genetic control of the day length response in buckwheat accessions.

**TABLE 22.2** Photoperiodic Sensitivity of *Fagopyrum* Species

| VIR Cat. No. | Species | "Sprouting to the Beginning of Flowering" Period Duration (days) | | $I_{PhS}$* |
| --- | --- | --- | --- | --- |
| | | 12 hours | 18 hours | |
| 108 | *Fagopyrum tataricum* (VIR-17, 4 n) | 29.9±0.35 | 44.2±1.54 | 1.48 |
| 4231 | *Fagopyrum cymosum* | 49.3±1.92 | 83.0±6.42 | 1.68 |
| 109 | *Fagopyrum giganteum* | 40.5±2.50 | 82.0±7.24 | 2.02 |

## PHOTOPERIODIC SENSITIVITY AND MORPHOLOGICAL CHARACTERS OF PLANTS

There is a point of view that early-ripening populations of different plant species are less sensitive to day length (Koshkin, 1997; Vrazhnov et al., 2012). It is not always true for *F. esculentum*. The instability of the correlation between the developmental rate and PS in populations is illustrated by Fig. 22.3A. Under SD, the "sprouting to early flowering" period averaged 21.1 days for the least day-length sensitive populations with $I_{PhS}$ below 1.20, and to 19.9 days for the most sensitive ones. Under LD, the "sprouting to early flowering" period significantly increased (up to 31.5 days) for the group of sensitive accessions. In this case, the number of vegetative nodes on the stem has increased 1.4 times in the least photoperiod sensitive populations, and 2.7 times in the most photoperiod sensitive ones, while the number of generative nodes on the stem has increased 1.1 and 1.6 times, respectively (Fig. 22.3B−D).

The tendency of buckwheat plants to a more vigorous growth under an increased day length is a well-known fact (Hao et al., 1998; Yang et al., 1998; Michiyama et al., 2004). In the mid-20th century, Russian taxonomists separated the accessions from Japan, China, and India into a subspecies *multifolium* Stol. with "vigorous growth" as its main distinctive feature. However, the author of the subspecies (Stoletova E.) has found two subspecies to be indistinguishable when grown under SD (Stoletova, 1958). However, there is no data on the differences in the measurable plant habitus features. In our test, the plants displayed a more vigorous growth of the vegetative organs under LD than under SD. Under LD, the stem length increased 1.2 times in the least photoperiod sensitive populations, and 1.7 times in the most sensitive ones.

It may be supposed that plants of the same accession that start flowering simultaneously under LD and SD are very close to each other morphologically. The plants with same flowering start dates under LD and SD are identified within one accession, and the size of the largest leaf and its node are analyzed. The largest leaf on plants grown under the natural LD was $10 \times 11-13$ cm, but under SD it was notably smaller ($6-8 \times 8-10$ cm). Location of the largest leaf on the stem follows no regularity (usually it is the 4th, more rarely the 5th node). Morphotypes from accessions most sensitive to day length and therefore starting to flower on different dates were also characterized by similar differences in the size of largest leaves under LD ($11-12 \times 11-12$ cm) and SD ($8 \times 9-10$ cm).

## CONCLUSION

The day length sensitivity variation obviously was a primary condition for buckwheat adaptation to regions with long summer days and expansion northwards. Common buckwheat from the European countries, Russia, and Kazakhstan is characterized in most cases by low photoperiodic response

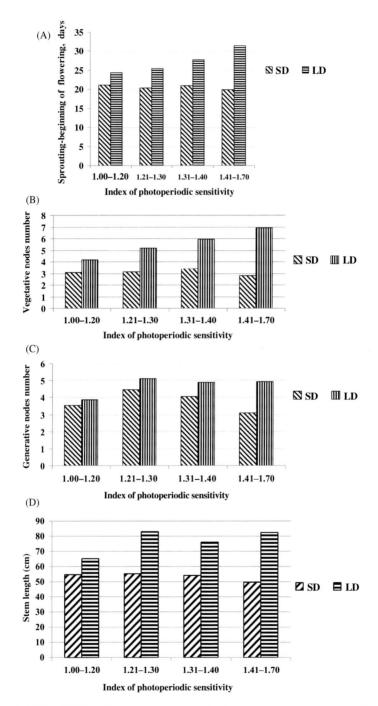

**FIGURE 22.3** (A) The effect of day length on the "sprouting to the beginning of flowering" period duration in common buckwheat accessions. (B) The effect of day length on the stem branching zone development in common buckwheat accessions. (C) The effect of day length on the stem fruiting zone development in common buckwheat accessions. (D) The effect of day length on stem development in common buckwheat accessions.

($I_{PhS}$ values of up to 1.20). However, only one accession with the least day-length sensitivity (VIR-2433 from Central Russia) has been found.

The further adaptation to specific local soil and climatic conditions was based on the polymorphism for the number of vegetative nodes on the stem, which facilitated prompt formation of populations with the optimal duration of the growing season (Fesenko et al., 2010; Fesenko et al., 2012; Fesenko and Romanova, 2010).

## REFERENCES

Baniya, B.K., Riley, K.W., Donghol, D.M.S., Sherchan, K.K., 1992. Characterization and evaluation of Nepalese buckwheat (*Fagopyrum* ssp.) landraces. In: Proceedings 5th International Symposium on Buckwheat. China, pp. 64−74.

Fesenko, A., Romanova, O., 2010. Studies on the stem branching zone development in local populations of common buckwheat in central and northwestern Russia. In: Proceedings of 11th International Symposium on Buckwheat. Orel, pp. 35−40.

Fesenko, A., Romanova, O., Fesenko, N.N., 2012. Peculiarities of common buckwheat adaptation to growing conditions. The European Journal of Plant Science and Biotechnology 6 (Special Issue 2). Global Science Books, Ltd, UK, pp. 75−79.

Fesenko, N.N., Martynenko, G.E., Funatsuki, H., Romanova, O.I., 1998. Express evaluation of Russian and Japanese varieties of buckwheat based on characteristics − indicators of duration of vegetation period. Proc. 7th Int. Symp. on Buckwheat. Canada, 185−192.

Fesenko, N.V., Fesenko, N.N., Romanova, O.I., Alexeeva, E.S., Suvorova, G.N., 2006. The Gene Bank and breeding of groat crops: buckwheat. In: Dragavtsev, VA (Ed.), Theoretical Basis of Plant Breeding, Vol V. VIR, St. Petersburg, Russia, 195 pp. (In Russian).

Fesenko, N.V., Fesenko, A.N., Romanova, O.I., 2010. Morphological structure of populations as the main element of the functional system of environmental adaptation of common buckwheat *Fagopyrum esculentum* Moench. Vestnik OrelGAU 4, 47−52 (In Russian).

Hao, X., Yang, W., Li, G., Zhou, N., Rufa, L., Zhou M. 1998. Relationship between vegetative growth of common buckwheat and light duration. In: Proceedings 7 International Symposium on Buckwheat. Canada, vol. 2, pp. 49−56.

Koshkin, V.A., 1997. Reaction of spring wheat varieties of different photoperiodic sensitivity to the influence of short day. Dokl. RASKhN [Reports of Russian Academy of AgriculturalSciences] 4, 13−15 (In Russian).

Koshkin, V.A., Koshkina, A.A., Matvienko, I.I., Pryadehina, A.K., 1994. Use of the original forms of spring wheat with weak photoperiod sensitivity to create productive early-maturing lines. Dokl. RASKhN [Reports of Russian Academy of AgriculturalSciences] 2, 8−10 (In Russian).

Lachmann, S., Adachi, T., 1992. Inheritance of photoperiod-induced flowering in common buckwheat. In: Proceedings of 5th International Symposium on Buckwheat. China, pp. 105−110.

Matano, T., Ujihara, A., 1981. Differentiation of agroecotypes of *Fagopyrum esculentum* Moench in Japan. In: Proceedings 1st International Symposium on Buckwheat. Ljubljana, 1980, pp. 7−12.

Michiyama, H., Tsuchimoto, K., Tani, K., Hirano, T., Hayasashi, H., Campbell, C., 2004. Influence of day length on the growth of stem, flowering, the morphology of flower clusters, and seed-set in buckwheat (*Fagopyrum esculentum* Moench). In: Proceedings of 9th International Symposium on Buckwheat. Prague, pp. 35−40.

Ohnishi, O., 1991. Discovery of wild ancestor of common buckwheat. Fagopyrum 11, 5−10.

Romanova, O., Koshkin, V., 2010. Photoperiod response of landraces and improved varieties of buckwheat from Russia and from the main buckwheat cultivating countries. The European Journal of Plant Science and Biotechnology 4 (Special Issue 1). Global Science Books, Ltd, UK, pp. 123−127.

Skripchinskiy, V.V., 1971. Physiology of Individual PlantDevelopment. Nauka, Moscow, Russia, 244 pp (In Russian).

Stoletova, E.A., 1958. Buckwheat, Moscow-Leningrad, Russia, 255 pp.

Vrazhnov, V.A., Koshkin, V.A., Rigin, B.V., Potokina, E.K., Tyunin, V.A., Schroeder, R.E., et al., 2012. Ecological test of ultra early-maturing forms of soft wheat in the conditions of different photoperiod. Dokl. RASKhN [Reports of Russian Academy of Argicultural Sciences] 2, 3−8 (In Russian).

Yang, W., Hao, X., Li, G., Zhou, N., Rufa, L., Zhou, M., 1998. Relationship between reproductive growth of common buckwheat and light duration. In: Proceedings of 7 International Symposium on Buckwheat. Canada, vol. 2, pp. 44−48.

## Chapter | Twenty Three

# Main Morphological Types of Cultivated Buckwheat Populations in Russia

**Aleksey N. Fesenko[1], Nikolay N. Fesenko[1], Olga I. Romanova[2], and Ivan N. Fesenko[1]**

[1]*All-Russia Research Institute of Legumes and Groat Crops, Orel, Russia,*
[2]*N. I. Vavilov Institute of Plant Genetic Resources (VIR), St. Petersburg, Russia*

## INTRODUCTION

The breeding of common buckwheat (*Fagopyrum esculentum* Moench.) in Russia started in the 1900s on Shatilov's Experimental Station in Orel region. The first variety, Bogatyr, was registered in 1938. This variety was bred by selection of more weighty grain fraction from the local population.

In the Institute of Grain Legumes and Groats Crops, buckwheat breeding and research has been conducted since the 1960s. During this time, varieties with very different morphologies and physiologies have been bred. In this chapter we review the essential differences between the different types of buckwheat varieties.

## VARIETIES OF TRADITIONAL TYPE

In the Institute of Grain Legumes and Groats Crops, over half a century, the buckwheat breeding for high yields has been based on marker traits which reflect vegetative development of plants, i.e., the number of vegetative nodes on the main stem and branches, which together constitute the number of vegetative nodes on a plant (Table 23.1). The number of vegetative nodes on the main stem is correlated with time interval from sowing to efflorescence (Fesenko et al., 1998); the number of vegetative nodes per whole plant

**235**

*Buckwheat Germplasm in the World.* DOI: https://doi.org/10.1016/B978-0-12-811006-5.00023-9
© 2018 Elsevier Inc. All rights reserved.

**TABLE 23.1** Vegetative Sphere Architecture

| Variety Type | Variety (population) | Number of Vegetative Nodes | | | Share of LSB Plants (%) |
|---|---|---|---|---|---|
| | | On Main Stem | On Branches | On Plant | |
| Local populations of Orel region | k-406 | 4.8 | 9.4 | 14.2 | 7.4 |
| | k-1709 | 4.8 | 8.9 | 13.7 | 6.5 |
| | Mean value | 4.8 | 9.2 | 13.9 | 6.9 |
| Traditional | Bogatyr (1938) | 5.1 | 10.1 | 15.3 | 6.3 |
| | Kalininskaya (1954) | 4.9 | 9.1 | 14.0 | 16.7 |
| | Shatilovskaya 5 (1967) | 4.9 | 9.3 | 14.2 | 8.5 |
| | Aromat (1985) | 4.8 | 9.0 | 13.8 | 10.3 |
| | Mean value | 4.9 | 9.4 | 14.3 | 10.4 |
| Krasnostreletskiy (Tatarstan) | Kazanka (1989) | 4.8 | 8.8 | 13.5 | 18.2 |
| | Karakityanka (1991) | 4.8 | 9.1 | 13.9 | 13.4 |
| | Kama (1993) | 4.6 | 8.6 | 13.1 | 11.1 |
| | Saulyk (1997) | 4.5 | 8.2 | 12.8 | 13.1 |
| | Cheremshanka (2001) | 4.7 | 8.7 | 13.4 | 9.9 |
| | Chatyr-Tau (2005) | 4.9 | 9.2 | 14.1 | 7.8 |
| | Batyr (2008) | 5.0 | 8.9 | 13.9 | 25.0 |
| | Nikolskaya (2013) | 5.0 | 10.0 | 15.0 | 13.0 |
| | Mean value | 4.8 | 8.9 | 13.7 | 14.0 |
| Krasnostreletskiy (Bashkortostan) | Agidel (2001) | 4.8 | 8.9 | 13.7 | 5.7 |
| | Inzerskaya (2002) | 4.6 | 8.0 | 12.6 | 10.8 |
| | Ilishevskaya (2008) | 5.0 | 9.2 | 14.2 | 17.5 |
| | Bashkirskaya krasnostebelnaya (2009) | 4.9 | 9.6 | 14.5 | 6.0 |
| | Zemlyachka (2013) | 4.9 | 9.2 | 14.1 | 14.1 |
| | Mean value | 4.8 | 9.0 | 13.8 | 10.8 |
| LSB | Ballada (1985) | 5.6 | 10.2 | 15.8 | 60.3 |
| | Esen (1993) | 5.5 | 10.4 | 15.9 | 31.3 |
| | Molva (1997) | 5.7 | 12.0 | 17.7 | 27.5 |
| | Mean value | 5.6 | 10.9 | 16.5 | 39.7 |
| Determinate growth habit | Sumchanka (1985) | 5.5 | 13.3 | 18.8 | 0.8 |
| | Demetra (1995) | 5.5 | 12.7 | 18.2 | 4.6 |
| | Dozhdik (1998) | 5.7 | 12.4 | 18.1 | 11.9 |
| | Dikul (1999) | 5.6 | 12.5 | 18.1 | 6.8 |
| | Devyatka (2004) | 5.8 | 13.3 | 19.1 | 17.3 |
| | Dialog (2008) | 5.8 | 12.5 | 18.3 | 26.0 |
| | Temp (2010) | 4.5 | 7.9 | 12.4 | 39.0 |
| | Design (2010) | 5.8 | 13.8 | 19.7 | 9.3 |
| | Druzhina (2014) | 5.6 | 12.3 | 17.9 | 10.8 |
| | Mean value | 5.5 | 12.3 | 17.8 | 14.1 |
| $LSD_{0.05}$ | | 0.42 | 1.12 | 1.57 | |

(branching potential) characterizes the vegetation period length (Table 23.2). In addition, these traits are correlated with the potential productivity of plants (Fesenko, 1983; Fesenko et al., 2006, 2010).

It was shown by morphological analysis of mid-ripening buckwheat varieties registered in Russia that growth of buckwheat grain productivity was mainly associated with the increasing of branching potential and improvement of vegetative sphere architecture (Fesenko et al., 2006).

**TABLE 23.2** Vegetation Period Characteristics

| Variety Type | Variety (population) | Cotyledons–Efflorescence | Period (days) Efflorescence–Harvest Ripening | Cotyledons–Harvest Ripening |
|---|---|---|---|---|
| Local populations of Orel region | k-406 | 22.8 | 52.3 | 75.0 |
| | k-1709 | 23.0 | 52.0 | 75.0 |
| | Mean value | 22.9 | 52.1 | 75.0 |
| Traditional | Bogatyr | 23.3 | 51.3 | 74.5 |
| | Kalininskaya | 23.0 | 51.9 | 74.9 |
| | Shatilovskaya 5 | 24.1 | 50.5 | 74.6 |
| | Aromat | 24.3 | 50.6 | 74.8 |
| | Mean value | 23.7 | 51.0 | 74.7 |
| Krasnostreletskiy (Tatarstan) | Kazanka | 23.3 | 47.1 | 70.3 |
| | Karakityanka | 23.3 | 46.6 | 69.8 |
| | Kama | 22.5 | 46.8 | 69.3 |
| | Saulyk | 23.3 | 46.6 | 69.8 |
| | Cheremshanka | 23.8 | 46.1 | 69.8 |
| | Chatyr-Tau | 24.0 | 46.5 | 70.5 |
| | Batyr | 24.3 | 47.2 | 71.5 |
| | Mean value | 23.5 | 46.7 | 70.1 |
| Krasnostreletskiy (Bashkortostan) | Agidel | 24.0 | 46.0 | 70.0 |
| | Inzerskaya | 23.0 | 46.5 | 69.5 |
| | Ilishevskaya | 24.0 | 47.5 | 71.5 |
| | Bashkirskaya krasnostebelnaya | 23.5 | 46.8 | 70.3 |
| | Mean value | 23.6 | 46.7 | 70.3 |

*Continued*

**TABLE 23.2** continued

| Variety Type | Variety (population) | Cotyledons–Efflorescence | Period (days) Efflorescence–Harvest Ripening | Cotyledons–Harvest Ripening |
|---|---|---|---|---|
| LSB | Ballada | 25.3 | 49.6 | 74.8 |
| | Esen | 25.5 | 49.3 | 74.8 |
| | Molva | 25.6 | 49.4 | 75.0 |
| | Mean value | 25.5 | 49.4 | 74.9 |
| Determinate growth habit | Sumchanka | 24.0 | 51.3 | 75.3 |
| | Demetra | 24.9 | 51.4 | 76.3 |
| | Dozhdik | 23.8 | 51.6 | 75.3 |
| | Dikul | 24.3 | 50.9 | 75.1 |
| | Devyatka | 25.5 | 51.5 | 77.0 |
| | Dialog | 24.5 | 49.5 | 74.0 |
| | Temp | 22.3 | 44.7 | 67.0 |
| | Design | 26.8 | 51.3 | 73.0 |
| | Druzhina | 25.0 | 51.8 | 76.8 |
| | Mean value | 24.5 | 50.4 | 75.0 |

**FIGURE 23.1** Buckwheat with (A) indeterminate (traditional) and (B) determinate growth habits.

First breeding generation varieties of the traditional morphotype (with indeterminate growth, see Fig. 23.1A) very slightly differed from local populations of the central region on the average number of vegetative nodes and branching potential (Table 23.1); for the duration of the growing season these groups of varieties are also very similar (Table 23.2).

Bred varieties of traditional morphotypes formed 17% fewer flowers than the local population, on average. Significant reduction in the average number of flowers per plant was observed only for the variety Bogatyr (Table 23.3). The average duration of blossoming and seed filling period decreased by 1.4 days (Table 23.2).

Varieties of traditional types show the same inflorescence architecture (Table 23.4) and main stem length (Table 23.5) in comparison to local populations.

## VARIETIES OF KRASNOSTRELETSKIY TYPE

In 1971 it was registered a first unusual variety Krasnostreletskaya with indeterminate morphology, markedly larger achenes and physiological limitation of growth. Later, such type of varieties was named "Krasnostreletskiy" − (KS).

Vegetative sphere architectures of the KS varieties from Tatarstan and Bashkortostan—where they have been mainly bred (Kadyrova, 2003)—and

**TABLE 23.3** Generative Sphere Architecture

| Variety Type | Variety (population) | Mean Number of Inflorescences on a Main Stem | Mean Number of Flowers Per Plant | Mean Share (%) of Fertile Flowers | Mean Share (%) of Developed Seeds |
|---|---|---|---|---|---|
| Local populations of Orel region | k-406 | 6.7 | 440 | 10.0 | 59.8 |
| | k-1709 | 6.6 | 442 | 11.1 | 55.8 |
| | Mean value | 6.7 | 441 | 10.5 | 57.8 |
| Traditional | Bogatyr | 5.6 | 325 | 14.7 | 67.2 |
| | Kalininskaya | 6.4 | 390 | 8.6 | 62.7 |
| | Shatilovskaya 5 | 6.1 | 380 | 10.6 | 64.2 |
| | Aromat | 5.7 | 369 | 13.8 | 55.4 |
| | Mean value | 6.0 | 366 | 11.9 | 62.4 |
| Krasnostreletskiy (Tatarstan) | Kazanka | 5.1 | 270 | 10.4 | 60.0 |
| | Karakityanka | 5.1 | 260 | 14.4 | 64.7 |
| | Kama | 4.9 | 223 | 16.4 | 61.4 |
| | Saulyk | 5.4 | 289 | 14.6 | 67.4 |
| | Cheremshanka | 5.3 | 299 | 15.8 | 68.1 |
| | Chatyr-Tau | 5.2 | 278 | 18.5 | 66.2 |
| | Batyr | 5.9 | 292 | 13.7 | 62.1 |
| | Mean value | 5.3 | 273 | 14.8 | 64.3 |
| Krasnostreletskiy (Bashkortostan) | Agidel | 5.3 | 294 | 20.2 | 69.7 |
| | Inzerskaya | 5.5 | 345 | 19.2 | 67.6 |
| | Ilishevskaya | 5.1 | 277 | 18.6 | 64.1 |
| | Bashkirskaya krasnostebelnaya | 5.6 | 291 | 16.7 | 70.5 |
| | Mean value | 5.4 | 302 | 18.7 | 68.0 |

| | | | | |
|---|---|---|---|---|
| **LSB** | | | | |
| Ballada | 6.7 | 443 | 12.3 | 61.8 |
| Esen | 6.3 | 413 | 14.4 | 72.5 |
| Molve | 7.0 | 460 | 12.3 | 59.9 |
| Mean value | 6.3 | 439 | 13.0 | 64.7 |
| **Determinate growth habit** | | | | |
| Sumchanka | 3.8 | 302 | 14.5 | 62.5 |
| Demetra | 3.8 | 409 | 14.4 | 54.9 |
| Dozhcik | 3.9 | 401 | 13.4 | 56.8 |
| Dikul | 3.9 | 396 | 12.8 | 56.1 |
| Devyatka | 3.8 | 396 | 15.4 | 59.1 |
| Dialog | 3.7 | 380 | 16.9 | 63.5 |
| Design | 3.7 | 316 | 16.3 | 58.7 |
| Druzhina | 3.9 | 457 | 11.5 | 50.4 |
| Mean value | 3.8 | 382 | 14.4 | 57.8 |
| $LSD_{0.05}$ | 1.14 | 128.4 | 3.22 | 12.42 |

**TABLE 23.4** Inflorescence Architecture

| Variety Type | Variety (population) | Mean Number of Elementary Clusters Per Inflorescence | Mean Number of Flowers Per Inflorescence | Mean Number of Flowers Per Elementary Cluster |
|---|---|---|---|---|
| Local populations of Orel region | k-406 | 10.1 | 51.1 | 5.0 |
| | k-1709 | 10.4 | 52.8 | 5.1 |
| | Mean value | 10.2 | 52.0 | 5.1 |
| Traditional | Bogatyr | 10.1 | 48.4 | 4.8 |
| | Kalininskaya | 9.7 | 49.4 | 5.0 |
| | Shatilovskaya 5 | 9.6 | 50.3 | 5.2 |
| | Aromat | 11.0 | 54.2 | 4.9 |
| | Mean value | 10.1 | 50.6 | 5.0 |
| Krasnostreletskiy (Tatarstan) | Kazanka | 9.4 | 44.0 | 4.6 |
| | Karakityanka | 9.4 | 42.2 | 4.5 |
| | Kama | 9.1 | 39.4 | 4.3 |
| | Saulyk | 9.4 | 44.0 | 4.7 |
| | Cheremshanka | 9.8 | 47.0 | 4.8 |
| | Chatyr-Tau | 9.9 | 43.9 | 4.5 |
| | Batyr | 9.7 | 41.8 | 4.3 |
| | Mean value | 9.5 | 43.2 | 4.5 |
| Krasnostreletskiy (Bashkortostan) | Agidel | 10.1 | 46.0 | 4.6 |
| | Inzerskaya | 10.5 | 50.8 | 4.9 |
| | Ilishevskaya | 10.0 | 46.2 | 4.6 |
| | Bashkirskaya krasnostebelnaya | 9.3 | 42.3 | 4.5 |
| | Mean value | 10.0 | 46.3 | 4.7 |

| | | | | |
|---|---|---|---|---|
| LSB | Ballada | 10.7 | 53.3 | 4.9 |
| | Esen | 11.0 | 54.7 | 4.9 |
| | Molva | 10.5 | 55.4 | 5.2 |
| | Mean value | 10.8 | 54.5 | 5.0 |
| Determinate growth habit | Sumchanka | 11.3 | 58.3 | 5.1 |
| | Demetra | 14.0 | 76.5 | 5.4 |
| | Dozhdik | 12.7 | 76.9 | 6.0 |
| | Dikul | 12.1 | 71.0 | 5.8 |
| | Devyatka | 13.2 | 74.8 | 5.6 |
| | Dialog | 12.4 | 72.4 | 5.8 |
| | Design | 11.5 | 60.5 | 5.3 |
| | Druzhina | 14.4 | 85.8 | 5.9 |
| | Mean value | 12.7 | 72.0 | 5.6 |
| $LSD_{0.05}$ | | 1.58 | 14.2 | 0.83 |

**TABLE 23.5** Length of Main Stem (Branching Zone + Flowering Zone)

| Variety Type | Variety (Population) | Branching Zone (cm) | Flowering Zone (cm) | Main Stem (cm) |
|---|---|---|---|---|
| Local populations of Orel region | k-406 | 45.9 | 48.2 | 94.1 |
| | k-1709 | 50.5 | 45.5 | 96.0 |
| | Mean value | 48.2 | 46.9 | 95.1 |
| Traditional | Bogatyr | 54.8 | 42.5 | 97.3 |
| | Kalininskaya | 50.9 | 44.6 | 95.5 |
| | Shatilovskaya 5 | 54.6 | 43.8 | 98.4 |
| | Aromat | 53.3 | 43.3 | 96.6 |
| | Mean value | 53.4 | 43.6 | 97.0 |
| Krasnostreletskiy (Tatarstan) | Kazanka | 43.6 | 39.5 | 83.2 |
| | Karakityanka | 45.1 | 40.3 | 85.4 |
| | Kama | 45.2 | 35.4 | 80.5 |
| | Saulyk | 44.4 | 41.0 | 85.4 |
| | Cheremshanka | 45.8 | 38.9 | 84.7 |
| | Chatyr-Tau | 44.5 | 35.8 | 80.3 |
| | Batyr | 48.4 | 40.6 | 89.0 |
| | Nikolskaya | 41.4 | 43.7 | 85.1 |
| | Mean value | 44.8 | 39.4 | 84.2 |
| Krasnostreletskiy (Bashkortostan) | Agidel | 43.4 | 40.0 | 83.4 |
| | Inzerskaya | 46.8 | 38.8 | 85.6 |
| | Ilishevskaya | 42.6 | 38.4 | 81.0 |
| | Bashkirskaya krasnostebelnaya | 44.5 | 37.4 | 81.9 |
| | Zemlyachka | 44.4 | 40.4 | 84.8 |
| | Mean value | 44.3 | 39.0 | 83.3 |
| LSB | Ballada | 52.9 | 47.0 | 99.9 |
| | Esen | 56.0 | 41.4 | 97.4 |
| | Molva | 57.2 | 44.1 | 101.2 |
| | Mean value | 55.4 | 44.2 | 99.5 |
| Determinate growth habit | Sumchanka | 53.3 | 31.0 | 84.2 |
| | Demetra | 56.6 | 30.8 | 87.4 |
| | Dozhdik | 54.6 | 30.8 | 85.4 |
| | Dikul | 55.2 | 26.3 | 81.5 |
| | Devyatka | 64.2 | 25.3 | 89.5 |
| | Dialog | 53.4 | 24.7 | 78.1 |
| | Temp | 38.7 | 32.7 | 71.4 |
| | Design | 60.2 | 28.6 | 88.9 |
| | Druzhina | 58.0 | 27.5 | 85.5 |
| | Mean value | 54.9 | 28.6 | 83.5 |
| $LSD_{0.05}$ | | 7.2 | 8.9 | 9.7 |

the local populations of traditional type, do not differ (Table 23.1), but vegetation period of the KS-type varieties is significantly shorter (Table 23.2). This is due to some peculiarities of their generative sphere development.

KS varieties produce minimal flower numbers per plant (31.5%−38.0% fewer in comparison to local populations) (Table 23.3). The duration of flowering and seed filling period of KS varieties was reduced by 5.4 days, on average (Table 23.2).

Reducing the number of flowers per plant of KS morphotype was due to the decrease in the number of inflorescences on the shoots, together with decreasing number of flowers per inflorescence (Table 23.4).

KS varieties develop a significantly shorter main stem compared to local populations (Table 23.5).

## VARIETIES WITH LIMITED SECONDARY BRANCHING

The indeterminate varieties with limited secondary branching (LSB) have been bred in the Institute of Grain Legumes and Groats Crops. Such varieties are distinguished by increasing (up to 30%−60%) share of plants with reduced branching zone on upper branches from 2−3 to 0−1 vegetative nodes (Fig. 23.2A, B) (Fesenko, 1983; Fesenko and Fesenko 2006; Fesenko et al., 2006). First such variety, Ballada, was registered in 1985.

LSB varieties have 5.5−5.7 vegetative nodes on the main stem (a later-ripening habit indicator). But branching potential of plants of these varieties have been increased to a relatively small degree (up to 15.8−17.7 nodes) due to the reduction of vegetative zones of the upper branches (Table 23.1). Therefore, the vegetation period of LSB varieties is almost no longer in comparison to varieties of the traditional type (Table 23.2).

The number of flowers per plant in LSB-varieties remained almost unchanged in comparison with the local populations (Table 23.3). At the same time, the duration of their blossoming and seed filling period reduced by 2.7 days, on average. This is one of the effects of the limited secondary branching mutation. Reduction in the number of vegetative nodes on the upper branches leads to reducing the time interval between the beginning of flowering on the main stem and on the branches. As a result, the flowering of these plants is more intense with a faster ending.

LSB varieties and local populations are similar in the main stem length (Table 23.5).

## VARIETIES WITH DETERMINATE GROWTH HABIT (DET)

A recessive mutant with a determinate growth habit was first published by N. V. Fesenko in 1968 (Fesenko, 1968). This mutation restricts shoot development by 3−5 inflorescences (Fig. 23.1B) (Fesenko et al., 2009). The first variety of this type was Sumchanka, which was registered in 1985. At the present time, more than a half of the area under buckwheat in Russia has been sown with determinate varieties. Extending their sowing area from 8.2% (in 1999) to 56.7% (in 2011) has led to buckwheat yield increasing in 1.5 times, on average (Fesenko et al., 2012).

**FIGURE 23.2** LSB-plants with (A) full reduction and (B) partial reduction (1 vegetative node) of branching zone on first branch.

The branching potential of determinate varieties has grown notably: the average number of vegetative nodes on the stem increased up to 5.5−5.8, and on the plant up to 18.1−19.7 (Table 23.1). However, the duration of their growing season was not significantly increased, due to almost simultaneous efflorescence of all inflorescences on a shoot (Fesenko et al., 2011).

Flower number per plant in determinate varieties was 13.3% fewer in comparison with the local populations, on average; inflorescence number in these varieties is reduced (Table 23.3), but an inflorescence is larger, with increased number of flowers (Table 23.4). It is accompanied by a 1-day shortening of flowering and seed filling period.

Varieties with determinate growth habits show significantly reduced stem length in comparison to local populations (Table 23.5).

## CONCLUSION

For varieties of all morphotypes a low efficiency of seeds formation is shown: the percentage of fertile flowers was 8.6%−20.2% (Table 23.3). Bred varieties manifest slightly higher fertility of flowers (11.9%−18.7% vs 10.5% in local varieties, on average) and proportion of developed seeds (62.4%−68.0% vs 57.8 in local varieties, on average).

Varieties of each morphotype represent sufficiently homogeneous groups in terms of the number of vegetative nodes both on the main stem and on the plant. So, KS varieties, which have been bred in the different soil and climate conditions of Tatarstan and Bashkortostan, manifest a similar architecture of vegetative sphere—and this similarity supports the idea that the main economical features of the KS type are tightly associated with the plant architecture. An additional support for this conclusion is that we failed to increase the length of the vegetation period (and potential productivity) without the loss of physiological determination of growth (the main feature of the KS type).

All varieties studied have kept the long flowering and seed filling period. It is somewhat shorter than that of the local populations, but with the decline of only 0.5−2.3 days.

## REFERENCES

Fesenko, A.N., Fesenko, N.N., 2006. Influence of the locus *LIMITED SECONDARY BRANCHING* (*LSB*) on development of reproductive sphere and productivity of buckwheat plants. Russ. Agricult. Sci. №3, 4−6.

Fesenko, A.N., Biryukova, O.V., Fesenko, I.N., Shipulin, O.A., Fesenko, M.A., 2011. Peculiarities of flowering dynamics of plants of mutant buckwheat morphotypes. Vestnik OrelGAU №3, 9−13 (in Russian).

Fesenko, A.N., Martynenko, G.E., Selikhov, S.N., 2012. Buckwheat production in Russia: current state and perspectives. Zemledelie № 5, 12−14 (in Russian).

Fesenko, I.N., Fesenko, A.N., Biryukova, O.V., Shipulin, O.A., 2009. Genes regulating inflorescences number in buckwheat with a determinate growth habit (homozygote at the recessive allele *det*). Fagopyrum 26, 21−24.

Fesenko, N.N., Martynenko, G.E., Funatsuki, H., Romanova, O.I., 1998. Express evaluation of Russian and Japanese varieties of buckwheat based on characteristics-indicator of duration of vegetative period. In: Proceedings of the Seventh International Symposium on Buckwheat in Winnipeg, Canada, part 1, pp. 185−192.

Fesenko, N.V., 1968. A genetic factor responsible for the determinant type of plants in buckwheat. Rus. J. Genet. 4, 165−166 (in Russian).

Fesenko, N.V., 1983. Breeding and seed farming of buckwheat. Kolos, Moskow, 191 p., (in Russian).

Fesenko, N.V., Fesenko, N.N., Romanova, O.I., Alekseeva, E.C., Suvorova, G.N., 2006. Theoretical basis of plant breeding, The Gene Bank and Breeding of Groats Crops, vol. V. Buckwheat. VIR, St, Petersburg (in Russian).

Fesenko, N.V., Fesenko, A.N., Shipulin, O.A., Savkin, V.I., Kolomeichenko, V.V., Martynenko, G.E., Mazalov, V.I., 2010. Metameric architecture of plants vegetative sphere as system criterion of adaptive and productive properties of plants and varieties of buckwheat. In: Proceedings of the 11th International Symposium on Buckwheat at Orel, Russia, pp. 425–428.

Kadyrova, F.Z., 2003. Buckwheat breeding in Tatarstan. PhD Thesis. Nemchinovka, 42 p., (in Russian).

## FURTHER READING

Fesenko, A.N., Fesenko, N.V., 2004. Productive characteristics of buckwheat morphobiotypes with various architectonics of branch vegetative zone. Russ. Agric. Sci. 3, 6–8.

# Chapter Twenty Four

# Interspecific Crosses Between *Fagopyrum cymosum* and Other Species Through Embryo Culture Techniques

**Sun Hee Woo[1], Swapan K. Roy[1], Soo J. Kwon[1], Seong-Woo Cho[2], and Hag H. Kim[3]**

[1]*College of Agriculture, Life and Environment Science, Chungbuk National University, Cheongju, South Korea,* [2]*Chonbuk National University, Jeonju, South Korea,* [3]*Woosong College, Daejeon, South Korea*

## INTRODUCTION

Wide hybridization has been and continues to be a useful method for plant breeders to create new plant forms or to introduce genes from related wild species into the crop plant of interest. Recently, vegetative reproduction of many economically important plants has been achieved through in vitro embryo rescue methods. Buckwheat is of nutritional importance as a pseudocereal because of its content of high quality seed protein (Eggum, 1980; Marshall and Pomeranz, 1982; Rufeng et al., 1992). However, for centuries buckwheat has remained a minor crop of unstable yield due to inherent genetic characters which prevent direct application of effective conventional breeding methods (Kreft, 1983). This has been shown to be due to a very high seed abortion rate (Morris, 1951), making the plant an inefficient user of available photosynthates.

Buckwheat breeding programs have looked at the development of self-compatible flower types resulting from mutations in the proposed

**249**

© 2018 Elsevier Inc. All rights reserved.

sporophytic, heteromorphic self-incompatibility system (Sharma and Boyes, 1961) of common buckwheat. These, however, have not been successful in increasing yield and therefore breeders are now attempting introgression of characters, including self-compatibility from other species. The family Polygonaceae, which includes the genus *Fagopyrum*, comprises many wild species that could form a large resource of genes imparting resistance/tolerance to a wide spectrum of biotic and abiotic stresses. The newly-found self-fertile species *Fagopyrum homotropicum* (Ohnishi, 1995) is more closely related to common buckwheat than is the species *Fagopyrum tataricum* (Adachi et al., 1992; Ohnishi, 1983) and therefore present efforts are being made to introgress the self-compatible allele found in *F. homotropicum* into common buckwheat.

Programs aimed at improving the strategies of buckwheat breeding are further fraught with numerous problems. Prominent among them is the self/cross-incompatibility phenomenon of the reproductive biology of this seed-propagated genus (Adachi, 1990; Guan and Adachi, 1992; Kreft, 1983; Marshall, 1969; Woo et al., 1995; Hirose et al., 1994). In buckwheat, these breeding barriers have in no small measure contributed to the recalcitrance of buckwheat to conventional improvement techniques. For instance, the transfer of valuable genetic traits such as self-fertility, frost tolerance, and important agronomic characters such as grain yield have so far proved to be very difficult due to the cross-incompatibility of common buckwheat. These phenomena pose a challenge to research capabilities in increasing buckwheat yields. In order to broaden the genetic basis of common buckwheat, several interspecific hybrids have already been attempted. A previous study reported that crosses between *Fagopyrum esculentum* and *F. tataricum* resulted in immature embryos if *F. tataricum* was used as the female parent (Morris, 1951), but no embryos developed if *F. esculentum* was used as the female parent. Nagatomo (1961) was unsuccessful in developing crosses between *F. esculentum* and *Fagopyrum cymosum* at the tetraploid level. Krotov and Dranenko (1975) produced a new species, *Fagopyrum giganteum*, from the hybrid between tetraploid *F. tataricum* and *F. cymosum*. Samimy et al. (1996) developed an infertile plant from the cross of diploid *F. esculentum* and *F. tataricum*. Wagatsuma and Un-no (1995) reported on the first fertile cross of diploid *F. esculentum* and *F. tataricum*. Several studies have reported on the first fertile cross of *F. homotropicum* and *F. esculentum* and further backcrosses to *F. esculentum* (Campbell, 1995; Wang and Campbell, 1998; Woo et al., 1999). With the aid of such interspecific crosses, genes for tolerance and/or resistance to specific stress may be transferred from wild species to cultivated forms. However, reports on successful interspecific hybrids that are fertile between common buckwheat *F. esculentum* and wild species are rare. Recently, the $F_1$ hybrid of common buckwheat and *F. tataricum* proved to be functionally sterile or had an embryo abortion (i.e., they did not yield viable seeds or viable pollen) (Wang and Campbell, 1998). The improvement of agricultural characteristics of this crop by using conventional breeding methods is in progress.

In vitro techniques together with conventional breeding methods could be used to speed up breeding programs and the overall improvement of buckwheat. In this aspect ovule culture has a very important role, because it could produce hybrid seedlings in interspecific hybridization.

Prior to this study, Ujihara et al. (1990) successfully obtained hybrid tetraploid ($2n = 32$) buckwheat between common buckwheat and *F. cymosum*, through ovule culture. The plants, however, proved to be sterile. Recently, hybrids with *F. cymosum* ($2n = 32$) and other species have been reported. Hybrid plants were characterized by sterility, according to Suvorova et al. (1994). Hirose et al. (1993) produced similar hybrids through embryo rescue. They also were successful in backcrossing to common buckwheat. The objective of the present investigation was to utilize embryo rescue procedures and tissue culture techniques to hybridize the perennial species, *F. cymosum*, with another diploid species in the cymosum group, to develop additional germplasm with potential usefulness for introgression of useful characteristics such as frost resistance and perennial growth habit into common or self-pollinating buckwheat.

## MATERIALS AND METHODS

In this study the diploid species *F. cymosum* was used as the male parent, and other diploid species, *F. esculentum*, *F. tataricum*, *F. homotropicum*, *Fagopyrum pilus*, self-pollinating pin and self-pollinating homostyle buckwheat and an unknown dipoid species (#3) were used as the female parent in the production of interspecific hybrids. The self-pollinating pin and homostyle plants were previously produced from crosses between *F. esculentum* and *F. homotropicum*. Both dimorphic self-incompatible and homomorphic self-compatible species were included as parents. Thrum (t) plants with tall anthers and short pistils, pin (p) plants with tall pistils and short anthers, and homomorphic (h) plants with anthers and pistil at the same level were used in the crosses, as shown in Table 24.1. The plants were grown in pots in a greenhouse. Maternal parental plants were emasculated one day before opening to prevent self-compatibility. *F. cymosum* has shortstyled (thrum) and long-styled (pin) plants. Homomorphic (*F. tataricum*, *F. homotropicum*, and selfpollinating (homostylc)) and heteromorphic (*F. esculentum*, *F. pilus*, and self-pollinating (pin)) were used as female parents, while heteromorphic (*F. cymosum*) were used as a pollen donor. Hand-pollination was conducted by rubbing the anther against the fresh stigma of the female parent under pollinator free conditions between other species and *F. cymosum* in the morning when the flowers were in full blossom. Enlarged ovaries resulting from the interspecific hybridization were removed 5–7 days after cross-pollination. After removal of the ovaries, the ovules are surface sterilized in 70% (v/v) ethanol for a few seconds and 10% (v/v) bleach solution for 2 min and subsequently rinsed three times in sterile distilled water. The ovules were removed

**TABLE 24.1** Results of Interspecific Crosses Between *Fagopyrum cymosum* and Other Species Through Embryo Rescue Techniques

| Cross Combination | Media | Days After Pollination | No. of Ovules | | | Germination B/AX100 | Regeneration B/CX100 | Seedlings 100 ovules |
|---|---|---|---|---|---|---|---|---|
| | | | Cultured (A) | Germinated (B) | Regenerated (C) | | | |
| F.e.(p) × F.c (t) | MS1 | 5 | 134 | 31 | 7 | 22.0 | 22.6 | 5.2 |
| | | 7 | 134 | 28 | 3 | 21.0 | 10.7 | 2.2 |
| | MS2 | 5 | 110 | 21 | 4 | 19.0 | 19.0 | 3.6 |
| | | 7 | 135 | 44 | 5 | 32.6 | 11.7 | 3.7 |
| F.t(h) × F.c.(t) | MS1 | 5 | 20 | 2 | 2 | 10.0 | 100.0 | 10.0 |
| | | 7 | 18 | 2 | 1 | 11.1 | 50.0 | 5.6 |
| | MS2 | 5 | 30 | 6 | 4 | 20.0 | 66.7 | 13.3 |
| | | 7 | 31 | 7 | 2 | 22.6 | 28.6 | 6.5 |
| F.h(h) × F.c.(t) | MS1 | 5 | 27 | 6 | 3 | 22.2 | 50.0 | 11.1 |
| | | 7 | 30 | 5 | 2 | 16.7 | 40.0 | 6.7 |
| | MS2 | 5 | 62 | 9 | 8 | 14.5 | 88.9 | 12.9 |
| | | 7 | 48 | 6 | 5 | 12.5 | 83.3 | 10.4 |
| SP(p) × F.C(t) | MS1 | 5 | 74 | 39 | 21 | 52.7 | 53.8 | 28.4 |
| | | 7 | 70 | 59 | 17 | 84.2 | 28.8 | 24.3 |
| | MS2 | 5 | 85 | 40 | 10 | 47.1 | 25.0 | 11.8 |
| | | 7 | 75 | 51 | 7 | 68.0 | 13.7 | 9.3 |
| SP(h) × F.c.(t) | MS1 | 5 | 93 | 4 | 2 | 4.3 | 80.0 | 2.2 |
| | | 7 | 98 | 21 | 5 | 21.4 | 23.8 | 5.1 |
| | MS2 | 5 | 147 | 8 | 4 | 5.4 | 50.0 | 2.7 |
| | | 7 | 132 | 23 | 5 | 17.4 | 0.0 | 3.8 |
| #3(h) × F.c(t) | MS1 | 5 | 8 | 0 | 0 | 0.0 | 0.0 | 0.0 |
| | | 7 | 11 | 0 | 0 | 0.0 | 0.0 | 0.0 |
| | MS2 | 5 | 12 | 0 | 0 | 0.0 | 0.0 | 0.0 |
| | | 7 | 12 | 0 | 0 | 0.0 | 0.0 | 0.0 |
| F.P.(t) × F.c(p) | MS2 | 5 | 52 | 49 | 42 | 94.2 | 85.7 | 80.8 |
| | | 7 | 50 | 45 | 40 | 90.0 | 88.9 | 80 |

under a stereo microscope and plated on a medium containing Murashige and Skoog basal media (Murashige and Skoog, 1962), MS1 = MS basal + Indole acetic acid (IAA) (0.2 mg/L) + 6-Benzylaminopurine (BA) (2 mg/L) or MS2 = MS basal + Zeatin (2 mg/L) + 3% sucrose + 0.8% agar. The pH was adjusted to 5.7 before autoclaving.

The cultures were kept at 25°C under fluorescent light (1500 Lux) with a 16-h photoperiod. The embryos that enlarged and formed calli or plantlets from the ovules were subsequently transferred to hormone-free MS media containing 3% sucrose medium for further development. After 1−2 months, the hybrid plants were transferred to pots filled with sterilized soil and grown in a growth chamber under 25°C temperature and 90% humidity, with a continued light.

## RESULTS

Hybrids between *F. cymosum* and five other species were successfully developed into mature plants. As presented in Table 24.1, *F. cymosum*, when used as a male parent, produced interspecific hybrids with 5 species, although the germination percentage varied between different cross combinations.

On the other hand, distinctive species differences were noted in regeneration. The present hybrids were germinated and grown successfully in vitro. As given in Table 24.1, many of the hybrids showed nearly normal growth. However, some hybrid plants did not produce a green callus, which did not produce healthy roots, and died. Some calluses formed somatic embryos, which developed roots when subcultured on hormone-free MS media.

However, as mentioned above, in the process of producing somatic embryos or roots, some of the plants/calluses died. The influence of hormone concentration during ovule culture showed that the percentage of germinated hybrid embryos were the same for both media. However, MS1 + IAA (0.2 mg/L) + BA (2.0 mg/L) produced a higher percentage of seedlings when compared to $MS_2$ + Zeatin (2.0 mg/L). The number of germinated embryos was also found to rise, with an increase from 5−7 days after pollination. Other data suggested that degeneration of the hybrid embryos had already begun at 7 days after pollination (data not shown) and therefore hybrid embryos should be rescued at that stage or earlier.

Ultimately, seven plants of *F. esculentum* × *F. cymosum*, four plants of *F. tataricum* × *F. cymosum*, six plants of *F. homotropicum* × *F. cymosum*, two plants of self-pollinating (pin) × *F. cymosum*, and two plants of self-pollinating (homostyle) × *F. cymosum* reached the flowering stage. The fertility of the plants was identified and backcrosses and/or selfings were made wherever possible.

The parent differed significantly in plant height, internode length, leaf number, and flower color, as well as flower morphology. All hybrid plants had morphologically intermediate traits between their parents and were vigorous in their growth habits (Fig. 24.1A−D). The $F_1$ hybrids were generally

**FIGURE 24.1** Hybrids of interspecific crosses between *Fagopyrum cymosum* and other species. (A) *F. esculentum* X *F. cymosum* interspecific $F_1$ hybrid. (B) *F. tutaricum* X *F. cymosum* interspecific $F_1$ hybrid. (C) *F. homotoropicum* X *F. cymosum* interspecific $F_1$ hybrid. (D) Self-pollinating (homomorphic) X *F. cymosum* interspecific $F_1$ hybrid.

intermediate between the parents in height, intermode length, and flower diameter, but resembled the male parent in leaf shape and flower color, and had thinner stems and fewer leaves than either parent.

The hybrid plants between *F. esculentum* (p) and *F. cymosum* (t) were vigorous in their growth and had a greater branch number than did their parents. They exhibited heteromorphic flowers with only pin plants being recovered. The flowers were normal, but were self-sterile. While fertilized ovules could be obtained, only two seeds were obtained with over 500 pollinations with self-pollinating homomorphic buckwheat. *F. tataricum* and *F. cymosum* (t) hybrids had flowers that were distinct enough to be differentiated from their parents. The hybrids exhibited heteromorphic flower types with the pin to thrum plants obtained being 4 and 2 respectively. They produced seed that was close to Tartary buckwheat in size. However, no $F_2$ hybrid plants have been obtained from over 200 embryo rescues. Normally embryo deterioration was observed 5−7 days after pollination.

When the hybrid plants were crossed using self-pollinating homomorphic pollen, embryos developed to the 5- and 7-day stage and over 200 are presently on culture media. The hybrids between *F. homotropicum* and *F. cymosum* (t) all showed homomorphic flowers. The flower development, however, appeared smaller than normal and no embryos have been recovered. The hybrid plants have a somewhat decumbent growth habit as compared to the erect habit of the parents. Self-pollinating homomorphic plants when crossed with *F. cymosum* (t) produced fairly uniform growth habits. All plants exhibited homomorphic flowers that appeared normal. The seed set was normal with seeds being close in size, shape, and color to the self-pollinating parent. Approximately 20% of the plants produced were albino, which quickly died. Self-pollinating pin plants crossed with *F. cymosum* (t) produced plants that were intermediate between the parents in growth habit. Growth, however, was slow and the plants produced flowers that were poorly developed. No embryos have been recovered. *F. pilus* (p) when crossed with *F. cymosum* (t) produced normal hybrid plants that were intermediate between the two parents in growth habit.

Over 80% of the attempted crosses were successful and could be obtained without embryo rescue. The hybrids could be distinguished morphologically by their pericarp. It is apparent that an in vitro culture method can serve as a powerful tool for producing hybrid plants to be used in plant improvement programs.

Based on the results of this study and previous ones, an effective in vitro culture method can make it possible to transfer useful genes between *F. tataricum, F. homotropicum, F. cymosum, F. pilus*, and self-pollinating (homostyle) buckwheat. The successful hybrids crosses are shown in Fig. 24.2.

## DISCUSSION

Buckwheat interspecific hybrids are often very difficult to obtain by conventional breeding methods, since strong pre-and postfertilization barriers exist in the genus. Postzygotic barriers appear to be a primary cause of the reproductive isolation, postfertilization failure commonly associated with endosperm abortion and consequent abnormal differentiation and starvation of the hybrid embryo (Adachi, 1990; Woo et al., 1995; Hirose et al., 1994). The method of in vitro culture of immature hybrid embryos prior to their abortion has been used to circumvent these barriers. Successful buckwheat ovule culture was reported by Ujihara et al. (1990). Subsequently embryo rescue methods have enabled several hybrids to be produced in different crosscombination.

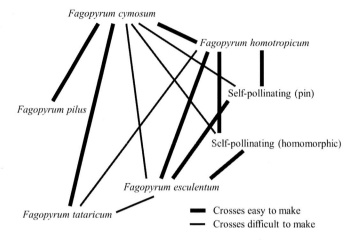

**FIGURE 24.2** Illustration of alien gene transfer path in the cymosum group of buckwheat.

*Fagopyrum cymosum* and other species are reproductively isolated from one another by a strong postzygotic mechanism, and these mechanisms seem to operate on two levels. The breakdown of interspecific embryos after 7 days could be attributed to endosperm abortion and consequent starvation of hybrid embryos. In nature this barrier is very efficient because no seeds are produced from ovules left in situ on the female parent. The excision of embryos 5–7 days old could be attributed to circumvention of this barrier, with the hybrid plants being developed through embryo rescue. The production of albino seedlings in hybrids, regardless of the cross, is another powerful isolation mechanism which operates on a second level, and can effectively eliminate plants that are produced. The cross *F. cymosum* by *F. pilus* produced seeds very readily, even without embryo rescue, while all other crosses required embryo rescue and were difficult to produce. Therefore, it is possible that these two species are either very closely related or are morphological variants of the same species.

The results of the present study seem to indicate that *F. cymosum* has a self/cross-incompatibility with the other species of the cymosum group. However, this barrier can be overcome through the use of embryo rescue. Bud pollination was regarded as the effective methods to transverse interspecific cross incompatibility caused by prezygotic barriers. To overcome postzygotic barriers, ovule culture has facilitated the production of interspecific hybrids. Several studies have been investigated to obtain interspecific hybrids using this technique (Adachi, 1997; Wang and Campbell, 1998). These results suggest that ovule culture is an easy and effective tool for obtaining wide interspecific hybrids in the genus *Fagopyrum*. We also found that the compatibility in interspecific crosses was influenced by the species of the maternal parent. Thus, attempts can be made to overcome incompatibility through using different species as the male or female parent. This work is still in progress.

Today, conventional breeding methods, such as embryo, ovule, and ovary culture for useful gene introgression from wild species to cultivated ones are being augmented by molecular genetic manipulation. However, because production of interspecific hybrids between *F. cymosum* and other species could be obtained quite readily through embryo rescue techniques, this simple and low-cost method can be used as an efficient way to overcome interspecific breeding barriers in the genus *Fagopyrum*. This broadening of the genetic resources that can be used in the breeding of buckwheat speaks well for the ongoing improvement of the domesticated species.

## REFERENCES

Adachi, T., 1990. How to combine the reproductive system with biotechnology in order to overcome the breeding barrier in buckwheat. Fagopyrum 10, 7−11.

Adachi, T., 1997. Production of interspecific hybrids between *Fagopyrum esculentum* and *F. homotropicum* through embryo rescue. Sabrao J. 29, 89−96.

Adachi, T., Mizugami, K., Ogura, K., Kishima, Y., 1992. Chloroplast genomes of genus Fagopyrum II. Phylo genetic relation among the main three species. Search Results Jpn. J. Breed. 42, 272−273.

Campbell, C., 1995. Inter-specific hybridization in the genus Fagopyrum, Proceedings of the 6th International Symposium on Buckwheat. Citeseer, pp. 255−263.

Eggum, B.O., 1980. Protein quality of buckwheat in comparison with other protein sources of plant or animal origin. Buckwheat: genetics, plant breeding, utilization, edited by Ivan Kreft, Branka Javornik, Blanka Dolinsek.

Guan, L., Adachi, T., 1992. Reproductive deterioration in buckwheat (*Fagopyrum esculentum*) under summer conditions. Plant Breed. 109, 304−312.

Hirose, T., Kitabayashi, H., Ujihara, A., Minami, M., 1993. Morphology and identification by isozyme analysis of interspecific hybrids in buckwheats. Fagopyrum (Slovenia).

Hirose, T., Ujihara., A., Kitabayashi, H., Minami, M., 1994. Interspecific cross-compatibility in Fagopyrum according to pollen tube growth. Jpn. J. Breed. 44, 307−314.

Kreft, I., 1983. Buckwheat breeding perspectives. Buckwheat Research, Proceedings of the 2nd International Symposium on Buckwheat, pp. 39−50.

Krotov, A., Dranenko, E., 1975. An amphidiploid buckwheat, *Fagopyrum giganteum*, Krot sp. Nova. Vavilva 30: 41−45. From Plant Breed. Abst44, 1722.

Marshall, H., Pomeranz, Y., 1982. Buckwheat: description, breeding, production, and utilization. Advances in Cereal Science and Technology (USA).

Marshall, H.G., 1969. Description and culture of buckwheat. Bull. 754 Pennsylvania Agric. Exp. Stn.

Morris, M.R., 1951. Cytogenetic studies on buckwheat. Genetic and cytological studies of compatibility in relation to heterostyly in common buckwheat, *Fagopyrum sagittatum*. J. Hered. 42, 85−89.

Murashige, T., Skoog, F., 1962. A revised medium for rapid growth and bio assays with tobacco tissue cultures. Physiol. Plant. 15, 473−497.

Nagatomo, T., 1961. Studies on physiology of reproduction and some cases of inheritance in buckwheat. Report of Breeding Science Laboratory, Faculty of Agriculture, Miyazaki University 1, 1−212.

Ohnishi, O., 1983. Isozyme variation in common buckwheat, *Fagopyrum esculentum* Moench, and its related species. Proceedings of the 2nd International Symposium. Buckwheat at Miyazaki, pp. 39−50.

Ohnishi, O., 1995. Discovery of new Fagopyrum species and its implication for the studies of evolution of Fagopyrum and of the origin of cultivated buckwheat. Proceedings of the 6th International Symposium. Buckwheat at Ina, pp. 175−190.

Rufeng, C.C.N., Bin, Z.Z.C., Rang, C., 1992. Preliminary study on the addition of buckwheat flour and reserve of nutrient elements in buckwheat health food.

Samimy, C., Bjorkman, T., Siritunga, D., Blanchard, L., 1996. Overcoming the barrier to interspecific hybridization of *Fagopyrum esculentum* with *Fagopyrum tataricum*. Euphytica 91, 323−330.

Sharma, K., Boyes, J., 1961. Modified incompatibility of buckwheat following irradiation. Can. J. Bot. 39, 1241−1246.

Suvorova, G., Fesenko, N., Kostrubin, M., 1994. Obtaining of interspecific buckwheat hybrid (*Fagopyrum esculentum* Moench × *Fagopyrum cymosum* Meissner). Fagopyrum 14, 13−16.

Ujihara, A., Nakamura, Y., Minami, M., 1990. Interspecific hybridization in genus Fagopyrum-properties of hybrids (*F. esculentum* Moench × *F. cymosum* Meissner) through ovule culture. Gamma Field Symp. 45−53.

Wagatsuma, T., Un-no, Y., 1995. In vitro culture of interspecific ovule between buckwheat (*F. esculentum*) and tartary (*F. tataricum*). Breed. Sci. 45, 312.

Wang, Y., Campbell, C., 1998. Interspecific hybridization in buckwheat among *Fagopyrum esculentum*, *F. homotropicum* and *F. tataricum*, Proceedings of the 7th International Symposium Buckwheat at Winnipeg, Canada. I, pp. 1−12.

Woo, S., Tsai, Q., Adachi, T., 1995. Possibility of interspecific hybridization by embryo rescue in the genus Fagopyrum. Curr. Adv. Buckwheat Res. 6, 225−237.

Woo, S.H., Adachi, T., Jong, S.K., Campbell, C.G., 1999. Inheritance of self-compatibility and flower morphology in an inter-specific buckwheat hybrid. Can. J. Plant Sci. 79, 483−490.

# Chapter Twenty Five

# Cell Cultures of *Fagopyrum tataricum* as a Source of Biologically Active Phenolic Compounds

**Anton N. Akulov[1,2], Elena A. Gumerova[1], and Natalya I. Rumyantseva[1,2]**

[1]*Kazan Institute of Biochemistry and Biophysics of Kazan Science Centre of the Russian Academy of Sciences, Kazan, Russia,* [2]*Kazan Federal University, Kazan, Russia*

## INTRODUCTION

In higher plants, phenolic compounds (PCs) are one of the most widespread and numerous classes of the secondary metabolites: they are synthesized by all plants and are present in all plant organs (Zaprometov, 1993). PCs act as defense (against herbivores, microbes, viruses, or competing plants) and signal compounds (to attract pollinating or seed dispersing animals), as well as compounds protecting the plant from ultraviolet radiation and oxidants (Lattanzio et al., 2006). PCs can also carry out a structural role (as components of plant cell walls) (Zaprometov, 1993) and a regulatory role (as inhibitors of auxin transport) (Kefeli, 1987, Brown et al., 2001, Peer and Murphy, 2014).

Practical interest in PCs is primarily due to their high antioxidant properties, which largely determine significant antimicrobial, antifungal, antiviral, antiinflammatory, and anticancer properties of these compounds (Dai and Mumper, 2010, Kumar and Pandey, 2013).

Two members of the genus *Fagopyrum*, common buckwheat *(Fagopyrum esculentum* Moench) and Tartary buckwheat *(Fagopyrum tataricum* (L.)

**259**

Buckwheat Germplasm in the World. DOI: https://doi.org/10.1016/B978-0-12-811006-5.00025-2
© 2018 Elsevier Inc. All rights reserved.

Gaertn.) are among the best sources of PCs. Rutin is the major flavonoid in both common and Tartary buckwheat; its content may reach 70%−90% of total flavonoids (Jiang et al., 2007). Buckwheat is the only plant among cereals and pseudocereals whose seeds contain rutin (Fabjan et al., 2003). It was shown that Tartary buckwheat produces much more PCs compared with common buckwheat; rutin content is also higher in the Tartary buckwheat than in common buckwheat which is typical of all tissues and at different growth stages (Gupta et al., 2011). Tartary buckwheat phenolics are powerful antioxidants that can be used to treat cardiovascular diseases, inflammation, hypertension, neurological disorders, diabetes, and obesity (Morishita et al., 2007, Chua, 2013; Kreft et al., 2016).

Rutin production from buckwheat plants can be promising from an economic and environmental point of view (Klykov et al., 2016). However, to a certain extent it is a poorly controlled process because the growth conditions can significantly affect both the plant growth and the secondary metabolite synthesis (Ramakrishna and Ravishankar, 2011). This problem can be solved by the method of plant cell and tissue cultures grown asceptically throughout the year under controlled conditions. The application of callus and suspension cultures is widely used for studying biosynthesis and the production of biologically active PCs in plants (Dias et al., 2016). The studies of total content and composition of PCs in cultured cells and tissues of common buckwheat were represented in several reports (Moumou et al., 1992a,b). It was found that the predominant PCs in buckwheat callus were proanthocyanidins and catechins; the maximum content of rutin ranged from 0.05 to 0.15 mg/g, which was nearly 20 times less than in leaf tissues. To date, the studies concerning the PCs in the callus and suspension cultures of Tartary buckwheat are not available. We have previously shown that long-term (over 10 years) morphogenic callus and suspension cultures of Tartary buckwheat maintain the ability to synthesize phenolic acids and flavonoids (Sibgatullina et al., 2012; Gumerova et al., 2015). Since the *in vitro* phenolic metabolism rate can be affected not only by physical and chemical factors but also by the degree of the cell differentiation in culture, the aim of this work was to study the PC composition in Tartary buckwheat calli with different types of cell differentiation and distinct morphogenic activity.

## MATERIALS AND METHODS
### Cell Culture

The lines of MC were established from immature embryos of *F. tataricum* (sample k-17 from N.I. Vavilov Institute of Plant Genetic Resources). NCs were obtained by the selection of single friable clones formed on the morphogenic calli, which were further subcultured on the medium RX. For the experiments, two lines of the MC (1-8 and K5) and 4 lines of the NC (1-8p, 6p1, 1-5p, and 10p2A) were used: NC 1-8p was derived from MC 1-8, NC 6p1—from MC K5, NC 1-10p2A—from MC 1-10 and NC 1-5p—from

**TABLE 25.1** General Characteristics of the Analyzed Callus Lines

| Line | Type of the Callus | Origin | Year of Generation |
|------|--------------------|--------|--------------------|
| 1-8 | Morphogenic | Immature embryo | 2006 |
| 1-8p | Nonmorphogenic | Line 1-8 | 2007 |
| K5 | Morphogenic | Immature embryo | 2010 |
| 6p1 | Nonmorphogenic | Line K5 | 2013 |
| 1-10p2A | Nonmorphogenic | Line 1-10 | 2007 |
| 1-5p | Nonmorphogenic | Line 1-5 | 2007 |

MC 1-5. General characteristics of the callus lines, including their origin, type, and year of generation, are shown in the Table 25.1. All callus lines were maintained on the medium RX (Rumyantseva et al., 1998) (mg/L): B5 macro- and microsalts (Gamborg et al., 1968), thiamin-HCl-2.0, pyridoxine-HCl-1.0, nicotinic acid—1.0, myo-inositol—100.0, casein hydrolyzate—2,000.0, 2.4 D—2.0, NAA—0.5, IAA—0.5, kinetin—0.2, sucrose—25,000.0, agar—0.8%, pH 5.5−5.6. Morphogenic lines were subcultured every 20 days, and the nonmorphogenic ones every 14 days.

## Preparation of Methanolic Extracts

The callus biomass was lyophilized and dry weight (d.w.) was calculated. Dry biomass was ground into powder, 25 mg samples were incubated with 1.5 mL of 80% methanol in a water bath at 80°C for 30 min. The obtained extracts were centrifuged at 12,000 g for 10 min. The supernatants were used to determine the total content of PCs. Total PC determination was performed by the spectrophotometric method (Folin and Ciocalteu, 1927; Gumerova et al., 2015).

## HPLC

For HPLC analysis of PCs, freeze-dehydrated biomass of callus was ground into powder, 25 mg of sample were incubated with 500 μL of 80% methanol in a water bath at 80°C for 30 min. The obtained extract was centrifuged at 12,000 g for 10 min. Chromatographic separation of PCs was performed using a high performance liquid chromatography Breeze ("Waters", USA). Original Symmetry C18 column, pore size 100 Å, particle size 5 μm, 3.9 mm × 150 mm ("Waters", USA) was used. Peak detection was performed by dual-wavelength UV HPLC Waters 2489 detector ("Waters", USA) at a wavelength of 280 nm. For PC separation solvents A and B were used. Solvent A contained 2.5% acetic acid, acetonitrile (40:1); solvent B contained 2.5% acetic acid, methanol, acetonitrile (40:40:1). Gradient solution B was formed as follows: at 0−2 min solution A/B ratio was 95:5; at 2−30 min solution A/B ratio decreased from 95:5 to 5:95); at 30−40 min solution A/B ratio was 5:95; at 40−43 min solution A/B ratio increased from 5:95 to 95:5); and at 43−45 min solution A/B ratio was 95:5. The flow rate was 0.25 mL/min. 20 µL of sample containing 1-5 µg of PCs was added to the column. Identification of the peaks was performed using a set of standard PCs (gallic, chlorogenic, gentisic, caffeic, *p*-coumaric, ferulic, anisic, salicylic, cinnamic acids, catechin, epicatechin, rutin, quercitrin, quercetin, naringenin). The calibration curves of PC content versus peak area from the chromatogram were built individually for each of the identified compounds. Statistical data processing was performed using the Analysis ToolPak in Microsoft Office Excel 2007. Each experiment was carried out in three biological replicates. Data are presented as mean ± standard error of the mean at a 95% confidence interval.

## RESULTS

Plant embryogenic cultures characterized by a high-frequency somatic embryo production are a widely used material in both basic research and biotechnics such as plant micropropagation and transformation (Jiménez, 2001). Morphogenic cultures of Tartary buckwheat have a nodular morphology which is typical of plant embryogenic calli (Williams and Maheswaran, 1986). Morphologically, these cultures are formed by two compartments: white or yellowish dense structures (nodules)—proembryonal cell complexes (PECCs) or, according to another definition, proembryogenic masses (PEMs), and cells of "soft" callus (Rumyantseva et al., 1989, 1992, 2003) (Fig. 25.1A). New PECCs arise from single cells of loosening PECCs, while the main mass of PECCs being disintegrated gives rise to a "soft" callus whose cells are not divided. The "soft" callus seems to be the nurse tissue

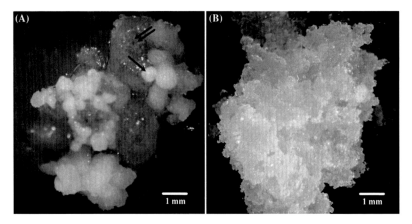

**FIGURE 25.1** Morphology of morphogenic (A) and nonmorphogenic (B) calli of Tartary buckwheat. PECCs are signed by single arrows; "soft" callus is signed by double arrows.

that supports growth of PECCs. The friable clones of NCs arise in MCs during the long-term cultivation with a very low frequency of about one clone within 30−40 passages and can be considered as a result of spontaneous point mutations and endopolyploidy (Kamalova et al., 2009). Such clones can be taken away from MC and grown separately as NCs allowing to compare the cytological, biochemical, and physiological traits of callus cultures arising from one explant, grown on the same medium, but having distinct morphogenic abilities. NCs of Tartary buckwheat have a morphology typical of nonmorphogenic cultures of most plant species and consist of large, high vacuolated cells having no ability for any type of morphogenesis (Fig. 25.1B).

Morphogenic and nonmorphogenic callus cultures of Tartary buckwheat significantly differed in the biomass growth; the biomass growth of NCs exceeded the growth of MCs by about 4 times (Kamalova et al., 2009). The study of total phenolic content showed that MCs accumulated more PCs than NCs (the difference was 2.9−7.5 times) (Fig. 25.2). It should be noted that the total phenolic content in two morphogenic calli slightly varied from 7.11 to 9.11 mg/g d.w. In the nonmorphogenic calli total phenolic content varied to a larger extent: if in lines 1-8p and 6p1 it was 4.3−6.04 mg/g d.w., then in lines 1-5p and 10p2A it was much smaller at 1.13 and 1.85 mg/g d.w., respectively.

HPLC analysis of the methanol extracts of the studied cultures revealed a number of PCs typical of Tartary buckwheat, including phenolic acids (gallic, chlorogenic acid, gentisic, caffeic, p-coumaric, ferulic, benzoic, cinnamic) and flavonoids (rutin, quercitrin, quercetin, catechin, epicatechin) (Table 25.2). Fig. 25.3 shows the HPLC patterns of two couples of MC and NC lines: 1-8 and 1-8p; K5 and 6p1 (chromatograms of two another NCs 1-5p and 10p2A are not presented here). Gentisic, p-coumaric, ferulic, and benzoic acid were minor PCs in both types of cultures, in NCs the content of

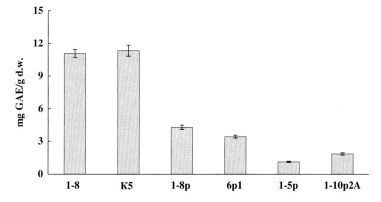

**FIGURE 25.2** Total content of phenolic compounds in various lines of morphogenic and nonmorphogenic calli of Tartary buckwheat.

**TABLE 25.2** Content of Phenolic Compounds Identified in Calli of Tartary Buckwheat, mg/g d.w.

| Penolic Compound Peak Number<br>Callus Line | Gallic Acid<br>1 | Cathechin<br>2 | Chlorogenic Acid<br>3 | Gentisinic Acid<br>4 | Coffeic Acid<br>5 | Epicathechin<br>6 | p-Coumaric Acid<br>7 | Ferulic Acid<br>8 | Benzoic Acid<br>9 | Rutin<br>10 | Quercitrin<br>11 | Cinnamic Acid<br>12 | Quercetin<br>13 |
|---|---|---|---|---|---|---|---|---|---|---|---|---|---|
| 1-8 | 0.26 | 4.35 | 0.19 | N/D | 0.14 | 0.1 | N/D | N/D | 0.11 | 8.71 | 0.34 | 0.03 | 0.1 |
| K5 | 0.17 | 1.73 | 0.02 | 0.01 | 0.15 | 1.14 | <0.01 | <0.01 | 0.01 | 5.77 | 0.91 | 0.24 | 0.98 |
| 1-8p | 0.1 | 0.45 | 0.06 | 0.04 | 0.03 | N/D | N/D | N/D | N/D | N/D | N/D | 0.01 | N/D |
| 6p1 | 0.01 | 0.43 | 0.01 | 0.04 | 0.03 | N/D | N/D | N/D | N/D | N/D | N/D | 0.01 | N/D |
| 1-5p | 0.01 | 0.47 | <0.01 | N/D | 0.01 | N/D | <0.01 | <0.01 | N/D | 0.01 | N/D | N/D | N/D |
| 1-10p2A | 0.01 | 0.47 | 0.01 | N/D | N/D | 0.016 | 0.01 | 0.01 | N/D | 0.01 | N/D | N/D | N/D |

N/D, not detected.

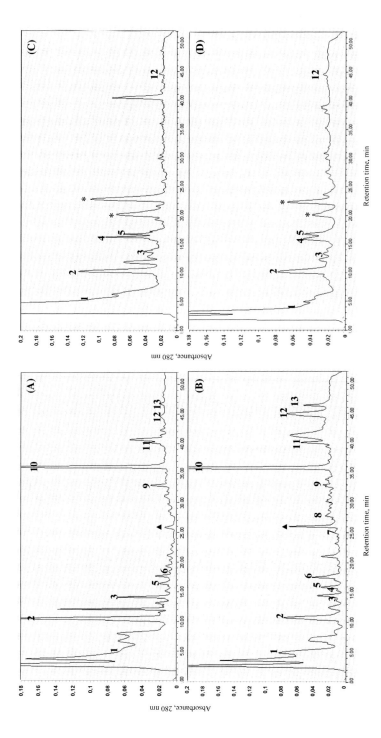

**FIGURE 25.3** HPLC patterns of methanol extracts of Tartary buckwheat calli. (A) morphogenic callus 1-8, (B) morphogenic callus K5, (C) nonmorphogenic callus 1-8β, (D) nonmorphogenic callus 6p1. 1, gallic acid; 2, catechin; 3, chlorogenic acid; 4, gentisinic acid; 5, coffeic acid; 6, epicathechin; 7, p-coumaric acid; 8, ferulic acid; 9, benzoic acid; 10, rutin; 11, quercitrin; 12, cinnamic acid; 13, quercetin; ▲, *, nonidentified peaks.

other identified compounds was significantly lower than in the morphogenic ones (Table 25.2). Catechin is the only one of the identified PCs whose content predominates in all studied NCs. Nevertheless, even in this case catechin content in NCs was several times lower than in the MCs. Fig. 25.3 and Table 25.2 showed that in all studied NC lines catechin was the most predominant PC, while in the MCs the main PC was rutin, which was 0.58%–0.87% of callus dry weight. In NCs, rutin was in trace amounts. The predominant PCs in both MC lines of Tartary buckwheat were rutin and catechin. Lines 1-8 and K5 differed in PC qualitative composition: the 1-8 line contained more catechin, chlorogenic and benzoic acids and rutin, while the K5 line contained more cinnamic acid, epicatechin, quercetin, and quercitrin. In all NCs, quercitrin, and quercetin as well as benzoic acid were not detected. The content of other phenolics varied in different lines of NCs, but generally did not exceed 0.01 mg/g d.w. Comparing the HPLC profiles of MC and NC lines we revealed two specific peaks of nonidentified PCs (marked by asterisks in Fig. 25.3C and D) for nonmorphogenic lines and one specific peak for morphogenic lines (signed by triangle in Fig. 25.3A and B). HPLC peaks specific for nonmorphogenic lines 1-8p and 6p1 were also detected on chromatograms of two another nonmorphogenic lines, 1-5p and 10p2A (data are not presented).

## DISCUSSION

PCs are usually considered as undesirable components in the plant tissue culture because the tissue aging and the explant death are often associated with phenolic oxidation to cytotoxic quinones able to form cross-links with various polymeric compounds—proteins, polysaccharides, and lipids (Pourcel et al., 2007). However, a significant body of literature now exists detailing that PCs are not only components of the antioxidant defense system of cultured cells but also participate in the regulation of morphogenesis (Sarkar and Naik, 2000; Lorenzo et al., 2001;. Debeaujon, 2003).

For the first time, we analyzed PC composition in callus cultures of Tartary buckwheat—a promising species for breeding and pharmacology. It is known that the majority of *in vitro* callus and suspension plant cultures are characterized by genetic, cytological, and morphological instability (D'Amato, 1991; Krishna et al., 2016), however the Tartary buckwheat MCs we have obtained maintain their morphology and morphogenic activity over years of cultivation (Rumyantseva et al., 1998). It is shown that usually MC cells have a diploid chromosome number (Rumyantseva et al., 1998; Kamalova et al., 2009). The highest biomass growth was typical of NCs (Kamalova et al., 2009), but the content of PCs was several times less compared to morphogenic cultures. Rutin, the main flavonoid of MCs, was not determined or was in trace amounts in NCs. The difference in phenolics content in morphogenic and nonmorphogenic cultures can be caused by a significant disturbance of the shikimate pathway and phenylpropanoid metabolism in NCs. Biosynthesis of phenolics is known to take place in three main cell compartments: the cytoplasm, the endoplasmic reticulum (ER), and plastids (Zaprometov, 1993; Vogt, 2010). Some stages of the flavonoid biosynthesis can occur in the nucleus (Saslowsky et al., 2005). In our study, we revealed (data not shown) that NCs cells, unlike MCs cells, either poorly synthesize chlorophyll or do not synthesize it at all. The anomalies in plastid development and their incapacity to form a thylakoid system and perform photosynthesis and certain biosynthetic processes, including biosynthesis of PCs, may underlie the observation mentioned above. Transformation of morphogenic cultures to the nonmorphogenic ones may be a result of various mutational or epigenetic DNA changes leading to the inactivation of genes involved in the biosynthesis of PCs. This is supported by the fact that NCs of Tartary buckwheat differ from MCs not only in chromosome numbers, but also in increased proportion of chromosomal aberrations (Rumyantseva et al., 1998; Kamalova et al., 2009). At the cellular level, NCs are formed by uniform parenchymal and highly vacuolated cells incapable of differentiation, whereas MCs consists of various cell types with a different ability to differentiation, as shown previously (Rumyantseva et al., 1992, 2004). It should be noted that different lines of NCs of Tartary buckwheat are markedly distinguished from each other in a total PC content. Among them, the lines 1-5p and 10p2A have the highest biomass growth rate and these lines also have the lowest PC concentration. Since metabolism pathways of phenolics, as well as of proteins and carbohydrates, have the same precursors, therefore the redirection of them to metabolic reactions involved in cell proliferation and cell expansion could in large part weaken the intensity of secondary metabolism pathways. Our data on the reduced phenolic metabolism in nonmorphogenic cultures are consistent with the results previously obtained on the cotton suspension cultures (Kouakou et al., 2007). Besides nonbound PCs, morphogenic and nonmorphogenic cultures have a different level of cell wall-bound phenolics. We have earlier shown that the morphogenic calli of corn and common and Tartary buckwheat, unlike nonmorphogenic calli, contain considerably more ferulic acid bound to the cell walls (Lozovaya et al., 1996). Similarly, Cvikrova et al. (1996) also found that the embryogenic culture of alfalfa contained more hydroxycinnamic acids associated with cell walls, compared to nonembryogenic culture.

In MCs, a higher phenolic and rutin contents compared to NCs can probably be considered as their biochemical feature caused by a certain degree of differentiation of calli. Souter and Lindsey (2000) supposed that PEMs (PECCs) seem to represent the preglobular-stage embryos, arrested in their further development by the presence of auxin in the culture medium. In this regard, MCs containing PECCs and maintaining a high ability for somatic embryogenesis over many years of culture also preserve the biochemical and metabolic features of embryonic tissues. Since rutin in Tartary buckwheat fruits is shown to be mainly localized in the embryo (Li et al., 2010; Suzuki et al., 2002), we can suppose that a high rutin content in MCs of Tartary buckwheat relates to its synthesis and accumulation in the PECCs. This assumption is fully confirmed by the fact that in NCs, which have no PECCs, rutin is detected in trace amounts, while quercitrin and quercetin are completely absent. In different lines of morphogenic calli, the variations in rutin contents are probably due to the change in the ratio of PECCs and "soft" callus. Moumou et al. (1992a,b) have shown that the main PC in common buckwheat callus cultures was catechin, while the maximum content of rutin did not exceed 0.15 mg/g d.w. These results may indicate that the calli they used were nonmorphogenic and poorly differentiated. As compared to calli of common buckwheat established by Moumou et al. (1992a), the rutin content in our 1-8 line of Tartary buckwheat was 60−170 times greater.

Accumulated data suggest that there is a positive correlation between the total phenol content in tissues and their antioxidant activity (Watanabe et al., 1995; Morishita et al., 2002; Gorinstein et al., 2007). Previously, we showed that the methanol extract of MC of Tartary buckwheat possessed higher antioxidant activity than methanol extract of NC (Sibgatullina et al., 2012). Low antioxidant activity of NC methanol extract may be primarily due to the loss of rutin synthesis in NC, since according to Morishita et al. (2007) rutin was the major contributor to the antioxidative activity of Tartary buckwheat.

Thus, we can conclude that nodular callus of Tartary buckwheat capable of somatic embryogenesis may be a suitable object to study buckwheat phenolic metabolism. Moreover, it may be a valuable alternative source for biotechnological production of rutin and other biologically-active phenolic substances. The comparative study of the composition of phenolic compounds in Tartary buckwheat calli with different morphogenic activity and distinct types of cell differentiation may be the proper approach to identify the biological functions of these compounds and the regulation of their synthesis.

## REFERENCES

Brown, D.E., Rashotte, A.M., Murphy, A.S., Normanly, J., Tague, B.W., Peer, W.A., et al., 2001. Flavonoids act as negative regulators of auxin transport in vivo in arabidopsis. Plant Physiol. 126, 524−535.

Chua, L.S., 2013. A review on plant-based rutin extraction methods and its pharmacological activities. J. Ethnopharmacol. 12, 805−817.

Cvikrova, M., Hrubcova, M., Eder, J., Binarova, P., 1996. Changes in the levels of endogenous phenolics, aromatic monoamines, phenylalanine ammonia-lyase, peroxidase and auxin oxidase activities during initiation of alfalfa embryogenic and nonembryogenic calli. Plant Physiol. Biochem. (Paris) 34, 853−861.

Dai, J., Mumper, R.J., 2010. Plant phenolics: extraction, analysis and their antioxidant and anticancer properties. Molecules 15, 7313−7352.

D'Amato, F., 1991. Nuclear changes in cultured plant cells. Caryologia 44, 217−224.

Debeaujon, I., 2003. Proanthocyanidin-accumulating cells in arabidopsis testa: regulation of differentiation and role in seed development. Plant Cell Online 15, 2514−2531.

Dias, M.I., Sousa, M.J., Alves, R.C., Ferreira, I.C.F.R., 2016. Exploring plant tissue culture to improve the production of phenolic compounds: a review. Ind. Crops Prod. 82, 9−22.

Fabjan, N., Rode, J., Košir, I.J., Wang, Z., Zhang, Z., Kreft, I., 2003. Tartary buckwheat (*Fagopyrum tataricum* Gaertn.) as a source of dietary rutin and quercitrin. J. Agric. Food Chem. 51, 6452−6455.

Folin, O., Ciocalteu, V., 1927. On tyrosine and tryptophane determinations in proteins. J. Biol. Chem. 73, 627−650.

Gamborg, O.L., Miller, R.A., Ojima, K., 1968. Plant cell cultures. Nutrient requirements of suspension cultures of soybean root cells. Exp. Cell Res. 50, 151−158.

Gorinstein, S., Vargas, O.J.M., Jaramillo, N.O., Salas, I.A., Ayala, A.L.M., Arancibia-Avila, P., et al., 2007. The total polyphenols and the antioxidant potentials of some selected cereals and pseudocereals. Eur. Food Res. Technol. 225, 321−328.

Gumerova, E.A., Akulov, A.N., Rumyantseva, N.I., 2015. Effect of methyl jasmonate on growth characteristics and accumulation of phenolic compounds in suspension culture of tartary buckwheat. Russ. J. Plant Physiol. 62, 195−203.

Gupta, N., Sharma, S.K., Rana, J.C., Chauhan, R.S., 2011. Expression of flavonoid biosynthesis genes vis-à-vis rutin content variation in different growth stages of Fagopyrum species. J. Plant Physiol. 168, 2117−2123.

Jiang, P., Burczynski, F., Campbell, C., Pierce, G., Austria, J.A., Briggs, C.J., 2007. Rutin and flavonoid contents in three buckwheat species *Fagopyrum esculentum, F. tataricum*, and *F. homotropicum* and their protective effects against lipid peroxidation. Food Res. Int. 40, 356−364.

Jiménez, V.M., 2001. Regulation of in vitro somatic embryogenesis with emphasis on to the role of endogenous hormones. Braz. J. Plant Physiol. 13, 196−223.

Kamalova, G.V., Akulov, A.N., Rumyantseva, N.I., 2009. Comparison of redox state of cells of tatar buckwheat morphogenic calluses and non-morphogenic calluses obtained from them. Biochemistry (Moscow) 74, 686−694.

Kefeli, V.I., 1987. Some phenolics as plant growth and morphogenesis regulators. In: Purohit, S.S. (Ed.), Hormonal Regulation of Plant Growth and Development. Martinus Nijhoff Publishers, Dordrecht and Agro Botanical Publishers, India, pp. 89−101.

Klykov, A.G., Moiseenko, L.M., Barsukova, Y.N., 2016. Biological resources and selection value of species of Fagopyrum Mill. genus in Far East Russia. In: Zhou, M., Kreft, I., Woo, S-H., Chrungoo, N., Wieslander, G. (Eds.), Molecular Breeding and Nutritional Aspects of Buckwheat. Academic Press, London, pp. 51−60.

Kouakou, T.H., Waffo-Téguo, P., Kouadio, Y.J., Valls, J., Richard, T., Decendit, A., et al., 2007. Phenolic compounds and somatic embryogenesis in cotton (*Gossypium hirsutum* L.). Plant Cell Tissue Organ Cult. 90, 25−29.

Kreft, I., Wieslander, G., Vombergar, B., 2016. Bioactive flavonoids in buckwheat grain and green parts. In: Zhou, M., Kreft, I., Woo, S-H., Chrungoo, N., Wieslander, G. (Eds.), Molecular Breeding and Nutritional Aspects of Buckwheat. Academic Press, London, pp. 161−168.

Krishna, H., Alizadeh, M., Singh, D., Singh, U., Chauhan, N., Eftekhari, M., et al., 2016. Somaclonal variations and their applications in horticultural crops improvement. 3 Biotech 6 (54), 1−18.

Kumar, S., Pandey, A.K., 2013. Chemistry and biological activities of flavonoids: an overview. Sci. World J. 2013, 1−16.

Lattanzio, V., Lattanzio, V.M.T., Cardinali, A., 2006. Role of phenolics in the resistance mechanisms of plants against fungal pathogens and insects. In: Imperato, F. (Ed.), Phytochemistry: Advances in Research. Research Signpost, Trivandrum, Kerala, pp. 23−67.

Li, D., Li, X., Ding, X., 2010. Composition and antioxidative properties of the flavonoid-rich fractions from tartary buckwheat grains. Food Sci. Biotechnol. 19 (3), 711−716.

Lorenzo, J.C., de los Angeles Blanco, M., Peláez, O., González, A., Cid, M., Iglesias, A., et al., 2001. Sugarcane micropropagation and phenolic excretion. Plant Cell Tissue Organ Cult. 65, 1−8.

Lozovaya, V., Waranyuwat, A., Widholm, J., Gorshkova, T., Yablokova, E, Zabotina, O., et al., 1996. Callus cell wall phenolics and plant regeneration ability. J. Plant Physiol. 148, 711−717.

Morishita, T., Hara, T., Suda, I., Tetsuka, T., 2002. Radicals-cavenging activity in common buckwheat (*Fagopyrum esculentum* Moench) harvested in the Kyushu region of Japan. Fagopyrum 19, 89−93.

Morishita, T., Yamaguchi, H., Degi, K., 2007. The contribution of polyphenols to antioxidative activity in common buckwheat and tartary buckwheat grain. Plant Prod. Sci. 10, 99−104.

Moumou, Y., Vasseur, J., Trotin, F., Dubois, J., 1992a. Catechin production by callus cultures of *Fagopyrum esculentum*. Phytochemistry 31, 1239−1241.

Moumou, Y., Trotin, F., Dubois, J., Vasseur, J., El-Boustani, E., 1992b. Influence of culture conditions on polyphenol production by *Fagopyrum esculentum* tissue cultures. J. Nat. Prod. 55, 33−38.

Peer, W.A., Murphy, A.S., 2014. Flavonoids and auxin transport: modulators or regulators? Trends Plant Sci. 12, 556−563.

Pourcel, L., Routaboul, J.M., Cheynier, V., Lepiniec, L., Debeaujon, I., 2007. Flavonoid oxidation in plants: from biochemical properties to physiological functions. Trends Plant Sci. 12, 9−36.

Ramakrishna, A., Ravishankar, G.A., 2011. Influence of abiotic stress signals on secondary metabolites in plants. Plant Signal Behav. 14, 1720−1731.

Rumyantseva, N.I., Sergeeva, N.V., Khakimova, L.E., Salnikov, V.V., Gumerova, E.A., Lozovaya, V.V., 1989. Organogenesis and somatic embryogenesis in tissue culture of two buckwheat species. Soviet J. Plant Physiol. 72, 152−158.

Rumyantseva, N.I., Salnikov, V.V., Fedoseeva, N.V., Lozovaya, V.V., 1992. Characteristic of morphogenesis in buckwheat calluses cultured for a long time. Soviet J. Plant Physiol. 39, 98−103.

Rumyantseva, N.I., Valieva, A.I., Samokhvalova, N.A., Mukhitov, A.R., Ageeva, M.V., Lozovaya, V.V., 1998. Peculiarities of lignification of cell walls of buckwheat calli with different morphogenic ability. Tsitologiya (in Russian) 391, 375−378.

Rumyantseva, N.I., Samaj, J., Ensikat, H.J., Sal'nikov, V.V., Kostyukova, Y.A., Baluska, F., et al., 2003. Changes in the extracellular matrix surface network during cyclic reproduction of proembryonic cell complexes in the *Fagopyrum tataricum* (L.) Gaertn. callus. Russ. J. Plant Physiol. 391, 375−378.

Rumyantseva, N.I., Akulov, A.N., Mukhitov, A.R., 2004. Extracellular polymers in callus cultures of *Fagopyrum tataricum* (L.) Gaertn. with different morphogenic activities: time courses during the culture cycle. Appl. Biochem. Microbiol. 40, 494−500.

Sarkar, D., Naik, P.S., 2000. Phloroglucinol enhances growth and rate of axillary shoot proliferation in potato shoot tip cultures in vitro. Plant Cell Tissue Organ Cult. 60, 139−149.

Saslowsky, D.E., Warek, U., Winkel, B.S.J., 2005. Nuclear localization of flavonoid enzymes in Arabidopsis. J. Biol. Chem. 280, 23735−23740.

Sibgatullina, G.V., Rumyantseva, N.I., Khaertdinova, L.R., Akulov, A.N., Tarasova, N.B., Gumerova, E.A., 2012. Establishment and characterization of the line of *Fagopyrum tataricum* morphogenic callus tolerant to aminotriazole. Russ. J. Plant Physiol. 59, 662−669.

Souter, M., Lindsey, K., 2000. Polarity and signaling in plant embryogenesis. J. Exp. Bot. 51, 971−983.

Suzuki, T., Honda, Y., Funatsuki, W., Nakatsuka, K., 2002. Purification and characterization of flavonol 3-glucosidase, and its activity during ripening in tartary buckwheat seeds. Plant Sci. 163, 417−423.

Vogt, T., 2010. Phenylpropanoid biosynthesis. Mol. Plant 3, 2−20.

Watanabe, M., Sato, A., Osawa, R., Terao, J., 1995. Antioxidative activity of buckwheat seed extracts and its rapid estimate for the evaluation of breeding materials. Nippon Shokuhin Kagaku Kogaku Kaishi 42, 649−655.

Williams, E.G., Maheswaran, G., 1986. Somatic embryogenesis: factors influencing coordinated behavior of cells as an embryogenic group. Ann. Bot. 57, 443−462.

Zaprometov, M.N., 1993. Phenolic Compounds. Nauka, Moscow.

# Chapter | Twenty Six

# Molecular Genetics of Buckwheat and Its Role in Crop Improvement

**Fayaz A. Dar, Tanveer B. Pirzadah, Bisma Malik, Inayatullah Tahir, and Reiaz U. Rehman**
*University of Kashmir, Srinagar, Jammu and Kashmir, India*

## INTRODUCTION

Buckwheat is a pseudo-cereal crop, which belongs to the family Polygonaceae and consists of about 19 species reported so far across the world. Additionally, the newly discovered species in the genus *Fagopyrum* include *Fagopyrum crispatifolium* (Liu et al., 2008), *Fagopyrum pugense* (Tang et al., 2010), *Fagopyrum wenchuanense* (Shao et al., 2011), *Fagopyrum qiangcai* (Shao et al., 2011), *Fagopyrum hailuogouense* (Zhou et al., 2015), and *Fagopyrum luojishanense* (Hou et al., 2015), and the taxonomic position and phylogenetic relationships among the first four species have been elucidated (Zhou et al., 2012; Shao et al., 2011). The crop is of Chinese origin and was introduced into Europe and America in the 15th and 17th centuries, respectively, and has been considered as a vital food source for humans of the present era (Ohnishi and Matsuoka, 1996; Guo et al., 2007). Its remarkable nutritional importance can be ascertained from the fact that it possesses highly valued biological proteins, starch, antioxidants, essential vitamins, minerals, flavonoids, and dietary fiber, besides exhibiting prebiotic activities (Krkoskova and Mrazova, 2005; Alamprese et al., 2007; Gimenz-Bastida and Zielinski, 2015). Several health promoting benefits attributed to buckwheat are mainly due to the presence of proteins and polyphenols in its grain, and it thus exhibits hypochlesterolemic, hypoglucemic, neuroprotective, antidiabetic, anticancerous, antiinflammatory, and antibacterial

**271**

*Buckwheat Germplasm in the World.* DOI: https://doi.org/10.1016/B978-0-12-811006-5.00026-4
© 2018 Elsevier Inc. All rights reserved.

properties (Mendler-Drienyovszki et al., 2013; Wang et al., 2013; Wloch et al., 2015; Gimenz-Bastida and Zielinski, 2015) (Fig. 26.1). Buckwheat thrives best in a cool, moist, and shady environment and grows in a wide range of soil types. The crop is sensitive to stress conditions like spring frost, extreme temperatures, drying winds, and drought. Its indeterminate growth is characterized by the presence of mature seeds at the bottom of the plant and at the same time, the uppermost part starts flowering and therefore making it difficult for the grower to determine appropriate time for harvesting. Amongst the varieties of buckwheat species, a few of them are known to have economic, agricultural, and nutritional importance. *Fagopyrum esculentum* (commonly known as sweet buckwheat) and *Fagopyrum tataricum* (commonly known as bitter buckwheat) are two important species in the buckwheat genepool, which are being cultivated predominantly in most parts of the world (Krkoskova and Mrazova, 2005).

*Fagopyrum esculentum* Moench (Common buckwheat) is widely cultivated in Asia, Europe, and America, especially in the cold and arid regions across the globe (Koyama et al., 2013; Yasui et al., 2016). Buckwheat is an underutilized crop that holds tremendous potential due to its short life cycle, its capability of growing at high altitudes, and the high quality protein content of its grains. The flour obtained is gluten-free and is an alternative to wheat

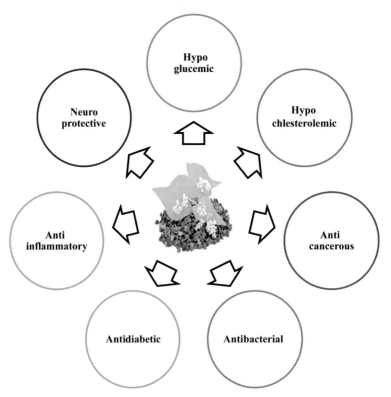

**FIGURE 26.1** Some of the important health promoting effects of buckwheat consumption.

flour for people suffering from celiac disease (Comino et al., 2013). These characteristics make it an important crop in the mountainous regions of India, Nepal, Bhutan, China, Ukraine, Kazakhstan, parts of Eastern Europe, Canada, Mongolia, North Korea, far Eastern Russia, and Japan (Campbell, 1997; Senthilkumaran et al., 2008; Farooq et al., 2016). The classification of buckwheat growing in the wild habitat was first given by Gross in 1913. *Fagopyrum tataricum* (popularly known as Tartary buckwheat) is a self-pollinating species of buckwheat and is found in both cultivated and wild forms. In Asia, Tartary buckwheat is mainly cultivated in China, Nepal, Pakistan, and India (Farooq et al., 2016). Tartary buckwheat is highly nutritious as it consists of proteins of high biological value, vitamins (B1, B2, and B6), balanced amino acids, starch and mineral compounds, and is fibrous (Fabjan et al., 2003; Lee et al., 2013; Wang et al., 2014). Its seeds are enriched with bioflavonoid rutin (quercetin 3-$O$-β-rutinoside) which is so far the highest in comparison to other buckwheat species and is thus receiving increased attention as an important pharmaceutical plant (Fabjan et al., 2003; Hou et al., 2016). Quercetin 3-$O$-β-rutinoside, commonly known as rutin, has great potential, being antioxidative, antihypertensive, and antiinflammatory. Another buckwheat species, known as *Fagopyrum cymosum* (popularly known as Perennial buckwheat), is an insect-pollinated heterostylous species. Moreover, studies on the basis of morphological, biochemical, and molecular characterization have helped in classifying buckwheat into two monoplyletic groups: the cymosum, which includes *F. esculentum, F. tataricum, Fagopyrum homotropicum, Fagopyrum megaspartnium, and Fagopyrum pilus*; and the urophyllum, which includes *Fagopyrum. urophyllum, Fagopyrum lineare, Fagopyrum gracilipedoides, Fagopyrum jinshaense,* and other wild species that have small lustrous achenes completely covered with a persistent perinath (Ohnishi and Matsuoka, 1996; Chen, 1999; Ohsako et al., 2002; Sangma and Chrungoo, 2010; Chrungoo et al., 2012). More than 2500 accessions belonging to different species of buckwheat have been documented from different regions of the world. These accessions could provide a vast repository of genes for various quality traits (International Plant Genetic Resources Institute and the Consultative Group on International Agriculture). The genus consists of both perennial and annual species, with most of the species being diploid ($2n = 2x = 16$), except *F. cymosum* and *Fagopyrum gracilipes* being tetraploid ($2n = 4x = 32$), which includes both self- as well as cross-pollinated species with the occurrence of dimorphic heterostyly rendering some of the species self-incompatible (Chrungoo et al., 2012; Farooq et al., 2016). Even though a large collection of buckwheat is currently available in different laboratories across the globe, there appears to be little agreement on important traits to be documented for characterization of different accessions of the crop. The occurrence of large genotype-environment (GE) interactions possesses a major problem of relating phenotypic performance to the genetic constitution and makes the selection of genotypes difficult (Chrungoo et al., 2012). Advancing the utilization of buckwheat would require an integrated approach, involving development of proper trait descriptors and marker assisted selection (MAS) of genotypes showing quality traits, and a mutagenic approach aimed at elimination of immune-dominant allergenic protein screening of the entire gene pool of buckwheat, including its wild species, for genes which could be used for improvement of the cultivated species through biotechnological approaches.

## IMPORTANCE OF BUCKWHEAT IN THE PRESENT SCENARIO

Buckwheat plays a pivotal role in sustaining the livelihoods of the poor and marginal farmers, especially those living in the food deficit and remote areas, particularly in the Himalayan regions around the globe. The crop has a short harvesting cycle and can be grown under adverse environmental conditions not suitable for the cultivation of other traditional crops. Buckwheat is often regarded as a neutraceutical, exportable, and industrial commodity, due to its potential health promoting benefits. Despite having high economic, agricultural, and nutritional status, the crop is considered as a minor crop and could not attain the status of a major cereal crop so far as its production, yield, and other growth parameters are concerned (Adachi, 2004).

## MAJOR DRAWBACKS

Seed shattering is a major drawback in some of the important species of buckwheat predominantly observed during harvesting period (Lee et al., 1996). It has been observed that both physiological and genetic factors are responsible for this shattering phenomenon in buckwheat. Moreover, extensive shattering has also been reported in few species of buckwheat (Tahir and Farooq, 1988; Farooq et al., 2016). Both conventional and nonconventional biotechnological approaches could be quite useful, especially in those species of buckwheat having an excessive shattering phenomenon.

In common buckwheat, flower abortion is one of the major limiting factors owing to various physiological and genetic factors and thus affecting its overall yield (Farooq et al., 2016). Abiotic stress conditions (drought, frost, temperature, water scarcity, and photoperiod) are also affecting the production yield in different species of buckwheat (Delperee et al., 2003; Jacquemart et al., 2012; Farooq et al., 2016).

Lodging is considered as a serious problem in buckwheat that affects its yield and quality and is characterized as the permanent dislocation of culms from their normal position, due to both internal and external factors (Pinthus, 1973; Murakami et al., 2012). Lodging has a severe impact on morphological features of the plant as it affects quality, yield, and contents of the grain besides reducing the efficiency of photosynthesis and mechanical harvesting in many important crop plants, including buckwheat (Setter et al., 1997; Kashiwagi and Ishimaru, 2004; Berry and Spink, 2012). It has been found that lodging has caused a significant loss in the grain yield of spring wheat, soybean, and maize (Tripathi et al., 2004; Noor and Caviness, 1980; Kang et al., 1999). Although extensive research has been performed worldwide to

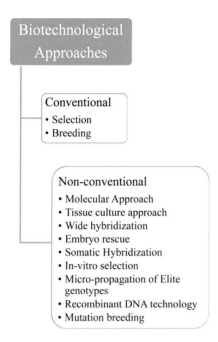

**FIGURE 26.2** Biotechnological methods for crop improvement.

counter the effect of lodging on crop plants like rice and wheat, little success has been achieved so far, as it is still one of the major limiting factors in yield and quality of grains (Crook and Ennos, 1995; Peng et al., 2014; Zhang et al., 2014). Lodging is still considered as one of the major drawbacks in the crop production of common buckwheat (Baniya, 1990). To overcome this problem, it has been proposed that increasing the physical strength of culm can be very useful in improving its lodging resistance.

The major objective in buckwheat crop improvement programs worldwide has been the improvement of seed yield. Moreover, various crop breeding programs have been initiated worldwide to address the other problems of buckwheat, which include: increased seed size, increased seed shattering resistance, early maturity, easier dehulling ability, determinate flowering, increased groat percentage, zero allergenic proteins, lodging resistance, and increased shelf life of its flour.

To achieve these objectives, various biotechnological tools are available and a few of them have been implemented to improve the crop productivity and resistance in buckwheat, a brief account of which will be discussed below (Fig. 26.2).

## DIFFERENT METHODS ADOPTED FOR CROP IMPROVEMENT OF BUCKWHEAT

The use of DNA techniques is rapidly expanding in crop improvement programs. Analysis of a limited number of individuals of different cultivars and populations of *F. esculentum* Moench. and populations of *F. tataricum* Gaertn. showed high polymorphism within cultivars and populations by using RAPD markers. RAPD analysis seems to be very promising in studying genetic variability in buckwheat (Javornik and Kump, 1993). Conventional breeding approaches have not proved useful in hybridizing *F. esculentum* with *F. tataricum* (Adachi et al., 1989; Samimy, 1991; Asaduzzaman et al., 2009). Efforts were being made to improve the buckwheat genus through interspecific hybridization between *F. esculentum* and tetraploid *F. cymosum* (Ujihara et al., 1990; Suvorova et al., 1994; Hirose et al., 1995; Woo et al., 1999; Asaduzzaman et al., 2009). However, the sterility of the resultant hybrids has limited its success. Campbell (1995) successfully hybridized *F. esculentum* with *F. homotropicum* in order to yield self-pollinated buckwheat is one of the major achievements in the buckwheat crop improvement program.

In order to overcome the breeding barriers, hybridization methods have been performed by crossing different buckwheat species with their wild relatives. In one such approach, in vitro embryo rescue techniques have been successfully employed in crossing between tetraploid *F. esculentum* and *F. cymosum* (Woo et al., 2008). Nonconventional methods adopted for the genetic improvement of buckwheat like plant tissue and cell culture have found that the hypocotyl segments of *F. esculentum* and *F. homotropicum* exhibit regenerative abilities through somatic embryogenesis and organogenesis techniques (Woo et al., 2012). The principle method of buckwheat crop improvement programs has been the mass and pure line selections within and between the different species of buckwheat accessions, especially in the developing nations. The pedigree method for breeding and backcrossing has not been so popular in buckwheat crop improvement programs (Woo et al., 2012).

On the basis of morphological characterization and breeding behavior, *F. tataricum* and *F. cymosum* were found to be unrelated. However, on the basis of chloroplast DNA analysis, it was found that *F. cymosum* and *F. tataricum* are closely related in comparison to *F. esculentum* (Kishima et al., 1995). On the other hand, studies on the basis of three chloroplast DNA regions revealed differences in the origin of *F. tataricum* and *F. cymosum* in the Tibet-Himalayan belt (Yamane et al., 2003). So far as the intraspecific genetic variation is concerned, *F. cymosum* was found to be more dominating as its populations have evolved independently to adapt under the variable environmental conditions in comparison to the populations of *F. tataricum*. Moreover, low specific genetic variations were observed in *F. tataricum* by making an analysis of the allozyme, cpDNA, and nucleotide sequence variation of the Adh gene. These findings clearly indicate the utilization of *F. cymosum* as an important source of genetic resource for the improvement of cultivated Tartary buckwheat. Additionally, embryo rescue technique was successfully performed to create interspecific hybrids between *F. cymosum* and *F. tataricum* (Woo et al., 1999). Therefore, the genetic potential of *F. cymosum* can be exploited to resolve the agronomic problems of cultivated Tartary buckwheat, which includes water resistance and enhanced grain yield. Moreover, the Tibet-Himalayan belt is a rich repository of the genetically divergent population of *F. cymosum* and therefore the conservation and utilization of these populations can help in the overall crop improvement programs of buckwheat germplasm. Although buckwheat is a hardy pseudocereal crop, its sensitivity to climatic conditions determines the overall yield of the crop. However, this problem can be addressed by assessing the buckwheat genus and identifying the species of buckwheat which exhibit climatic resilience, so that their genetic potential can be effectively harnessed and can play an important role in the crop improvement program of the *Fagopyrum* genus.

Presently buckwheat crop improvement programs are receiving increased attention for the development of haploids. Anthers of three Polish-bred common buckwheat cultivars were cultivated in vitro. The most prevalent class of regenerates were tetraploids, regardless of cultivar. No haploids were found among the regenerates derived from the diploid cultivars, likewise the tetraploid cultivar yielded no diploid regenerates (Berbec and Doroszewska, 1999). Mesophyll protoplasts of *F. esculentum* have also been fused by PEG-mediated fusion with hypocotyl protoplasts of *F. tataricum* serving as the carrier. A simple two-step preselection method using the intolerance of mesophyll protoplasts to the fusion procedure and the appearance of *F. tataricum* calli could be established and its hybrid nature could be confirmed by RFLP analysis (Samimy et al., 1996).

In recent times molecular genetics has emerged as an applied research discipline which promises to offer solutions to many agro-economic problems. For instance transgenesis/CRISPR techniques has made it possible to develop specific genotypes based on transfer of specific genes as compared to the mixing of two complete genomes of parental lines followed by back-crossing for elimination of undesireable genes.

## MARKER SYSTEMS ADOPTED FOR ELUCIDATING GENETICS OF BUCKWHEAT

There are a number of marker systems available for elucidating the genetics of crop plants, and several marker systems have already been applied for studying the buckwheat genetics. Buckwheat (*F.* esculentum Moench; $2n = 2x = 16$) is a widely cultivated cross-pollinated annual crop in the temperate zones (Yasui et al., 2016). *Fagopyrum esculentum*, however, encounters two major problems as a crop, first, its outcrossing nature, due to heteromorphic self-incompatibility (SI), that makes it difficult to produce pure cultivars of buckwheat, or to fix useful traits. Second, buckwheat grains contain allergens, which induce anaphylactic reactions in some people due to sensitivity (Heffler et al., 2014). Another area receiving attention is the identification of allergenic proteins in buckwheat and their removal. The major allergenic protein with a molecular mass of 22 kDa was identified and sequenced. In this regard, it has been considered that the nucleotide information generated can be effectively used to develop hypoallergenic buckwheat (Nair and Adachi, 1999).

Therefore, improving the nutritional quality of the grain and removing the genes responsible for self-incompatibility and allergens are some of the important breeding objectives in buckwheat and various genetic molecular marker systems have been developed for this purpose [e.g., Amplified Fragment Length Polymorphism (AFLP) markers (Yasui et al., 2004), Simple Sequence Repeats (SSRs) markers (Konishi and Ohnishi, 2006), Expressed Sequence Tag (ESTs) markers (Hara et al., 2011), and Array based markers (Yabe et al., 2014)]. However, AFLP markers have not yet been converted to single locus markers in the buckwheat genome. SSR markers have limited utility in the buckwheat genome, due to difficulty in amplifying specific loci, because of the high level of genetic diversity between buckwheat cultivars and EST marker systems that do not span the entire genome. The newest genome map of buckwheat constructed using array-based markers has sufficient markers to cover the entire genome; however, it requires a specialized instrument to interpret the fluorescence signals of the arrays (Yabe et al., 2014).

## MARKER ASSISTED SELECTION AND THEIR ROLE IN GENETIC IMPROVEMENT OF BUCKWHEAT

MAS is based on selecting plants carrying genomic regions that are involved in the expression of traits of interest through the use of DNA-based molecular markers (Fig. 26.3). With the development and availability of an array of marker techniques and dense molecular genetic maps in crop plants, MAS has become possible for traits both governed by major genes as well as quantitative trait loci (QTLs). There are potential benefits of using molecular markers linked to genes of interest in breeding programs, thus moving from phenotype-based

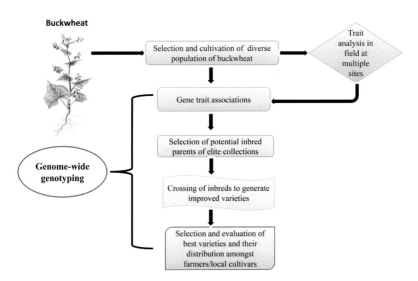

**FIGURE 26.3** Strategy of Marker Assisted Selection (MAS) for buckwheat crop improvement.

selection towards genotype-based selection. QTL mapping is a process of locating genes with effects quantitative traits using molecular markers. DNA based marker systems arc ideal to study QTLs and to map QTLs, which can be effectively used in MAS (Khan, 2015). The construction of the first *F. tataricum* genetic linkage map for elucidating the evolution, genomics, QTL mapping, MAS, and map-based cloning of important genes provided the basis for the research on gene location and molecular breeding (Xiaolei et al., 2013). Simple sequence repeat (SSR) markers were used for the construction of molecular genetic linkage maps in wild species of Tartary buckwheat (Xiaolei et al., 2013). A genomic survey on *F. tataricum* was carried out using simple sequence repeat (SSR) markers for the analysis of genetic diversity (Hou et al., 2016). In one such study, about 21.9 GB raw sequence reads were generated, which were assembled into 348.34 Mb genome sequences including 204,340 contigs. SSR marker analysis was performed on 64 accessions of *F. tataricum* collected from different regions across the world, and the clustering revealed two separate subgroups. In one group, there were Nepal, Bhutan, and the Yunnan-Guizhou Plateau regions, and the other group consisted of Loess Plateau regions, Hunan and Hubei of China, and USA. This clearly indicates that there is a significant amount of genetic diversity present among the accessions of *F. tataricum* and can be very useful in utilizing its germplasm resources in improving Tartary buckwheat breeding programs (Hou et al., 2016).

For genetic analysis of minor crops like buckwheat, MAS for breeding is preferred as there is no need for the availability of prior genomic information. The potential of using a novel array based genotyping system for the construction of high density linkage maps and QTL mapping has been implemented for the analysis of buckwheat genome (Yabe et al., 2014). So far this system has been successfully used for the construction of the high density linkage map of *F. esculentum* (common buckwheat). The map consisted of 751 loci and 8,884 markers and the average distance between adjacent loci was 2.13 cM. This linkage map has found to be very useful for analyzing the genome of other buckwheat populations. QTL based analysis has been utilized for MAS of minor crop plants including buckwheat. Although MAS has been very significant in development of genomic resources for buckwheat, so far little attention has been paid in this direction, as it is only a minor crop species within the Polygonaceae family that is phylogenetically distinct to other major crops and model plant species (Yabe et al., 2014). The linkage map of buckwheat is already available, but it has limited applicability due to its large genetic variation (Konishi and Ohnishi, 2006; Pan and Chen, 2010; Yasui et al., 2004).

## ROLE OF NGS IN MOLECULAR GENETIC STUDIES OF BUCKWHEAT

Next-generation Sequencing (NGS) has emerged as the most promising and powerful technique in analyzing the genomes of both model and nonmodel plant species. The sequences obtained using NGS technology have been found to be very useful in the creation of databases that contain information about genes of both model and nonmodel crops and their comparative analysis (Yasui et al., 2016). Therefore, NGS technology has enabled us to understand the molecular mechanism that controls the agronomically important traits in the genomes of nonmodel crops, including buckwheat. Recently, a versatile NGS-based genotyping method with a low-cost, genotyping by sequencing (GBS) marker system was developed (Elshire et al., 2011). The GBS system utilizes redundant libraries constructed with PCR fragments that have recognition sites of two kinds of restriction enzymes on both ends. The PCR fragments sequenced using NGS technologies are mapped to reference sequences for genome wide genotyping. The GBS system has been used to genotype various crop species to date (He et al., 2014). A draft genome of buckwheat could be used as a reference sequence for developing GBS markers to identify genes that control desirable breeding traits. The NGS based technology has been developed to sequence the genome of buckwheat in the name of the Buckwheat Genome Database (BGDB; http://buckwheat.kazusa. or.jp). This database can be used to rapidly access the sequences of homologous genes previously deposited in the database of other plants, for instance, the buckwheat genes identified using this approach include genes that control flavonoid biosynthesis and genes that encode 2 S albumin-type allergens and granule-bound starch synthases (GBSSs) (Yasui et al., 2016). Furthermore, to illustrate that the draft genome can be used as a reference sequence for NGS-based genotyping, GBS-based technology has been incorporated to identify novel candidate genes for controlling heteromorphic SI of buckwheat (Fig. 26.4).

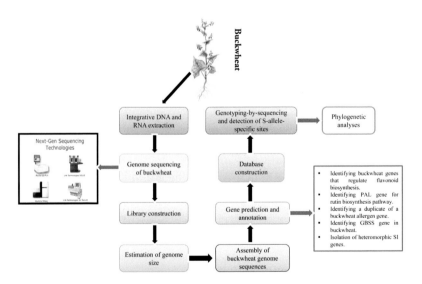

**FIGURE 26.4** Schematic diagram of Next generation Sequencing (NGS) for the identification of agronomically important genes in buckwheat genome.

In buckwheat, phenylalanine ammonia lyase (PAL) gene plays a key role in the biosynthesis of neutraceutically important bioactive compounds like rutin/ quercetin, and its sequencing could be very significant in understanding its genomics. PAL genes were sequenced from three *Fagopyrum* spp. (*F. tataricum*, *F. esculentum*, and *Fagopyrum dibotrys*), which revealed the presence of three single nucleotide polymorphisms (SNPs) and four insertion/deletions at intra and interspecific levels. In this study, it was observed that self-pollinated species (*F. tataricum*) has lower frequency of SNPs identified as compared to cross-pollinated species (*F. esculentum* and *F. dibotrys*). The identified SNPs were not having a significant change in amino acids in the case of *F. tataricum*, and both conservative and nonconservative variations were observed in the case of *F. esculentum* and *F. dibotrys* (Thiyagarajan et al., 2016). The information generated from the sequence characterization unveils the evolutionary significance of the PAL gene, which can be effectively utilized for the genetic improvement of genus *Fagopyrum* with respect to its medicinal relevance.

## FUTURE STRATEGIES AND PROSPECTS

Underutilized crops like buckwheat can contribute to the economic and agricultural sustainability, as it grows in a wide range of environments, requires relatively low inputs, and forms an essential part of the human diet. However, insufficient research efforts in developed countries have not significantly contributed to the production of improved cultivars of these underutilized crops like that of other major crops and therefore the economic potential of these crops has not been sufficiently explored. Some of the major constraints in the development and exploration of these underutilized crops include inadequate knowledge of genetics, quality traits, and agronomic practices, and lack of sufficient crop improvement programs. Other constraints that have contributed in the underutilization of these traditional crops include the low interest of farmers in their cultivation practices, lack of marketing management, experience, and inadequate financial resources. Moreover, molecular intervention in this crop has a vast scope that will offer vital cues to the factors involved in its low yield. It can also be developed for cultivation on marginal lands because of its tolerance toward abiotic stresses, and this strategy would diminish its competition for land with other high-yielding staple crops (Fig. 26.5).

The buckwheat crop improvement program has its emphasis on the development of high yielding varieties, determinate growth habits, enhanced rutin concentration, seed shattering resistance, early maturity, lodging resistance, zero allergic proteins, and so on. Moreover, the crop improvement program should also focus on some important agronomic traits such as improving the nutritional quality, self-compatibility, and homostyly in some species of buckwheat. Both conventional and nonconventional approaches should be adopted to improve the genetic potential of this very important crop plant. The need of the hour is to identify the promising species amongst the buckwheat gene pool so that the potential genes could be identified to address the

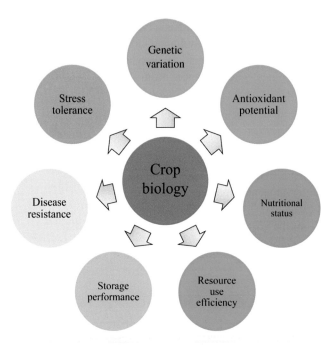

**FIGURE 26.5** Studying and assessing diverse aspects of crop biology.

major problems associated with this crop. Nonconventional approaches in this direction have proved to be very useful from the past few decades in providing solutions to the problems associated with the economically and agriculturally important crop plants including buckwheat, but the need of the hour is to intensify our research in this direction. Therefore, research efforts are needed through progressive collaborations, especially amongst the research institutes, agriculturists, and local growers, in order make buckwheat crop improvement a real success in the near future.

## REFERENCES

Adachi, T., 2004. Recent advances in overcoming breeding barriers in buckwheat. In: Proceedings of the 9th International Symposium on Buckwheat. Prague. pp. 22–25.

Adachi, T., Yamaguchi, A., Miike, Y., Hoffman, F., 1989. Plant regeneration from protoplasts of common buckwheat (*Fagopyrum esculentum*). Plant Cell Rep. 8, 247–250.

Alamprese, C., Casiraghi, E., Pagani, M.A., 2007. Development of gluten-free fresh egg pasta analogues containing buckwheat. Eur. Food Res. Technol. 225, 205–213.

Asaduzzaman, M., Minami, M., Matsushima, K., Nemoto, K., 2009. An in-vitro ovule culture technique for producing interspecific hybrid between tartary buckwheat and common buckwheat. J. Biol. Sci. 9 (1), 1–11.

Baniya, B.K., 1990. Buckwheat in Nepal. Fagopyrum 10, 86–94.

Berbec, A., Doroszewska, T., 1999. Investigations on androgenetic induction in three varieties of buckwheat (*Fagopyrum esculentum* Moench.). Fagopyrum 16, 43–47.

Berry, P.M., Spink, J., 2012. Predicting yield losses caused by lodging in wheat. Field Crops Res. 137, 19–26.

Campbell, G.C., 1995. Inter-specific hybridization in the genus *Fagopyrum*. Curr. Adv. Buckwheat Res. 6, 255–263.

Campbell, G.C., 1997. Buckwheat *Fagopyrum esculentum* Moench. Promoting the conservation and use of underutilized and neglected crops 19. Institute of Plant Genetics and Crop Plant Research, Gatersleben, Germany and the IPGRI, Rome, Italy.

Chen, Qing-Fu, 1999. A study of resources of *Fagopyrum* (Polygonaceae) native to China. Bot. J. Linnean Soc. 130, 53−64.

Chrungoo, N.K., Kreft, I., Sangma, S.C., Devadasan, N., Dohtdong, L., Chetri, U., 2012. Genetic diversity in Himalayan buckwheats: a perspective for use in crop improvement programmes. In: Proceedings of the 12th International Symposium on Buckwheat, Laško, Aug. 21−25. Pernica: Fagopyrum, pp. 198−211.

Comino, I., de Lourdes Moreno, M., Real, A., Rodríguez-Herrera, A., Barro, F., Sousa, C., 2013. The gluten-free diet: testing alternative cereals tolerated by celiac patients. Nutrients 5, 4250−4268.

Crook, M.J., Ennos, A.R., 1995. The effect of nitrogen and growth regulators on stem and root characteristics associated with lodging in two cultivars of winter wheat. J. Exp. Bot. 46, 931−938.

Delperee, C., Kinet, J.M., Lutts, S., 2003. Low irradiance modifies the effect of water stress on survival and growth related parameters during the early developmental stages of buckwheat (*Fagopyrum esculentum* Moench). Physiol. Plant. 119, 211−220.

Elshire, R.J., Glaubitz, J.C., Sun, Q., 2011. A robust, simple Genotyping-by-Sequencing (GBS) Approach for high diversity species. PLoS One 6, e19379.

Fabjan, N., Rode, J., Kosir, I.J., Wang, Z., Zhang, Z., Kreft, I., 2003. Tartary buckwheat (*Fagopyrum tataricum* Gaertn.) as a source of dietary rutin and quercitrin. J. Agric. Food Chem. 51, 6452−6455.

Farooq, S., Rehman, R.U., Pirzadah, T.B., Malik, B., Dar, F.A., Tahir, I., 2016. Cultivation, Agronomic Practices, and Growth Performance of Buckwheat. Academic Press, Oxford, pp. 299−320.

Gimenz-Bastida, J.A., Zielinski, H., 2015. Buckwheat as a functional food and its effects on health. J. Agric. Food Chem. 63, 7896−7913. Available from: http://dx.doi.org/10.1021/acs.jafc.5b02498.

Gross, M.H., 1913. Remarques sur les Polygonees de l'Asie Orientale. Bulletin del Académie internationale de Géographie Botanique 23, 7−32.

Guo, Y.Z., Chen, Q.F., Yang, L.Y., Huang, Y.H., 2007. Analyses of the seed protein contents on the cultivated and wild buckwheat *Fagopyrum esculentum* resources. Genet. Resour. Crop. Evol. 54, 1465−1472.

Hara, T., Iwata, H., Okuno, K., Matsui, K., Ohsawa, R., 2011. QTL analysis of photoperiod sensitivity in common buckwheat by using markers for expressed sequence tags and photoperiod-sensitivity candidate genes. Breed. Sci. 61, 394−404.

He, J., Zhao, X., Laroche, A., Lu, Z-X., Liu, H., Li, Z., 2014. Genotyping-by-sequencing (GBS), an ultimate marker-assisted selection (MAS) tool to accelerate plant breeding. Front. Plant Sci. 5, 484.

Heffler, E., Pizzimenti, S., Badiu, I., Guida, G., Rolla, G., 2014. Buckwheat allergy: an emerging clinical problem in. Europe, J. Allergy Ther. 5, 168.

Hirose, T., Ujihara, H., Kitabayashi, Minami, M., 1995. Pollen tube behavior related to self-incompatibilty interspecific crosses of *Fagopyrum*. Breed. Sci. 45, 65−70.

Hou, L.L., Zhou, M.L., Zhang, Q., Qi, L.P., Yang, X.B., Tang, Y., et al., 2015. *Fagopyrum luojishanense*, a new species of polygonaceae from Sichuan, China. Novon: A J. Bot. Nomenclature 24 (1), 22−26.

Hou, S., Sun, Z., Linghu, B., Xu, D., Wu, B., Zhang, B., et al., 2016. Genetic diversity of buckwheat cultivars (*Fagopyrum tartaricum* Gaertn.) assessed with SSR markers developed from genome survey sequences. Plant Mol. Biol. Rep. 34, 233−241.

Jacquemart, A.L., Cawoy, V., Kinet, J.M., Ledent, J.F., Quinet, M., 2012. Is buckwheat (*Fagopyrum esculentum* Moench) still a valuable crop today. Eur. J. Plant Sci. Biotechnol. 6 (Special Issue 2), 1−10.

Javornik, B., Kump, B., 1993. Random amplified polymorphic DNA (RAPD) markers in buckwheat. Fagopyrum 13, 35−39.

Kang, M.S., Din, A.K., Zhang, Y.D., Magari, R., 1999. Combining ability for rind puncture resistance in maize. Crop Sci. 39, 368−371.

Kashiwagi, T., Ishimaru, K., 2004. Identification and functional analysis of a locus for improvement of lodging resistance in rice. Plant Physiol. 134, 676−683.

Khan, S., 2015. QTL mapping: a tool for improvement in crop plants. Res. J. Recent Sci. 4, 7−12.

Kishima, Y.K., Ogura, K.M., Mikami, T., Adachi, T., 1995. Chloroplast DNA analysis in buckwheat species: phylogenetic relationships, origin of the reproductive systems and extended inverted repeats. Plant Sci. 108, 173−179.

Konishi, T., Ohnishi, O., 2006. A linkage map for common buckwheat based on microsatellite and AFLP markers. Fagopyrum 23, 1−6.

Koyama, M., Nakamura, C., Nakamura, K., 2013. Changes in phenols contents from buckwheat sprouts during growth stage. J. Food Sci. Technol. 50, 86−93.

Krkoskova, B., Mrazova, Z., 2005. Prophylactic components of buckwheat. Food Res. Int. 38, 561−568.

Lee, C.C., Shen, S.R., Lai, Y.J., Wu, S.C., 2013. Rutin and quercetin, bioactive compounds from tatary buckwheat, prevent liver inflammatory injury. Food Funct. 4, 794−802.

Lee, J.H., Aufhammer, W., Kubler, E., 1996. Produced, harvested and utilizable grain yield of the pseudocereals buckwheat (*Fagopyrum esculentum*, Moench), quinoa (*Chenopodium quinoa*, Wild) and amaranth (*Amaranthus hypochondriacus*, L × A. hybridus*, L.) as affected by production techniques. Bodenkultur 47, 5−14.

Liu, J.L., Tang, Y., Xia, Z.M., Shao, J.R., Cai, G.Z., Luo, Q., 2008. *Fagopyrum crispatifolium* Liu, J L., a new species of Polygonaceae from Sichuan, China. J. Syst. Evol. Res. 46, 929−932.

Mendler-Drienyovszki, N., Cal, A.J., Dobránszki, J., 2013. Progress and prospects for interspecific hybridization in buckwheat and the genus *Fagopyrum*. Biotechnol.Adv. 31, 1768−1775.

Murakami, T., Yui, M., Amaha, K., 2012. Canopy height measurement by photogrammetric analysis of aerial images: Application to buckwheat (*Fagopyrum esculentum* Moench) lodging evaluation. Comput. Electron. Agric. 89, 70−75.

Nair, A., Adachi, T., 1999. Immunoblotting and characterization of allergenic proteins in common buckwheat (*Fagopyrum esculentum*). Plant Biotech 16, 219−224.

Noor, R.B.M., Caviness, C.E., 1980. Influence of induced lodging on pod distribution and seed yield in soybeans. Agron. J. 72, 904−906.

Ohnishi, O., Matsuoka, Y., 1996. Search for the wild ancestor of buckwheat II. Taxonomy of *Fagopyrum* (Polygonaceae) species based on morphology, isozymes and cpDNA variability. Genes Genetic Syst. 72, 383−390.

Ohsako, T., Yamane, K., Ohnishi, O., 2002. Two new *Fagopyrum* (Polygonaceae) species, *F. gracilipedoides* and *F. jinshaense* from Yunnan, China. Genes Genet. Syst. 77, 399−408.

Pan, S.J., Chen, Q.F., 2010. Genetic mapping of common buckwheat using DNA, protein and morphological markers. Hereditas 147, 27−33.

Peng, D.L., Chen, X.G., Yin, Y.P., Lu, K.L., Yang, W.B., Tang, Y.H., et al., 2014. Lodging resistance of winter wheat (*Triticum aestivum* L.): Lignin accumulation and its related enzymes activities due to the application of paclobutrazol or gibberellin acid. Field Crops Res. 157, 1−7.

Pinthus, M.J., 1973. Lodging in wheat, barley, and oats: the phenomenon, its causes, and preventive measures. Adv. Agron. 25, 209−263.

Samimy, C., 1991. Barrier tointerspecific crossing of *Fagopyrum esculentum* with *Fagopyrum tataricum*: I. Site of pollen tube arrest: II. Organogenesis from immature embryos of *F. tataricum*. Euphytica 54, 215−219.

Samimy, C., Bjorkman, T., Siritunga, D., Blanchard, L., 1996. Overcoming the barrier to interspecific hybridization of *Fagopyrum esculentum* with *Fagopyrum tataricum*. Euphytica 91, 323−330.

Sangma, S.C., Chrungoo, N.K., 2010. Buckwheat gene pool: potentialities and drawbacks for use in crop improvement programmes. Eur. J. Plant Sci. Biotechnol. 4, 45−50.

Senthilkumaran, R., Bisht, I.S, Bhat, K.V., Rana, J.C., 2008. Diversity in buckwheat (*Fagopyrum* spp.) landrace populations from north-western Indian Himalayas. Genet. Resour. Crop Evol. 55, 287−302.

Setter, T.L., Laureles, E.V., Mazaredo, A.M., 1997. Lodging reduces yield of rice by self-shading and reductions in canopy photosynthesis. Field Crops Res. 49, 95−106.

Shao, J.R., Zhou, M.L., Zhu, X.M., Wang, D.Z., Bai, D.Q., 2011. *Fagopyrum wenchuanense* and *Fagopyrum qiangcai*, two new species of Polygonaceae from Sichuan. China. Novon 21, 256−261.

Suvorova, G.N., Fesenko, N.N., Kosturbin, M.M., 1994. Obtaining of interspecific buckwheat hybrid (*Fagopyrum esculentum* Moench × *Fagopyrum cymosum* Meissn.). Fagopyrum 14, 13−16.

Tahir, I., Farooq, S., 1988. Review article on buckwheat. Buckwheat Newsletter. Fagopyrum 8, 33−53.

Tang, Y., Zhou, M.L., Bai, D.Q., Shao, J.R., Zhu, X.M., Wang, D.Z., et al., 2010. *Fagopyrum pugense* (Polygonaceae), a new species from Sichuan. China. Novon 20, 239−242.

Thiyagarajan, K., Vitali, F., Tolaini, V., Galeffi, P., Cantale, C., Vikram, P., et al., 2016. Genomic characterization of phenylalanine ammonia lyase gene in buckwheat. PLoS ONE 11 (3), e0151187. Available from: http://dx.doi.org/10.1371/journal.pone.0151187.

Tripathi, S.C., Sayre, K.D., Kaul, J.N., Narang, R.S., 2004. Lodging behavior and yield potential of spring wheat (*Triticum aestivum* L.): effects of ethephon and genotypes. Field Crops Res. 87, 207−220.

Ujihara, A., Nakamura, Y., Minami, M., 1990. Interspecific hybridization in genus Fagopyrum-properties of hybrids (*F. esculentum* Moench × *F. cymosum* Meissner) through ovule culture. Gamma Field Radiation Breeding, 1990, NIAR, MAFF, Japan, pp, 45−51.

Wang, L., Yang, X., Qin, P., Shan, F., Ren, G., 2013. Flavonoid composition, antibacterial and antioxidant properties of Tartary buckwheat bran extract. Ind. Crop Prod. 49, 312−317.

Wang, X., Feng, B., Xu, Z., Sestili, F., Zhao, G., Xiang, C., et al., 2014. Identification and characterization of granule bound starch synthase I (GBSSI) gene of tatary buckwheat (*Fagopyrum tataricum* Gaertn.). Gene 534, 229−235.

Wloch, A., Strugala, P., Pruchnik, H., Zylka, R., Oszmianski, J., Kleszczynska, H., 2015. Physical effects of buckwheat extract on biological membrane *in vitro* and its protective properties. J Membrane Biol DOI 10.1007/s00232-015-9857-y.

Woo, S.H., Wang, Y.J., Campbell, C., 1999. Interspecific hybrids with *Fagopyrum cymosum* in the genus Fagopyrum. Fagopyrum 16, 13−18.

Woo, S.H., Tsai, K.S., Adachi, T., Jong, S.K., Choi, J.S., 2008. Pollen-tube behavior and embryo development of interspecific crosses among genus *Fagopyrum*. J. Plant Biol. 52, 302−310.

Woo, S.H., Suzuki, T., Mukasa, Y., Morishita, T., Yun, H.Y., Park, H.C., 2012. Present status, future breeding strategy and prospects for buckwheat. In: Proceedings of the 12th International Symposium on Buckwheat, Laško, Aug. 21−25. Pernica: *Fagopyrum*, pp, 25−26.

Xiaolei, D., Zongwen, Z., Bin, W., Yanqin, L., Anhu, W., 2013. Construction and analysis of genetic linkage map in tartary buckwheat (*Fagopyrum tataricum*) using SSR. Chin. Agric. Sci. Bull. 29 (21), 61−65.

Yabe, S., Hara, T., Ueno, M., Enoki, H., Kimura, T., Nishimura, S., et al., 2014. Rapid genotyping with DNA micro-arrays for high-density linkage mapping and QTL mapping in common buckwheat (*Fagopyrum esculentum*). Breed. Sci. 64, 291−299.

Yamane, K., Yasui, Y., Ohnishi, O., 2003. Interspecific cpDNA variations of diploid and tetraploid perennial buckwheat, *Fagopyrum cymosum* (Polygonaceae). Am. J. Bot. 90, 339−346.

Yasui, Y., Wang, Y., Ohnishi, O., Campbell, C.G., 2004. Amplified fragment length polymorphism linkage analysis of common buckwheat (*Fagopyrum esculentum*) and its wild self-pollinated relative *Fagopyrum homotropicum*. Genome 47, 345−351.

Yasui, Y., Hirakawa, H., Ueno, M., Matsui, K., Katsube-Tanaka, T., Yang, S.J., et al., 2016. Assembly of the draft genome of buckwheat and its applications in identifying agronomically useful genes. DNA Research 1−10. Available from: http://dx.doi.org/10.1093/dnares/dsw012.

Zhang, W.J., Li, G.H., Yang, Y.M., Li, Q., Zhang, J., Liu, J.Y., et al., 2014. Effects of nitrogen application rate and ratio on lodging resistance of super rice with different genotypes. J. Integr. Agri. 13, 63−72.

Zhou, M.L., Bai, D.Q., Tang, Y., Zhu, X.M., Shao, J.R., 2012. Genetic diversity of four new species related to southwestern Sichuan buckwheats as revealed by karyotype, ISSR and allozyme characterization. Plant System. Evol. 298, 751−759.

Zhou, M.L., Zhang, Q., Zheng, Y.D., Tang, Y., Li, F.L., Zhu, X.M., et al., 2015. *Fagopyrum hailuogouense* (Polygonaceae), one new species from Sichuan, China. Novon: A J. Bot. Nomenclature 24 (2), 222−224.

# Chapter Twenty Seven

# Complete Chloroplast Genome Sequence of Buckwheat (*Fagopyrum* spp.) and Its Application in Genome Evolution and Authentication

**Su-Young Hong and Kwang-Soo Cho**

*Highland Agriculture Research Institute, National Institute of Crop Science, Rural Development Administration, Pyeongchang, South Korea*

## INTRODUCTION

Chloroplasts are essential organelles in plant cells that perform photosynthesis, in addition to other functions including synthesizing sugars, pigments, and certain amino acids. The chloroplast (cp) is considered to have originated from an ancestral endosymbiotic cyanobacteria. In addition to the larger dominant genome located in the nucleus of the plant cell, chloroplasts contain their own independent genome encoding a specific set of proteins. The nonrecombinant nature of the cp genome makes it a potentially useful tool in genomics and evolutionary studies. Although the cp genome is highly conserved in vascular plants, evolutionary hotspots such as single nucleotide polymorphisms [SNPs] and insertion/deletions [In/Dels] resulting from inversions, translocations, rearrangements, and copy number variation of tandem repeats have been found in many plants (Jheng et al., 2012). As such, these SNPs and In/Dels are useful as cp genome molecular markers as the cp genome is highly conserved within the species. Further, cp DNA can be easily extracted from samples because of the high copy number. The small size of the cp

**287**

Buckwheat Germplasm in the World. DOI: https://doi.org/10.1016/B978-0-12-811006-5.00027-6
© 2018 Elsevier Inc. All rights reserved.

genome makes it suitable for complete sequencing and the data can be further applied to phylogeny construction (De las Rivas et al., 2002), DNA bar coding (Hollingsworth et al., 2011), and transplastomic studies (Bock and Khan, 2004). Complete cp DNA sequencing began in 1991 (Taberlet et al., 1991) and to date cp genomes of various algae and plants, including crop species, have been reported (CpBase: http://chloroplasr.ocean.washington.edu).

Until recently, cp genome sequencing was a costly and time-consuming process. The majority of such research, therefore, has been limited to sequencing a small portion of the cp genome, which in many cases is insufficient for determining evolutionary relationships, thereby limiting its utility for plant evolutionary and genomic studies. As complete cp genome sequences harbor sufficient information, sequencing of whole cp genomes is essential for the comparison and analyses of diversifications among plant species. The advent of next-generation sequencing (NGS) has made it considerably cheaper and easier to sequence complete cp genomes. NGS is advantageous as it provides extremely high yield and the opportunity for multiplexing when investigating whole-cp genomes, rather than targeting individual regions (Nock et al., 2011; Straub et al., 2012). NGS allows potentially hundreds of flowering plant cp genomes to be sequenced simultaneously, significantly reducing the per-sample cost of cp genome sequencing (McPherson et al., 2013). The complete cp genome of *Fagopyrum* may provide useful information for phylogenetic comparisons with the related species.

*Fagopyrum esculentum* (common buckwheat) and *Fagopyrum tataricum* (tartary buckwheat) are the two cultivated species with high economic importance due to their nutritional value and their usage in human consumption as both greens and grains (Li et al., 2001). Buckwheat is an important functional food as it contains various polyphenols, proteins with high biological value, high contents of available minerals, and a relatively higher fiber content (Liu et al., 2001; Bonafaccia et al., 2003). Especially, Tartary buckwheat has superior nutritional benefits compared to common buckwheat, due to the presence of higher levels of rutin which is a major component of flavonoids (Fabjan et al., 2003).

As tartary buckwheat products are highly preferred for consuming due to their nutritional properties, food products made of tartary buckwheat are expensive compared to common buckwheat products. Often, tartary buckwheat food products are adulterated with traces of common buckwheat. In addition to food products, buckwheat seeds used for cultivation also contain a mixture of seeds from both species. Hence, a method to distinguish products from different buckwheat species is required.

## COMPARATIVE ANALYSIS OF CHLOROPLAST GENOME

The tartary buckwheat complete cp genome has a total sequence length of 159,272 bp, which is 327 bp shorter than that of the common buckwheat genome (159,599 bp). The cp genome of both of these species share the common feature of containing two inverted repeats, which divide the whole genome into a large single copy region (LSC) and a small single copy region (SSC). The LSC is comprised of 84,397 bp in tartary buckwheat and 84,888 bp in common buckwheat, whereas the SSC is 13,343 bp and the inverted repeat region (IR) is 61,532 bp and 61,368 bp in tartary and common buckwheat, respectively (Table 27.1). The gene content, order, and orientation of the *F. tataricum* cp genome were similar to those of common buckwheat. The *F. tataricum* cp genome has a total of 104 genes including 82 protein coding genes, 29 transfer RNA (tRNA) genes and 4 ribosomal RNA (rRNA) genes. Protein coding genes include photosynthesis related genes (the majority), in addition to transcription and translation related genes. The LSC region of the *F. tataricum* cp genome has 62 protein coding genes and 22 tRNA genes, whereas the SSC region contains 11 protein coding genes and one tRNA gene. Nine cp protein coding genes and six tRNA genes contain introns in *F. tataricum*. Among the tRNA genes, *trnK*-UUU has the largest intron in both *F. tataricum* (2,460 bp) and *F. esculentum* (2,458 bp).

The total size variation between *F. tataricum* and *F. esculentum* cp genomes can be accounted for an 82 bp shorter IR region in *F. esculentum*. The border regions of the *F. tataricum* and *F. esculentum* cp genomes were compared to analyze the expansion variation in junction regions. *Rps19, ycf1, ndhF, rps15,* and *trnH* were found in the junctions of LSC/IR and SSC/IR regions. The *rps19* gene of the LSC in *F. tataricum* extended into the IRb region, which created a short pseudo gene of 108 bp at the LSC/IRb junction. This *rps19* pseudo gene is 104 bp in *F. esculentum*. The *ndhF* gene of SSC in *F. esculentum* extends into the IRb with the initial 71 bp 5′ portion of the gene initiating in the IRb region, whereas this is located at the beginning of the SSC region in *F. tataricum*. Similarly, the SSC region of *F. esculentum* extended exactly within the *rps15* gene, whereas in *F. tataricum* the SSC region extended to 2 bp beyond the *rps15* gene. The location of other genes (e.g., Ψ*rps19*, *trnH*, and *ycf1* pseudogene) are similar in both cp genomes.

## DIVERGENCE HOTSPOT

The complete cp genomes of *F. tataricum* and *F. esculentum* were compared and plotted using the mVISTA program to elucidate the level of sequence divergence. The comparison shows that the coding regions of both cp genomes are highly conserved compared to noncoding regions. However, the intergenic region showed the greatest sequence divergence between the two cp genomes. More divergence was found in the sequences of the *trnL*-UAA, *ndhF, trnM*-CAU *ndhK, petN, rpoB, trnS*-GCU, and *trnR*-UCU regions, compared to others. The nucleotide and amino acid sequences of protein coding genes of *F. tataricum* and *F. esculentum* are highly similar with an average sequence similarity of 98.8% and 98.3%, respectively. Between the two species, the nucleotide sequence identity of the LSC, SSC, and IR are 96%, 99.5%, and 99%, respectively. The most conserved genes include the four rRNA genes, along with genes from photosystem I, cytochrome b/f complex, and ATP synthase.

**TABLE 27.1** Comparison of the Complete Chloroplast Genome Contents of *Fagopyrum* Species

| Species | Total Sequence Length (bp) | Large Single Copy (bp) | Inverted Repeat Region (bp) | Small Single Copy (bp) | GC Content (%) | Gene Bank Accession No. | References |
|---|---|---|---|---|---|---|---|
| *Fagopyrum tartaricum* (cv. Daekwan 3-3) | 159,272 | 84,397 | 61,532 | 13,343 | 37.9 | NC_027161 | Cho et al. (2015) |
| *Fagopyrum tartaricum* (cv. Miqiao) | 159,272 | 84,397 | 61,634 | 13,241 | 37.9 | KM201427 | Liu et al. (2016) |
| *Fagopyrum esculentum* | 159,599 | 84,888 | 61,368 | 13,343 | 38.0 | NC_010776 | Logacheva et al. (2008) |
| *Fagopyrum cymosum* | 160,546 | 84,237 | 32,598 | 11,014 | 36.9 | KP404630 | Yang et al. (2016) |

## DIVERGENCE OF CODING GENE SEQUENCE

The average Ks values between the two buckwheat species 0.1237, 0.0725, and 0.0088 in the LSC, SSC, and IR regions respectively, with a total average ratio of 0.0683 across all regions. Although the coding region is highly conserved, we observed a slight variation in the divergence of the coding region. Based on the comparison of Ks values among the regions, higher Ks values were observed for some genes, including *rpoC2, ycf3, accD*, and *clpP*.

## DISTRIBUTION OF TANDEM REPEATS

Within the cp genome, tandem repeats were compared between *F. tataricum* and *F. esculentum*. A total of 19 tandem repeats were identified in *F. tataricum* and *F. esculentum* combined, with varying sizes of repeat units. Of these, 15 were found within intergenic sequences (IGS) and four within coding sequences, all of which shared a similar sequence identity between the two species. These repeating units are repeated from one to four times in both species. Among these repeats, eleven repeats are located in the LSC, seven within in the IR and one in the SSC region. Those tandem repeats located in the IR region are highly diverged and the copy number variations of the repeats within the TR15 region account for the 63 bp InDel #7. Similarly, palindromic repeats were also compared between the two species, identifying three and four repeats in *F. tataricum* and *F. esculentum* respectively. All of these repeats are located in the IGS of the LSC region. Among these, both species have palindromic repeats at two similar locations, namely the IGS of *rbcL* and *accD* and the IGS of *psbT* and *psbN*. Despite sharing similar locations in the two species, two of the palindromes varied in their loop size.

## EVOLUTION OF *FAGOPYRUM TATARICUM* AND *FAGOPYRUM ESCULENTUM*

Variation in the divergence of the coding region was observed between tartary and common buckwheat species. Although the coding region exhibited a highly conserved nature, the *rpoC2, ycf3, accD*, and *clpP* genes of the LSC region of tartary buckwheat showed a higher evolution rate compared to other genes. Yamane et al. (2003) found that the *accD* gene had a high evolution rate in *Fagopyrum* and proposed that this gene was under a weak selection constraint. This is consistent with the high Ks value (2.4538) obtained for *accD* in this study. Other genes we identified as having an unexpectedly high evolution rate between the two studied *Fagopyrum* species include *rpoC2, ycf3*, and *clpP*. Cuénoud et al. (2002) reported that the *matK* gene has a higher Ks ratio than *accD*, but in this study, we found this value to be lower in *matK* than *accD*. These Ks values indicate that the LSC region is under greater selection pressure than the rest of the cp genome and our data confirm a positive selection pressure and neutral evolution of the protein coding genes. Based on the sequence similarity among the three regions, the IR

**TABLE 27.2** Estimation of Evolutionary Divergence Between Sequences Based on the matK Coding Sequences

| Species | Fagopyrum tartaricum | Fagopyrum cymosum | Fagopyrum esculentum |
|---|---|---|---|
| F. tartaricum | | | |
| F. cymosum | 0.00726 ± 0.00214 | | |
| F. esculentum | 0.02538 ± 0.00409 | 0.02402 ± 0.00408 | |

The number of base substitutions per site from between sequences are shown. Standard error estimate(s) are shown above the diagonal and were obtained by a bootstrap procedure (1000 replicates). Analyses were conducted using the Kimura 2-parameter model (Kimura, 1980). The analysis involved 3 nucleotide sequences. Codon positions included were 1st + 2nd + 3rd + Noncoding. All positions containing gaps and missing data were eliminated. There were a total of 1523 positions in the final dataset. Evolutionary analyses were conducted in MEGA6 (Tamura et al., 2013).

region is more conserved than the LSC and SSC regions. This is in agreement with earlier reports that hypothesized that the frequent recombinant events occurring in the IR result in selective constraints on sequence homogeneity, resulting in the IR region diverging at a slower rate than single copy regions (Huang et al., 2013; Wolfe et al., 1989). Based on a previous report (Qian et al., 2013), we calculated the number of base substitutions per site from between sequences, as well as the divergence time (Table 27.2). The divergence time ($T$) was given by $T = D_A/2\lambda$ (Qian et al. 2013), where $\lambda$ is approximately $2.1 \times 10^{-9}$ substitutions per site per year and $D_A$ is the number of base substitutions per site. We can assume a putative divergence time between *F. tataricum* and *Fagopyrum cymosum* and between *F. tataricum* and *F. esculentum* of 0.17 and 0.59 Mya, respectively.

## VALIDATION OF INDEL MARKERS IN BUCKWHEAT GERMPLASM

PCR was carried out to amplify these InDel regions in 75 tartary buckwheat accessions originated from China, Russia, Bhutan, Nepal, Japan, and Pakistan, and 21 common buckwheat accessions originated from China, Russia, Japan, and Pakistan. All the buckwheat accessions from both tartary and common buckwheat confirmed the presence of InDel regions which can be observed by the variation in the amplicon size in both buckwheat species. Hence, all the accessions in both species concurrently showed clear variation in the InDels, indicating that these InDels could be reliably used as biomarkers for the authentication of tartary and common buckwheat traces. The InDel evolutionary hot spot region contributes to the sequence variation and results in the amplicon size variation and hence this region is useful in developing species-specific PCR methods. PCR analyses of the all the selected InDels in this study showed variation in the amplicon size which is specific for each buckwheat species and hence can be utilized in the easy discrimination of common and tartary buckwheat.

## APPLICATION OF INDEL MARKERS TO IDENTIFY THE TRACE AMOUNTS IN PROCESSED FOOD

Genomic DNA was extracted from the mixed flour samples and PCR was performed using InDel marker. It was observed that all the mixed flour samples from 1:9 to 9:1 mixtures showed expected amplicon with a variation in the product size indicating the amplification of InDel region from both tartary and common buckwheat. This suggests that the contamination of buckwheat flour as low as 10% can be easily detected and differentiated using InDel markers.

Further, the applicability of the InDel marker detection method was checked in commercial food products for their authentication and to identify any mixture of trace amounts. Initially, buckwheat noodles made from common and tartary buckwheat were purchased from market and tested for the detect ability of InDel markers. This showed an amplicon of respective product size in both common and tartary buckwheat noodles corresponding to the amplicon obtained from the leaf genomic DNA of common and tartary buckwheat respectively, which was used as a reference. This indicates that InDel markers can be reliably used in the authentication of buckwheat noodles. In addition, to determine any trace amounts in buckwheat tea, around six types of buckwheat tea that were labeled as "made from 100% bitter buckwheat" were purchased from local markets and tested. Genomic DNA was isolated from the tea by a similar method as used for noodles, and InDel marker was detected by PCR. PCR was also performed with genomic DNA made from the leaves of common and tartary buckwheat and used as a reference to compare the amplicon sizes. It was observed that out of six types of bitter buckwheat tea, five showed the amplicon at a similar amplicon size of tartary buckwheat amplicon, whereas the sixth type of bitter buckwheat tea showed amplification product similar to both common and tartary buckwheat amplicons, indicating the possibility of cross contamination with common buckwheat in 100% bitter buckwheat tea. PCR analysis of buckwheat DNA made from common and tartary buckwheat noodles showed amplification with different product size, suggesting that adulteration of buckwheat noodles can be easily detected by this method. Buckwheat DNA was detected to identify the allergens in commercial food products through the PCR-based detection method (Yamakawa et al., 2008).

## REFERENCES

Bock, R., Khan, M.S., 2004. Taming plastids for a green future. Trends Biotechnol. 22, 311–318.

Bonafaccia, G., Marocchini, M., Kreft, I., 2003. Composition and technological properties of the flour and bran from common and tartary buckwheat. Food Chem. 80, 9–15.

Cho, K.S., Yun, B.K., Yoon, Y.H., Hong, S.Y., Mekapogu, M., Kim, K.H., et al., 2015. Complete chloroplast genome sequence of tartary buckwheat (*Fagopyrum tataricum*) and comparative analysis with common buckwheat (*F. esculentum*). Plos One 10.

Cuénoud, P., Savolainen, V., Chatrou, L.W., Powell, M., Grayer, R.J., Chase, M.W., 2002. Molecular phylogenetics of Caryophyllales based on nuclear 18S rDNA and plastid rbcL, atpB, and matK DNA sequences. Am. J. Bot. 89, 132–144.

De las Rivas, J., Lozano, J.J., Ortiz, A.R., 2002. Comparative analysis of chloroplast genomes: functional annotation, genome-based phylogeny, and deduced evolutionary patterns. Genome Res. 12, 567−583.

Fabjan, N., Rode, J., Kosir, I.J., Wang, Z.H., Zhang, Z., Kreft, I., 2003. Tartary buckwheat (*Fagopyrum tataricum* Gaertn.) as a source of dietary rutin and quercitrin. J. Agric. Food Chem. 51, 6452−6455.

Hollingsworth, P.M., Graham, S.W., Little, D.P., 2011. Choosing and using a plant DNA barcode. Plos One 6.

Huang, Y.Y., Matzke, A.J.M., Matzke, M., 2013. Complete sequence and comparative analysis of the chloroplast genome of coconut palm (*Cocos nucifera*). Plos One 8.

Jheng, C.-F., Chen, T.-C., Lin, J.-Y., Chen, T.-C., Wu, W.-L., Chang, C.-C., 2012. The comparative chloroplast genomic analysis of photosynthetic orchids and developing DNA markers to distinguish *Phalaenopsis orchids*. Plant Sci. 190, 62−73.

Kimura, M., 1980. A simple method for estimating evlutionanry rate of base substitutions through comparative studies of nucleotide sequences. J. Mol. Evol. 16, 111−120.

Liu, M., Zheng, T., Ma, Z., Wang, D., Wang, T., Sun, R., et al., 2016. The complete chloroplast genome sequence of Tartary Buckwheat Cultivar Miqiao 1(*Fagopyrum tataricum* Gaertn.). Mitochondrial DNA Part B 1, 577−578.

Liu, Z., Ishikawa, W., Huang, X., Tomotake, H., Kayashita, J., Watanabe, H., et al., 2001. A buckwheat protein product suppresses 1,2-dimethylhydrazine-induced colon carcinogenesis in rats by reducing cell proliferation. J. Nutr. 131, 1850−1853.

Logacheva, M.D., Samigullin, T.H., Dhingra, A., Penin, A.A., 2008. Comparative chloroplast genomics and phylogenetics of *Fagopyrum esculentum* ssp. ancestrale -a wild ancestor of cultivated buckwheat. BMC Plant Biol. 8, 59.

McPherson, H., van der Merwe, M., Delaney, S., Edwards, M., Henry, R., McIntosh, E., et al., 2013. Capturing chloroplast variation for molecular ecology studies: a simple next generation sequencing approach applied to a rainforest tree. BMC Ecol. 13, 8.

Nock, C., Waters, D., Edwards, M., Bowen, S., Rice, N., Cordeiro, G., et al., 2011. Chloroplast genome sequences from total DNA for plant identification. Plant Biotechnol. J. 9, 328−333.

Qian, J., Song, J., Gao, H., Zhu, Y., Xu, J., Pang, X., et al., 2013. The complete chloroplast genome sequence of the medicinal plant *Salvia miltiorrhiza*. Plos One 8, e57607.

Straub, S., Parks, M., Weitemier, K., Fishbein, M., Cronn, R., Liston, A., 2012. Navigating the tip of the genomic iceberg: next-generation sequencing for plant systematics. Am. J. Bot. 99, 349−364.

Taberlet, P., Gielly, L., Pautou, G., Bouvet, J., 1991. Universal primers for amplification of three non-coding regions of chloroplast DNA. Plant Mol. Biol. 17, 1105−1109.

Tamura, K., Stecher, G., Peterson, D., Filipski, A., Kumar, S., 2013. MEGA6: molecular evolutionary genetics analysis version 6.0. Mol. Biol. Evol. 30, 2725−2729.

Wolfe, K.H., Gouy, M.L., Yang, Y.W., Sharp, P.M., Li, W.H., 1989. Date of the monocot-dicot divergence estimated from chloroplast DNA sequence data. Proc. Natl. Acad. Sci. U.S.A. 86, 6201−6205.

Yamakawa, H., Akiyama, H., Endo, Y., Miyatake, K., Sakai, S., Kondo, K., et al., 2008. Specific detection of buckwheat residues in processed foods by polymerase chain reaction. Biosci. Biotechnol. Biochem. 72, 2228−2231.

Yamane, K., Yasui, Y, Ohnishi, O., 2003. Intraspecific cpDNA variations of diploid and tetraploid perennial buckwheat, *Fagopyrum cymosum* (Polygonaceae). Am. J. Bot. 90, 339−346.

Yang, J., Lu, C., Shen, Q., Yan, Y., Xu, C., Song, C., 2016. The complete chloroplast genome sequence of *Fagopyrum cymosum*. Mitochondrial DNA Part A 27, 2410−2411.

# Chapter Twenty Eight

# Distribution of Amino Acids in Buckwheat

Sun Hee Woo[1], Swapan K. Roy[1], Soo J. Kwon[1], Abu Hena Mostafa Kamal[2], Sang Un Park[3], Keun-Yook Chung[1], Moon-Soon Lee[1] and Jong-Soon Choi[3,4]

[1]College of Agriculture, Life and Environment Science, Chungbuk National University, Cheongju, South Korea, [2]University of Texas at Arlington, Arlington, TX, United States, [3]Chungnam National University, Daejeon, South Korea, [4]Korea Basic Science Institute, Daejeon, South Korea

## INTRODUCTION

Buckwheat has been widely used in eastern Asian countries as a non-cereal crop that is considered as a nutritional source as well as a medicinal one for promoting human health. Buckwheat has been also used as vegetables, endowed with the health benefits of their nutritionally-important substances (Shin et al., 2010).

The health promoting effects of buckwheat are related to the strong antioxidant activity during the protein digestion. Besides these pharmacological issues, little has been known about the nutritional aspect of buckwheat. Buckwheat protein improves health in various ways, notably reducing serum cholesterol (Kayashita et al., 1995), suppressing gallstones and tumors (Liu et al., 2001), and inhibiting the angiotensin I-converting enzyme (Ma et al., 2006). This nutritional characteristic results from high concentration of all essential amino acids, in particular including lysine, threonine, and tryptophan (Bonafaccia et al., 2003). However, there have not been any studies on the chemical composition of free amino acids (FAA) from buckwheat. Therefore, we examined the distribution of FAA in the tissues such as sprouting leaf, stem, and root from 2 species of buckwheats, i.e., *Fagopyrum esculentum* and *Fagopyrum tataricum*.

The mobilization of free amino acids happens in the plant tissues for anabolic reaction of macromolecule storage and catabolic metabolism into bioactive small molecules. The plant exhibits absorbing the metabolites from root and transporting to leaves via xylem and phloem of stem (Herschbach et al., 2012). Thus, studies on the distribution and utilization of FAAs in buckwheat

**295**

© 2018 Elsevier Inc. All rights reserved.

will give a primary understanding of amino acid metabolism for the synthesis of biomolecules and the mobilization of FAAs between the inter-tissues.

## MATERIALS AND METHODS

Common buckwheat (*F. esculentum*) and Tartary buckwheat (*F. tataricum*) were cultivated in the farm of Chungbuk National University, Korea. The seeds were germinated under dark conditions at $25°C \pm 2°C$ in a culture room by supplying spray water at regular intervals. Seven-day-old buckwheat sprouting leaves, stems, and roots were dissected and frozen at $-80°C$ prior to biochemical analysis.

## EXTRACTION OF FREE AMINO ACIDS (FAA)

The buckwheat tissues of each 10 mg were ground in a mortar jar with liquid nitrogen and the crude extracts were homogenized with 10 volumes of 70% (v/v) ethanol. The soluble fraction of extract was separated by the partition of the same volume (each 100 mL) with petroleum ether and water. The resulting aqueous phase was fractionated and used as the source of FAA. Protein concentration was measured by Bio-Rad protein assay (Bio-Rad Lab., Hercules, CA, USA) using bovine serum albumin (BSA) as a standard according to the manufacturer's instructions.

## FAA ANALYSIS

Buckwheat FAAs were analyzed by the Picotag method (Heinrikson and Meredith, 1984). FAAs in buckwheat sprouting leaves, stems, and roots were calibrated with norleucine and quantified with the standard physiological PTC-amino acid standards (Edman's reagent; Thermo Fisher Scientific, Rockford, IL, USA). The statistical significance of the composition of the FAAs in common buckwheat (CB) and Tartary buckwheat (TB) was assessed by the Student *t*-test. Pattern analysis of FAAs distribution was conducted using Excel 2007 and DAnTE version 1.0, R-based bioinformatic tool (Polpitiya et al., 2008).

## RESULTS AND DISCUSSION
### General Features of Amino Acids

The ingestion of protein is required to supply amino acids for the formation of body protein. The amino acids of proteins can be divided into 2 nutritional categories: essential and nonessential. The essential amino acids are amino acids that can not be synthesized by humans. Additionally, the amino acids such as arginine, cysteine, glycine, glutamine, proline, serine, and tyrosine are considered conditionally essential, meaning they are not normally required in the diet, but must be supplied exogenously when not being synthesized it in adequate amounts in body (Fürst and Stehle, 2004). In humans, the essential amino acid needs for net protein deposition are a minor portion of the total amino acid requirement (Dewey et al., 1996), and $>90\%$ of the total amino acid requirement, even of the young child, is connected with the maintenance of body protein stores. Formulating the amino acids required for 'maintenance' is tricky and is still an issue of controversy (Young and Borgonha, 2000).

## COMPARISON OF CB AND TB

The developmental changes of buckwheat sprouts significantly varied between CB and TB. Plant height and dry weights were slightly higher in CB than in TB, whereas root length was higher in TB. In particular, the fresh weight of CB was significantly higher than that of TB (Fig. 28.1).

The protein contents of dried masses of sprouting leaves, stems, and roots from CB and TB were compared. The abundance of protein quantity per dried mass in two buckwheat was commonly shown as a sequence of leaf > root > stem (data not shown).

Whereas the contents of leaf proteins were similarly observed in both buckwheat species, the quantities of stem and root were differentially obtained from CB and TB. The protein amount of TB stem was 51% higher than that of CB. However, the amount of TB root was 40% lower than that of CB. The difference of local protein amounts was simply due to the native characteristics of buckwheat species.

The total of amino acids existing as a free form in three tissues were similarly measured irrespective of buckwheat species (data not shown). In particular, the quantity of essential amino acids in TB leaf was 53% larger than that of CB leaves, whereas the amount of nonessential amino acids of TB leaf was reversely 38% smaller than that of CB leaves. Thus, the TB leaf with higher amounts of essential FAAs can be used as a diet for the benefit of human nutrition.

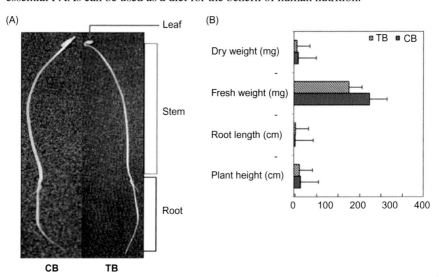

**FIGURE 28.1** Morphology of sprouting common buckwheat (CB) and Tartary buckwheat (TB) (A) and comparative growth characteristics of buckwheat (B).

## FREE AMINO ACID COMPOSITION IN 2 BUCKWHEAT SPECIES

For the composition of 20 FAAs, the overall distribution and relative amount of FAAs are listed according to 3 tissues and 2 species (Table 28.1). In general, Asn is known as the most abundant free amino acid in plant parts, followed by Gln and Ala (Choi et al., 2011). In the buckwheat, however, Asn was a minor amino acid and Gln was predominant in the stem and root in CB and TB. In sprouting leaves, Val was most abundant in both buckwheat and Tyr was second most abundant in CB. The comparison of leaf FAAs between CB and TB revealed the higher rates of His and Arg ($P < 0.01$) and of Thr, Leu, and Phe ($P < 0.05$) in TB. Surprisingly, the amount of Tyr in TB leaf was 15-fold lower compared to CB. Most of the phenolic compounds in plants are derived from phenylalanine and tyrosine, and the main step of the biosynthesis is the conversion of phenylalanine to cinnamic acid by the abolition of an ammonia molecule, which is catalyzed by phenylalanine ammonia lyase (PAL) as regulatory enzyme of secondary metabolism (Yao et al., 1995). A total of 14 different amino acids including Asp, Glu, and Lys were markedly increased during sprouting, but Arg and Cys were decreased (Sun Lim et al., 2001). TB has been known to contain higher amounts of phenolic compounds than CB (Kalinova and Vrchotova, 2009). Gln was measured as the most abundant amino acid in stem in both CB and TB. This finding was in good agreement with the previous report about the abundant amino acids containing Gln in the wheat phloem (Wang and Tsay, 2011). Higher FAAs in TB compared to CB were Asp, Glu, Ser, Thr, Tyr, Cys, and Phe ($P < 0.01$). Contrarily, His and Ile were significantly lower in TB ($P < 0.05$) and Arg and Leu were more significantly lower in TB ($P < 0.01$). As for buckwheat root, several TB FAAs such as Asp, Gly, Ala, Pro, and Val were more significantly lower at $P < 0.01$ compared to CB. Likewise, TB FAAs such as Glu, Met, and Lys were significantly lower at $P < 0.05$. Here, Pro content was known to increase 40-fold in response to drought stress (Pospisilova et al., 2011). Thus, it is notable whether the difference of Pro content between CB and TB links to the different response to abiotic stress. Pro accumulation is a common metabolic response of higher plants to water deficits and salinity stress (Verslues and Sharp, 1998), which is primarily localized in the cytosol in the plant. Pro protects membranes and proteins against the adverse effects of high concentrations of inorganic ions and temperature extremes (Stein et al., 2011), and also functions as a protein-compatible hydrotrope (Wang et al., 2011), and hydroxyl radical scavenger (Kaul et al., 2008). In contrast to down-regulated TB FAAs, Tyr, Trp, and Gln were more significantly larger in TB FAAs, and the content of Tyr, Trp ($P < 0.01$), and Gln ($P < 0.05$) were more significantly larger in TB.

The distributional pattern of free amino acids was determined using heat map analysis. The distribution of the FAAs between CB and TB was overall similar except for Gln in stem and root of both CB and TB, Tyr in leaves of

**TABLE 28.1** Free Amino Acid (FAA) Composition of Sprouting Leaves, Stems, and Roots Between Common Buckwheat (CB) and Tartary Buckwheat (TB)

| FAA | CB | | | TB | | | Fold Change (TB/CB) | | |
| --- | --- | --- | --- | --- | --- | --- | --- | --- | --- |
| | Sprout | Stem | Root | Sprout | Stem | Root | Sprout | Stem | Root |
| Asp | 1.6 ± 0.3 | 2.2 ± 0.4 | 2.9 ± 0.3 | 2.2 ± 0.2 | 3.7 ± 0.2 | 2.0 ± 0.1 | 1.4 | 1.7** | 0.7** |
| Glu | 5.1 ± 0.7 | 10.5 ± 0.6 | 6.4 ± 0.9 | 5.5 ± 0.5 | 16.4 ± 0.5 | 4.7 ± 0.2 | 1.1 | 1.6** | 0.7* |
| Asn | 1.7 ± 0.2 | 6.6 ± 0.8 | 2.2 ± 0.4 | 2.1 ± 0.2 | 7.2 ± 0.3 | 2.7 ± 0.2 | 1.2 | 1.1 | 1.2 |
| Ser | 2.7 ± 0.4 | 8.0 ± 0.8 | 7.0 ± 1.2 | 3.3 ± 0.3 | 12.0 ± 0.7 | 8.5 ± 0.4 | 1.2 | 1.5** | 1.2 |
| Gln | 15.5 ± 2.5 | 102.0 ± 10.5 | 31.3 ± 4.5 | 14.4 ± 1.4 | 103.6 ± 4.1 | 40.4 ± 2.7 | 0.9 | 1.0 | 1.3* |
| Gly | 1.4 ± 0.3 | 6.7 ± 0.4 | 5.1 ± 0.7 | 1.3 ± 0.1 | 6.0 ± 0.2 | 2.9 ± 0.2 | 1.0 | 0.9 | 0.6** |
| His | 3.8 ± 0.7 | 15.5 ± 1.3 | 6.6 ± 0.7 | 6.6 ± 0.6 | 12.9 ± 0.4 | 8.5 ± 0.3 | 1.7** | 0.8* | 1.3 |
| Arg | 4.7 ± 0.7 | 8.0 ± 1.1 | 5.3 ± 0.6 | 7.2 ± 0.6 | 4.5 ± 0.6 | 4.5 ± 0.1 | 1.5** | 0.6** | 0.8 |
| Thr | 2.4 ± 0.3 | 9.7 ± 0.4 | 9.9 ± 1.2 | 3.3 ± 0.3 | 17.4 ± 0.6 | 11.9 ± 0.7 | 1.3* | 1.8** | 1.2 |
| Ala | 1.4 ± 0.2 | 5.8 ± 0.8 | 4.4 ± 0.8 | 1.6 ± 0.2 | 5.6 ± 0.5 | 2.0 ± 0.1 | 1.2 | 1.0 | 0.5** |
| Pro | 1.3 ± 0.2 | 2.0 ± 0.2 | 2.7 ± 0.0 | 1.1 ± 0.1 | 1.9 ± 0.2 | 1.5 ± 0.2 | 0.8 | 1.0 | 0.5** |
| Tyr | 48.0 ± 4.8 | 8.7 ± 0.5 | 2.7 ± 0.1 | 3.2 ± 0.1 | 11.5 ± 0.5 | 3.7 ± 0.4 | 0.1** | 1.3** | 1.4** |
| Val | 68.9 ± 10.4 | 19.0 ± 1.0 | 7.6 ± 0.6 | 110.8 ± 46.6 | 17.9 ± 1.1 | 4.8 ± 0.3 | 1.6 | 0.9 | 0.6** |
| Met | 2.4 ± 0.3 | 3.6 ± 0.1 | 0.9 ± 0.1 | 3.2 ± 0.8 | 3.7 ± 0.1 | 0.8 ± 0.0 | 1.3 | 1.0 | 0.8* |
| Cys | 1.1 ± 0.1 | 0.1 ± 0.0 | 0.0 ± 0.0 | 0.9 ± 0.4 | 0.3 ± 0.0 | 0.1 ± 0.1 | 0.8 | 3.3** | 2.1 |

*Continued*

**TABLE 28.1** continued

| FAA | CB | | | TB | | | Fold Change (TB/CB) | | |
|---|---|---|---|---|---|---|---|---|---|
| | Sprout | Stem | Root | Sprout | Stem | Root | Sprout | Stem | Root |
| Ile | 1.2±0.2 | 9.1±0.4 | 2.4±0.2 | 1.9±0.5 | 8.0±0.3 | 2.0±0.2 | 1.6 | 0.9* | 0.8 |
| Leu | 1.1±0.2 | 7.4±0.6 | 2.1±0.2 | 2.1±0.4 | 5.2±0.1 | 2.0±0.2 | 1.8* | 0.7** | 1.0 |
| Phe | 1.2±0.2 | 3.9±0.3 | 1.2±0.0 | 1.8±0.2 | 4.8±0.2 | 1.4±0.2 | 1.5* | 1.3** | 1.1 |
| Trp | 7.3±1.0 | 14.7±0.7 | 2.7±0.1 | 7.0±0.4 | 14.9±0.6 | 5.1±0.5 | 1.0 | 1.0 | 1.9** |
| Lys | 0.9±0.2 | 2.3±0.4 | 1.3±0.2 | 0.7±0.1 | 1.9±0.2 | 0.9±0.1 | 0.7 | 0.8 | 0.7* |
| SUM | 173.7±23.7 | 245.7±21.4 | 104.9±12.7 | 180.1±54.0 | 259.4±11.5 | 110.5±7.1 | 1.0 | 1.1 | 1.1 |

Content of free amino acid was expressed as mean ± SD (mg/100 g d.w); Statistical significance was performed by Student's $t$-test at *$P < 0.05$ and **$P < 0.01$.

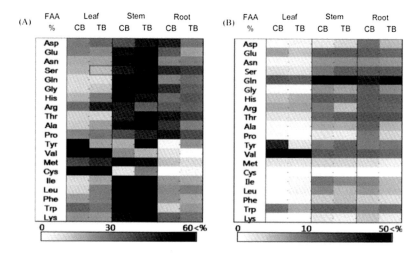

**FIGURE 28.2** Heat map analysis of free amino acids (FAA) distribution. Distribution of the whole FAA in individual buckwheat organs (A) and FAA in 3 different organs (B). Gray values of each bar represent the relative intensities as the mol% shown at the bottom.

CB, and Val in the leaf of both CB and TB (Table 28.1, Fig. 28.2A). Among the 20 FAAs, Tyr was revealed to be a distinct component to differentiate the leaf and stem in CB and TB. When the distribution of each FAA was analyzed according to the occupancy of 3 tissues, Tyr was clearly more abundant in CB leaf compared to other tissue residence of TB (Fig. 28.2B). This suggests that they have common properties in FAA utilization; however, the specific Tyr metabolism was differently evolved in both CB and TB.

In summary, several studies were carried out about amino acid composition of proteins, but information about the FAA pool and the role of these substances is very deficient. The main intention of this study was to contribute to the insufficient knowledge concerning the composition of the FAAs in buckwheat.

The results acquired highlighted an extreme variability in the FAA composition examined in root, stem, and leaf of CB and TB. Nevertheless, the most abundant amino acids were Gln, Tyr, and Val, which are involved in important metabolic pathways in humans. In light of the potential importance of the FAAs in a physiological setting, further studies are needed to evaluate the opportunity to include them in vegetables and dietary supplements in the near future. Finally, the basic study of free amino acids in buckwheat gives clear evidence for different inter-conversion and metabolism of amino acids in 2 buckwheat species.

# REFERENCES

Bonafaccia, G., Marocchini, M., Kreft, I., 2003. Composition and technological properties of the flour and bran from common and tartary buckwheat. Food Chem. 80, 9–15.

Choi, S.-H., Ahn, J.-B., Kozukue, N., Levin, C.E., Friedman, M., 2011. Distribution of free amino acids, flavonoids, total phenolics, and antioxidative activities of jujube (*Ziziphus jujuba*) fruits and seeds harvested from plants grown in Korea. J. Agric. Food Chem. 59, 6594–6604.

Dewey, K., Beaton, G., Fjeld, C., Lonnerdal, B., Reeds, P., Brown, K., et al., 1996. Protein requirements of infants and children. Eur. J. Clin. Nutr. 50.

Fürst, P., Stehle, P., 2004. What are the essential elements needed for the determination of amino acid requirements in humans? J. Nutr. 134, 1558S–1565S.

Heinrikson, R.L., Meredith, S.C., 1984. Amino acid analysis by reverse-phase high-performance liquid chromatography: precolumn derivatization with phenylisothiocyanate. Anal. Biochem. 136, 65–74.

Herschbach, C., Gessler, A., Rennenberg, H., 2012. Long-Distance Transport and Plant Internal Cycling of N-and S-Compounds, Progress in Botany,, vol. 73. Springer, pp. 161–188.

Kalinova, J., Vrchotova, N., 2009. Level of catechin, myricetin, quercetin and isoquercitrin in buckwheat (*Fagopyrum esculentum* Moench), changes of their levels during vegetation and their effect on the growth of selected weeds. J. Agric. Food Chem. 57, 2719–2725.

Kaul, S., Sharma, S., Mehta, I., 2008. Free radical scavenging potential of L-proline: evidence from *in vitro* assays. Amino Acids 34, 315–320.

Kayashita, J., Shimoka, I., Nakajyoh, M., 1995. Hypocholesterolemic effect of buckwheat protein extract in rats fed cholesterol enriched diets. Nutr. Res. 15, 691–698.

Liu, Z., Ishikawa, W., Huang, X., Tomotake, H., Kayashita, J., Watanabe, H., et al., 2001. A buckwheat protein product suppresses 1, 2-Dimethylhydrazine–induced colon carcinogenesis in rats by reducing cell proliferation. J. Nutr. 131, 1850–1853.

Ma, M.-S., Bae, I.Y., Lee, H.G., Yang, C.-B., 2006. Purification and identification of angiotensin I-converting enzyme inhibitory peptide from buckwheat (*Fagopyrum esculentum* Moench). Food Chem. 96, 36–42.

Polpitiya, A.D., Qian, W.-J., Jaitly, N., Petyuk, V.A., Adkins, J.N., Camp, D.G., et al., 2008. DAnTE: a statistical tool for quantitative analysis of-omics data. Bioinformatics 24, 1556–1558.

Pospisilova, J., Haisel, D., Vankova, R., 2011. Responses of transgenic tobacco plants with increased proline content to drought and/or heat stress. Am. J. Plant Sci. 2, 318.

Shin, D.-H., Kamal, A.H.M., Suzuki, T., Yun, Y.-H., Lee, M.-S., Chung, K.-Y., et al., 2010. Reference proteome map of buckwheat (*Fagopyrum esculentum* and *Fagopyrum tataricum*) leaf and stem cultured under light or dark. Aust. J. Crop Sci. 4, 633.

Stein, H., Honig, A., Miller, G., Erster, O., Eilenberg, H., Csonka, L.N., et al., 2011. Elevation of free proline and proline-rich protein levels by simultaneous manipulations of proline biosynthesis and degradation in plants. Plant Sci. 181, 140–150.

Sun Lim, K., Young Koo, S., Jong Jin Hwangi, S., Han Sun, H., Park, C.H., 2001. Development and utilization of buckwheat sprouts as functional vegetables. Fagopyrum 18, 49–54.

Verslues, P.E., Sharp, R.E., 1998. Role of amino acids in abiotic stress resistance. Plant Amino Acids Biochem. Biotechnol. 319.

Wang, W.-G., Li, R., Liu, B., Li, L., Wang, S.-H., Chen, F., 2011. Effects of low nitrogen and drought stresses on proline synthesis of *Jatropha curcas* seedling. Acta Physiol. Plantarum 33, 1591–1595.

Wang, Y.-Y., Tsay, Y.-F., 2011. Arabidopsis nitrate transporter NRT1. 9 is important in phloem nitrate transport. Plant Cell 23, 1945–1957.

Yao, K., De Luca, V., Brisson, N., 1995. Creation of a metabolic sink for tryptophan alters the phenylpropanoid pathway and the susceptibility of potato to *Phytophthora infestans*. Plant Cell 7, 1787–1799.

Young, V.R., Borgonha, S., 2000. Nitrogen and amino acid requirements: the Massachusetts Institute of Technology amino acid requirement pattern. J. Nutr. 130, 1841S–1849S.

# Chapter Twenty Nine

# Inheritance of Self-Compatibility in a Buckwheat Hybrid

**Sun Hee Woo[1], Swapan K. Roy[1], Seong-Woo Cho[2], Soo J. Kwon[1], Cheol-Ho Park[3], and Taiji Adachi[4]**

[1]*College of Agriculture, Life and Environment Science, Chungbuk National University, Cheongju, South Korea,* [2]*Chonbuk National University, Jeonju, South Korea,* [3]*Kangwon National University, Chuncheon, South Korea,* [4]*Institute for Plant Biotechnology R & D, Ltd., Osaka, Japan*

## INTRODUCTION

Buckwheat is a crop species of considerable importance and its use may become more widespread in the future due to its high nutritive value (Eggum, 1980; Marshall and Pomeranz, 1982; Rufeng et al., 1992). Consumption has already increased over the past few years. However, production is often troubled by persistent low and unstable yields that affect markets as well as producers. One of the major constraints to buckwheat production is low percentage seed set, which often can be 12% or lower with resulting low grain yield. This has been shown to be due to a very high abortion rate (MORRIS, 1951), making the plant an inefficient user of available photosynthates.

Buckwheat breeding programs have looked at the development of self-compatible flower types resulting from mutations in the proposed sporophytic, heteromorphic self-incompatibility system (Sharma and Boyes, 1961) of common buckwheat. These, however, have not been successful in increasing yield and therefore breeders are now attempting introgression of characters, including self-compatibility from other species. Wide hybridization has been considered to be an important approach for the improvement of many crop species. The family Polygonaceae, which includes the genus *Fagopyrum*, comprises many wild species that could form a large resource of genes imparting resistance/tolerance to a wide spectrum of biotic and abiotic stresses. The newly found species *Fagopyrum homotropicum* is more closely related to common buckwheat than

**303**

*Buckwheat Germplasm in the World.* DOI: https://doi.org/10.1016/B978-0-12-811006-5.00029-X
© 2018 Elsevier Inc. All rights reserved.

is the species *Fagopyrum tataricum* (Ohnishi, 1995), and therefore efforts are being made to introgress the self-compatible allele found in *F. homotropicum* into common buckwheat. Programs aimed at improving the strategies of buckwheat breeding are fraught with numerous problems. Prominent among them is the self/cross-incompatibility phenomenon of the reproductive biology of this seed-propagated genus (Adachi, 1990; Guan and Adachi, 1992; Kreft, 1983; Marshall, 1969; Woo et al., 1995). In buckwheat, these breeding barriers have in no small measure contributed to the recalcitrance of buckwheat to conventional improvement techniques. For instance, the transfer of valuable genetic traits such as self-fertility and frost tolerance, and important agronomic characters, such as grain yield, have so far proved to be very difficult due to the cross-incompatibility of common buckwheat.

These phenomena pose a challenge to research directed at increasing buckwheat yields. In order to broaden the genetic basis of common buckwheat, several interspecific hybrids have been attempted. A previous study reported that that crosses between *Fagopyrum esculentum* and *F. tataricum* resulted in immature embryos if *F. tataricum* was used as the female parent (MORRIS, 1951), while no embryos developed if *F. esculentum* was used as the female parent. Nagatomo (1961) was unsuccessful in developing crosses between *F. esculentum* and *Fagopyrum cymosum* at the tetraploid level (Krotov and Dranenko, 1973). Krotov and Dranenko (1975) produced a new species *Fagopyrum giganteum* from the hybrid between tetraploid *F. tataricum* and *F. cymosum*. Samimy (1991) and Samimy et al. (1996) developed an infertile plant from the cross of diploid *F. esculentum* and *F. tataricum*. Wagatsuma and Un-no (1995) reported on the first fertile cross of diploid *F. esculentum* and *F. tataricum*. Several studies reported on the first fertile cross of *F. homotropicum* and *F. esculentum* and further backcrosses to *F. esculentum* (Campbell, 1995; Wang and Campbell, 1998). With the aid of such interspecific crosses, genes for tolerance and/or resistance to specific stresses may be transferred from wild species to cultivated forms. However, reports on fertile interspecific hybrids between common buckwheat *F. esculentum* and wild species are rare. The recent reports of successful hybridization between *F. esculentum*, *F. homotropicum*, and *F. tataricum* at the diploid level have demonstrated that interspecific crosses can be utilized in the improvement of common buckwheat (Campbell, 1995; Wagatsuma and Un-no, 1995; Wang and Campbell, 1998). The purpose of this study was to develop the methodology for the transfer of desirable agronomic traits from a wild annual species (*F. homotropicum*) into elite lines of cultivated common buckwheat. Therefore, attempts were made to develop autogamous buckwheat by combining conventional breeding methods with tissue culture techniques.

## MATERIALS AND METHODS
### Plant Materials

Common buckwheat and *F. homotropicum* used for interspecific hybridization were raised in a glasshouse at Miyazaki University. Crosses were made

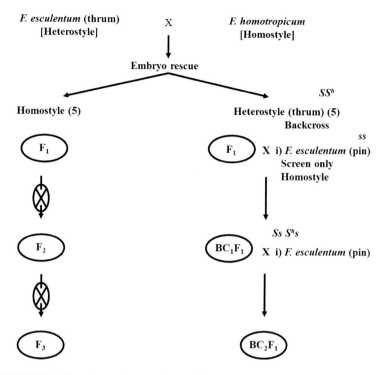

**FIGURE 29.1** Genealogy of restored material crossed to produce successive progenies in hybrids between *Fagopyrum esculentum* and *Fagopyrum homotropicum*.

as outlined in Fig. 29.1. All interspecific crosses were made in a phytotron by hand-pollination to prevent selfing or contamination from foreign pollen. Plants were bud emasculated during the morning hours and the emasculated flowers were pollinated using freshly collected pollen the following morning.

## OVULE CULTURE

Ovaries were excised at 3, 5, 7, or 11 days after pollination when the embryos were at the following stages: early globular, globular-early heart, late heart-early torpedo, and late torpedo. The ovules were surface-sterilized by dipping them for 1 min in 70% ethanol, followed by immersion in a 2% solution of sodium hypochloride together with one drop of detergent for 3 min. The ovaries were then rinsed three times in sterile distilled water. Ovules were removed from the ovaries with the tip of a scalpel under a dissecting microscope. The ovules were placed on the surface of the culture medium as described previously (Adachi, 1997). The ovules were cultured at 25°C under 16 h of light and 8 h of dark. After 4 weeks, the embryos were transferred to MS basal medium without hormones for germination and shoot development (Murashige and Skoog, 1962). The germinating seedlings were subcultured two or three times. When the plantlets had well-developed roots they were transplanted to pots containing an autoclaved mixture of vermiculite and soil (1:1) and grown in a growth chamber. $F_1$ plants were sib-mated and/or backcrossed to common buckwheat.

## GENETIC ANALYSIS

In order to identify the genetic constitution of the progenies derived from the $F_1$ hybrids, $F_2$ and $F_3$ populations were raised. The observed segregation ratios for the hetrostyle and the homostyle genes were tested for goodness-of-fit to the expected ratios by chi-square analysis. The $F_1$-hybrids (heterostyly, thrum-type) were also backcrossed with common buckwheat (pin).

## SEED PROTEIN ANALYSIS

The seed protein profile of the autogamous buckwheat obtained from the progenies of interspecific $F_1$ hybrid lines coded from H1 to H5 were examined. The water soluble proteins were extracted under reducing conditions with 0.125 M Tris-HCL, 10% SDS, 1% 2-mercaptoethanol, 30% glycerol, 0.2% bromophenol blue (sample buffer). The extracted proteins were then boiled in sample buffer in a water bath for 5 min, cooled, and stored at $-20°C$ until they were used. Ten microlitres of protein extract were used for electrophoresis, which was carried out on a SDS-PAGE (sodium dodecyl sulfate-poly acrylamide gel electrophoresis) discontinuous system following the procedure of previously described protocol (Laemmli, 1970). The proteins were separated on a 10% acrylamide and 0.1% SDS. The electrophoretic buffer contained 0.025 M Tris, 0.192 M glycine, pH 8.3, 0.1% wt/vol SDS. The electrophoresis was carried out for approximately 3 h at a constant current of 25 mA. The gels were stained in 0.25% coomassie blue. The gels were dried on a gel drying processor (Atto, Japan) at a constant temperature of 60°C for 3 h.

## RESULTS AND DISCUSSION

Application of in vitro techniques for overcoming breeding barriers in interspecific crosses in the genus *Fagopyrum* has been restricted by the lack of embryo-rescue culture techniques. Samimy et al. (1996) and Wagatsuma and Un-no (1995) have exploited possibilities for ovule culture. In our ovule culture experiments, more than 18% of the excised ovules germinated after 30−35 days in vitro, but only 25.6% of the germinated ovules regenerated into transplantable seedlings (Table 29.1; Fig. 29.2).

Of all the media tested, the best response was observed on the media supplemented with casein hydrolysateW1 and MS1 (White, 1964). Most of the ovules that failed to germinate developed calli and later formed embryoids.

However, the plantlets regenerated from these embryoids were spindly, rootless, and albino. Samimy et al. (1996) reported the development of a hybrid through embryo rescue in a cross between *F. esculentum* and *F. tataricum*. We were successful in developing hybrids between common

TABLE 29.1 Response in Sterile Culture Conditions of Ovules Containing *Fagopyrum esculentum* X *Fagopyrum homotropicum* Hybrid Embryos Excised on Different Days After Pollination (DAP)

| Flower Type of Female Parent | DAP | Cultured (A) | Number of Ovules Germinate (B) | Regenerated (C) | Germination (B)/(A)X100 | Regeneration (C)/(B)X100 | Seedlings/ 100 Ovules |
|---|---|---|---|---|---|---|---|
| Thrum | 3 | 65 | 13 | 2 | 20 | 15.4 | 3.2 |
| | 5 | 58 | 16 | 5 | 28 | 31.3 | 8.6 |
| | 7 | 62 | 13 | 3 | 21 | 23.1 | 4.8 |
| | 11 | 82 | 15 | 4 | 18.3 | 26.7 | 4.9 |
| Total | | 267 | 57 | 14 | 21.8 | 24.1 | 5.4 |
| Pin | 3 | 54 | 5 | 1 | 9.3 | 20 | 1.9 |
| | 5 | 62 | 11 | 3 | 17.7 | 27.3 | 4.8 |
| | 7 | 58 | 8 | 2 | 13.8 | 25 | 3.4 |
| | 11 | 78 | 14 | 5 | 17.8 | 35.7 | 6.4 |
| Total | | 252 | 38 | 11 | 14.7 | 27 | 4.1 |
| Grand Total | | 519 | 95 | 25 | 18.3 | 25.6 | 4.8 |
| P X Th | 5 | 85 | 70 | 43 | 82.4 | 61.4 | 9.2 |
| Th X P | 5 | 48 | 41 | 25 | 85.4 | 61 | 10.5 |
| Total | | 133 | 111 | 68 | 83.9 | 61.2 | 9.85 |

**FIGURE 29.2** Plant development following interspecific hybridization between *Fagopyrum esculentum* (thrum) and *Fagopyrum homotropicum* through embryo rescue. (A) A plantlet obtained from the hybrid embryo after 3 weeks of ovule culture on MS1 medium. (B) A hybrid plantlet from ovule culture on MS3 medium. (C) A hybrid plant growing in soil. (D) $F_1$ plants flowering at 12 weeks of age.

buckwheat and *F. homotropicum*. Heterogeneity 2 analysis showed that the germination rate of ovules from thrum parents was higher than that from pin parents ($P = 0.06$, Table 29.1). Therefore, use of thrum plants of *F. esculentum* in future crosses would be expected to increase the efficiency of production of hybrids between these two species. There were no significant effects of excision date at 3, 5, 7, and 11 days after pollination or from pin or thrum female flower type on plantlet regeneration rate. The morphological characters and the hybrid nature of the $F_3$ generation are shown in Table 29.2.

In growth habit, the hybrid plants were somewhat decumbent rather than having the erect habit of the female common buckwheat parent. However, they possessed an intermediate flower and plantlet shape, clearly indicating the hybrid nature of the plants. $F_3$ generations had thin, bright green, and ovate leaves. The inflorescences had the hairy rachis and pink-colored flowers of the wild parents. The hybrid had a seed size and weight closer to that

**TABLE 29.2** Some Characteristics of Interspecific Hybrids of Buckwheat Between *Fagopyrum esculentum* and *Fagopyrum homotropicum* in the $F_3$ Generation (Recorded for 40-Day Old Seedlings)

| Characters | *F. esculentum* | *F. homotropicum* | $F_3$ Hybrid |
|---|---|---|---|
| Plant height (cm) | 120 ± 20 | 70 ± 10 | 60 ± 20 |
| Width X length of leaf (cm) | 8.5 × 10.5 | 7.5 × 8.5 | 3.5 × 5.5 |
| Leaf shape | Sagittate | Ovate | Ovate |
| Anthers | Well developed | Small | Well developed |
| Number of primary branches | 5 | 4 | 5 |
| Flower color | White | Pink | Pale pink |
| Style shape | Heterostyle | Homostyle | Homo/heterostyle |
| Number of floret buds | 18 ± 4 | 15 ± 4 | 20 ± 4 |
| Number of seeds/ pollinated flowers | 1/10 | 2/10 | 3/10 |

**TABLE 29.3** Inheritance of Heterostylar and Homostylar Genes in Interspecific Hybrids Between *Fagopyrum esculentum* and *Fagopyrum homotropicum* at Different Generations

| Generation | Strain Number | Genetic Segregation | | | | | | $\chi_2$ | P |
|---|---|---|---|---|---|---|---|---|---|
| | | Observed | | | Expected | | | | |
| | | Ho[a] | Th[b] | P[c] | Ho | Th | P | | |
| $F_1$ | | 5 | 5 | 0 | 1 | 1 | 0 | 0.1 | 0.75 |
| $F_2$ | H1 | 6 | 0 | 4 | 3 | 0 | 1 | 0.533 | 0.50–0.75 |
| | H2 | 8 | 0 | 4 | 3 | 0 | 1 | 0.111 | 0.75–0.90 |
| | H3 | 12 | 0 | 2 | 3 | 0 | 1 | 0.381 | 0.50–0.75 |
| | H4 | 7 | 0 | 3 | 3 | 0 | 1 | 0 | >0.90 |
| | H5 | 9 | 0 | 2 | 3 | 0 | 1 | 0.03 | 0.75–0.90 |
| | Total | 42 | 0 | 15 | 3 | 0 | 1 | 0.006 | >0.90 |
| $BC_1F_1$ ($F_1$ × pin) | | 5 | 5 | 1 | 1 | 1 | 0 | 0.1 | 0.75 |

[a]*Homomorphic plant.*
[b]*Thrum plant.*
[c]*Pin plant.*

of common buckwheat. The findings had earlier reported the first successful interspecific hybridization of common buckwheat using a pin plant as the female parent and *F. homotropicum* (Campbell, 1995). The segregation pattern was studied in the $F_1$, $F_2$, and $BC_1F_1$ generations (Table 29.3).

**TABLE 29.4** Segregation Patterns of Heterostylar and Homostylar Genes From the Cross *Fagopyrum esculentum* and *Fagopyrum homotropicum* in the F$_3$ Generation

| Generation | F$_1$ Strain | F$_2$ Strain | Observed Ho[a] | Observed Th[b] | Observed P[c] | Expected Ho | Expected Th | Expected P | $\chi^2$ | P |
|---|---|---|---|---|---|---|---|---|---|---|
| F$_3$ | H1 | 1 | 5 | 0 | 0 | 1 | 0 | 0 | | |
| | | 5 | 5 | 0 | 0 | 1 | 0 | 0 | | |
| | | 6 | 7 | 0 | 0 | 1 | 0 | 0 | | |
| | H2 | 2 | 11 | 0 | 0 | 1 | 0 | 0 | | |
| | | 8 | 11 | 0 | 0 | 1 | 0 | 0 | | |
| | | 12 | 10 | 0 | 0 | 1 | 0 | 0 | 0.026 | 0.75–0.90 |
| | H3 | 3 | 14 | 0 | 0 | 1 | 0 | 0 | | |
| | | 4 | 11 | 0 | 0 | 1 | 0 | 0 | | |
| | | 7 | 10 | 0 | 3 | 3 | 0 | 1 | | |
| | H4 | 1 | 16 | 0 | 2 | 3 | 0 | 1 | 1.185 | 0.10–0.25 |
| | | 3 | 10 | 0 | 3 | 3 | 0 | 1 | 0.111 | 0.50–0.75 |
| | | 8 | 10 | 0 | 0 | 1 | 0 | 0 | | |
| | H5 | 1 | 10 | 0 | 0 | 1 | 0 | 0 | | |
| | | 8 | 5 | 0 | 1 | 1 | 0 | 0 | | |

[a]Homomorphic plant.
[b]Thrum plant.
[c]Pin plant.

The $F_1$ hybrids segregated 5:5, which fits a ratio of 1:1 for homomorphic: thrum type flower types, indicating a single dominant gene for homomorphism. The $F_2$ plants were produced to define the homomorphic gene and to examine the genetic basis of it.

Homostyly appears to be controlled by a single dominant gene as the $F_2$ progeny segregated 42:15, which fits a 3:1 (homomorphic:pin type flower types) ratio (Table 29.3). This confirms the results obtained earlier whereas the findings focused on a single dominate gene controlling homostyly (Campbell, 1995; Wang and Campbell, 1998). Only progeny of two of the lines (H3 and H4) were shown to be heteromorphic by segregating in the $F_3$ generation with a ratio of 3:1 (Table 29.4). These results suggest that homostyly genes may be controlled by a single locus with multiple alleles. The $BC_1F_1$ lines segregated at the expected 1:1 ratio with the exception of one pin-type plant. The chi-square test showed that the observed ratio of 42 to 15 did not differ significantly ($P > 0.90$) from the expected ratio (Table 29.3). From these results we deducted a tentative genotype for heterostyly and homostyly flower types.

Homomorphism was controlled by a single allele $S^h$, while the pin/thrum-complex was governed by a single genetic locus $S$, with two alleles, $S$ and $s$, which control $Ss$ (thrum-type) as well as the $ss$ (pin-type), respectively. Corresponding to the incompatibility mechanism in *Fagopyrum*, this represents the case of a single locus $S$ with three alleles, $S^h$, $S$, and $s$, and the phenotypes, homomorphic, pin and thrum. It can be characterized by relationship of dominance, $S > S^h > s$ (Table 29.5).

Electrophoretic analyses of water soluble proteins of different lines (H1 to H5) of the $F_3$ generation are shown in Fig. 29.3. Although variations were observed among different lines, the separation pattern shows specificity with respect to each strain. The quantitative variation between lines can be attributed to the relative production of proteins due to allelic differences. Single seed analysis showed similar banding patterns within the lines indicating strong autogamy and homozygosity.

In contrast, the common buckwheat seeds show strong polymorphism for water soluble proteins. Further studies on agronomic characters (Woo et al., 1997), allergenic proteins (Urisu et al., 1995; Yano et al., 1989) and other components such as rutin are also important to further improve the currently existing interspecific hybrid progenies (Ohsawa and Tsutsumi, 1995;

**TABLE 29.5** Tentative Scheme of the Relationship of Heterostyly and Homostyly Genes in Buckwheat

| Genotype | Ss | SS$^h$ | S$^h$S$^h$ | ss |
|---|---|---|---|---|
| Phenotype | Thrum | Thrum | Homo | Pin |
| Relationship of dominance | | $S > S^h > s$ | | |

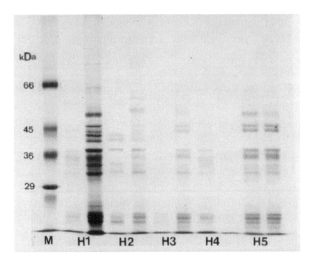

**FIGURE 29.3** Water soluble proteins of self-pollinating buckwheat $F_3$ plants from the interspecific hybrid between *Fagopyrum esculentum* and *Fagopyrum homotropicum.* Each line (H1—H5) came from a single $F_1$ plant. Each lane represents a single seed of the respective lines.

Kitabayashi et al., 1995). The autogamous nature of these progenies also simplifies the systematic RAPD, SCAR, AFLP, and SNP analysis of buckwheat (Aii et al., 1998; Nagano et al., 1998) and studies on allergenic proteins that are already underway.

## REFERENCES

Adachi, T., 1990. How to combine the reproductive system with biotechnology in order to overcome the breeding barrier in buckwheat. Fagopyrum 10, 7−11.

Adachi, T., 1997. Production of interspecific hybrids between *Fagopyrum esculentum* and *F. homotropicum* through embryo rescue. Sabrao J. 29, 89−96.

Aii, J., Nagano, M., Woo, S.H., Campbell, C., Adachi, T., 1998. Molecular marker linked to the S locus in buckwheat. Adv. Buckwheat Res. 7, V65−71.

Campbell, C., 1995. Inter-specific hybridization in the genus *Fagopyrum*. In: Proceedings of the 6th International Symposium on Buckwheat. Citeseer, pp. 255−263.

Eggum, B.O., 1980. The protein quality of buckwheat in comparison with other protein source of plant or animal origin. In: Proceedings of the 1st International Symposium. Buckwheat Research, pp. 115−120.

Guan, L., Adachi, T., 1992. Reproductive deterioration in buckwheat (*Fagopyrum esculentum*) under summer conditions. Plant Breed. 109, 304−312.

Kitabayashi, H., Ujihara, A., Hirose, T., Minami, M., 1995. Varietal differences and heritability for rutin content in common buckwheat, *Fagopyrum esculentum* Moench. Breed. Sci. 45, 75−79.

Kreft, I., 1983. Buckwheat breeding perspectives, Buckwheat Research. In: Proceedings of the 2nd International Symposium on Buckwheat. Miyazaki University, Japan, pp. 3−12.

Krotov, A., Dranenko, E., 1973. Amphidiploid buckwheat, *Fagopyrum giganteum* Krot. sp. nova. Biull Vses Ord Lenina Inst Rastenievod Im NI Vavilova.

Krotov, A.S., Dranenko, E., 1975. Buckwheat *Fagopyrum* Mill. In: Flora of Cultivated Plants. 3. Grout crops. Leningrad, Kolos, p. 7−118 (in Russian).

Laemmli, U.K., 1970. Cleavage of structural proteins during the assembly of the head of bacteriophage T4. Nature 227, 680–685.

Marshall, H., Pomeranz, Y., 1982. Buckwheat: description, breeding, production, and utilization. Advances in Cereal Science and Technology (USA).

Marshall, H.G., 1969. Description and culture of buckwheat. Bull. 754 Pennsylvania Agric. Exp. Stn.

Morris, M.R., 1951. Cytogenetics studies on buckwheat. J. Hered. 42, 85–89.

Murashige, T., Skoog, F., 1962. A revised medium for rapid growth and bioassays with tobacco tissue cultures. Physiol. Plant. 15, 473–497.

Nagano, M., Aii, J., Campbell, C., Adachi, T., 1998. Development of the amplified fragment length polymorphysims (AFLP) technology in buckwheat. Adv. Buckwheat Res. 7, V58–64.

Nagatomo, T., 1961. Studies on physiology of reproduction and some cases of inheritance in buckwheat. Report of Breeding Science Laboratory, Faculty of Agriculture, Miyazaki University 1, 1-212.

Ohnishi, O., 1995. Discovery of new Fagopyrum species and its implication for the studies of evolution of *Fagopyrum* and of the origin of cultivated buckwheat. In: Proceedings of the 6th International Symposium. Buckwheat at Ina, pp. 175–190.

Ohsawa, R., Tsutsumi, T., 1995. Improvement of rutin content in buckwheat flour. Curr. Adv. Buckwheat Res. 1, 365–372.

Rufeng, C.C.N., Bin, Z.Z.C., Rang, C., 1992. Preliminary study on the addition of buckwheat flour and reserve of nutrient elements in buckwheat health food.

Samimy, C., 1991. Barrier to interspecific crossing of *Fagopyrum esculentum* with *Fagopyrum tataricum*: I. Site of pollen-tube arrest. II. Organogenesis from immature embryos of F. tataricum. Euphytica 54, 215–219.

Samimy, C., Bjorkman, T., Siritunga, D., Blanchard, L., 1996. Overcoming the barrier to interspecific hybridization of *Fagopyrum esculentum* with *Fagopyrum tataricum*. Euphytica 91, 323–330.

Sharma, K., Boyes, J., 1961. Modified incompatibility of buckwheat following irradiation. Can. J. Bot. 39, 1241–1246.

Urisu, A., Kondo, Y., Morita, Y., Yagi, E., Tsuruta, M., Yasaki, T., Yamada, K., Kuzuya, H., Suzuki, M., Titani, K., 1995. Isolation and characterization of a major allergen in buckwheat seeds. Curr. Adv. Buckwheat Res. 965–974.

Wagatsuma, T., Un-no, Y., 1995. In vitro culture of interspecific ovule between buckwheat (*F. esculentum*) and tartary (*F. tataricum*). Breed. Sci. 45, 312.

Wang, Y., Campbell, C., 1998. Interspecific hybridization in buckwheat among *Fagopyrum esculentum*, F. homotropicum and F. Tataricum. In: Proceedings of the 7th International Symposium. Buckwheat at Winnipeg, Canada. I, pp. 1–12.

White, P.R., 1964. The cultivation of animal and plant cells. Soil Sci. 97, 74.

Woo, S., Tsai, Q., Adachi, T., 1995. Possibility of interspecific hybridization by embryo rescue in the genus *Fagopyrum*. Curr. Adv. Buckwheat Res. 6, 225–237.

Yano, M., Nakamura, R., Hayakawa, S., Torii, S., 1989. Purification and properties of allergenic proteins in buckwheat seeds. Agric. Biol. Chem. 53, 2387–2392.

# Chapter Thirty

# Analysis of Genetic Diversity and Population Structure of Buckwheat (*Fagopyrum esculentum* Moench.) Landraces of Korea Using SSR Markers

**Jae Young Song, Gi-An Lee, Mun-Sup Yoon, Kyung-Ho Ma, Yu-Mi Choi, Jung-Ro Lee, Yeon-Ju Jung, Hong-Jae Park, Chung-Kon Kim, and Myung-Chul Lee**

*National Academy of Agricultural Science, Suwon, South Korea*

## INTRODUCTION

Common buckwheat (*Fagopyrum esculentum* Moench) is an outcrossing, self-incompatible species belonging to the Polygonaceae family (Sharma and Boyes, 1961). Common buckwheat has been widely distributed and is a cultivated crop of considerable importance in many countries around the world in Asia, America, and Europe, although the cultivation of this crop has not increased in recent years (Alekseeva, 1986). The important component of buckwheat seeds has a well-balanced amount of essential amino acids and excellent nutritional value (Javornik et al., 1981). In addition, common

**315**

*Buckwheat Germplasm in the World*. DOI: https://doi.org/10.1016/B978-0-12-811006-5.00030-6
© 2018 Elsevier Inc. All rights reserved.

buckwheat is also important as a nectariferous and pharmaceutical plant (Alekseeva, 1986). Buckwheat produces grains and fodder and is also a source of succulent green leafy vegetable (Narain, 1979). Most of the varieties of common buckwheat grown are local populations adapted to their environmental conditions through cultivation. For crop variety protection, information on genetic distances among inbreds is important for the identification of essential derivation as well as legal protection of the germplasm (Smith et al., 1995). Therefore, information about the genetic diversity and population structure in breeding material is of fundamental importance for the improvement of crops (Hallauer and Miranda, 1988). So, the evaluation of germplasm diversity and relationships among present cultivated and wild varieties and populations is important both for future breeding and for the study of buckwheat evolution (Kump and Javornik, 1996). Genetic diversity among and within common buckwheat cultivars has been studied using allozyme analysis and the origin of cultivated common buckwheat has been studied by the diffusion routes analysis using RAPD markers (Murai and Ohnishi 1996).

Recent advances in molecular biology have offered more suitable molecular markers for assessing genetic diversity than RAPD markers. Among the PCR-based techniques, amplified fragment length polymorphisms (AFLP) (Vos et al., 1995) and simple sequence repeat (SSR) markers are widely used for studies of genetic diversity in crop species. The advantages of SSR markers are their codominant modes of inheritance and hypervariability, which make them ideal for a wide range of applications (Goldstein and Schlötterer, 1999). Simple sequence repeats (SSRs, also called microsatellites) are abundantly distributed throughout eukaryotic genomes (Litt and Luty, 1989). Microsatellite markers are powerful tools for the analysis of wide genetic variations within or among populations (Tautz, 1989). In many crops, several recent studies have used SSR markers to assess the genetic diversity, phylogenetic relationships, and population structures of various crops, for example in durum wheat (Thuillet et al., 2005), maize (Vigouroux et al., 2005), and rice (Li et al., 2010). The aim of the present study was to evaluate the genetic diversity, population structure, and genetic relationships among geographically diverse accessions of buckwheat landraces of Korea maintaining or conserving in National Agrobiodiversity Center of RDA using SSR markers.

## MATERIALS AND METHODS
### Plant Materials and DNA Extraction

A list of common buckwheat accessions used in this study is given in Table 30.1. A total of 179 accessions of common buckwheat were obtained from the National Agrobiodiversity Center of the Rural Development Administration (RDA) (http://genebank.rda.go.kr/), Korea (GW 19, GG 3, GN 24, JN 14, JB 43, CN 4, and CB 12 accessions). For the DNA extraction,

**TABLE 30.1** List of 179 Buckwheat Accessions of the Collection in the RDA

| Sample Number | IT or Tem. IT | Region | Country of Origin | Sample Number | IT or Tem. IT | Region | Country of Origin |
|---|---|---|---|---|---|---|---|
| 1 | 709851 | GB | KOR | 221 | 108889 | GW | KOR |
| 27 | 910167 | GN | KOR | 222 | 108892 | GW | KOR |
| 51 | K002646 | GN | KOR | 223 | 108934 | JB | KOR |
| 53 | K002648 | GN | KOR | 224 | 108957 | GB | KOR |
| 54 | K003292 | GN | KOR | 225 | 108968 | GB | KOR |
| 58 | K011766 | GW | KOR | 226 | 109053 | GB | KOR |
| 141 | 100906 | JN | KOR | 228 | 109078 | GB | KOR |
| 142 | 100973 | GB | KOR | 229 | 109095 | GB | KOR |
| 144 | 101006 | GB | KOR | 230 | 109106 | GB | KOR |
| 145 | 101022 | JB | KOR | 233 | 109175 | GB | KOR |
| 146 | 101091 | JB | KOR | 237 | 109601 | JN | KOR |
| 147 | 101120 | JB | KOR | 238 | 110977 | GB | KOR |
| 148 | 101271 | GW | KOR | 239 | 110978 | GB | KOR |
| 149 | 101282 | GW | KOR | 241 | 111123 | CN | KOR |
| 150 | 101389 | JB | KOR | 244 | 112812 | JB | KOR |
| 151 | 101391 | JB | KOR | 247 | 112911 | GG | KOR |
| 153 | 101431 | JB | KOR | 249 | 112949 | JB | KOR |

*Continued*

**TABLE 30.1** continued

| Sample Number | IT or Tem. IT | Region | Country of Origin | Sample Number | IT or Tem. IT | Region | Country of Origin |
|---|---|---|---|---|---|---|---|
| 154 | 102359 | GB | KOR | 250 | 112957 | JB | KOR |
| 155 | 102780 | GB | KOR | 252 | 112982 | JB | KOR |
| 157 | 103026 | GB | KOR | 254 | 113033 | GB | KOR |
| 158 | 103069 | JB | KOR | 255 | 113051 | GB | KOR |
| 159 | 103093 | JB | KOR | 256 | 113066 | GB | KOR |
| 160 | 103119 | GN | KOR | 258 | 113083 | GB | KOR |
| 163 | 103569 | GN | KOR | 260 | 113086 | GB | KOR |
| 165 | 103633 | GN | KOR | 261 | 113087 | GB | KOR |
| 167 | 103710 | GN | KOR | 262 | 113088 | GB | KOR |
| 169 | 103836 | JB | KOR | 263 | 113123 | CB | KOR |
| 170 | 103881 | JB | KOR | 264 | 113126 | CB | KOR |
| 173 | 104133 | GB | KOR | 266 | 113200 | GB | KOR |
| 174 | 104139 | GB | KOR | 268 | 113250 | JB | KOR |
| 175 | 104236 | GB | KOR | 269 | 113266 | JB | KOR |
| 177 | 104328 | GB | KOR | 270 | 113276 | JB | KOR |
| 178 | 104429 | GN | KOR | 271 | 113296 | JB | KOR |
| 179 | 104461 | GW | KOR | 272 | 113306 | JB | KOR |

Continued

| | | | | | | | | |
|---|---|---|---|---|---|---|---|---|
| 181 | 104526 | GW | KOR | | 274 | 113347 | JB | KOR |
| 182 | 104551 | GW | KOR | | 275 | 113353 | JB | KOR |
| 183 | 104769 | GN | KOR | | 276 | 113358 | JB | KOR |
| 187 | 105304 | GW | KOR | | 277 | 113371 | JB | KOR |
| 190 | 105398 | JB | KOR | | 278 | 113392 | JB | KOR |
| 194 | 105473 | GB | KOR | | 279 | 113406 | JB | KOR |
| 198 | 105523 | GB | KOR | | 280 | 113413 | JB | KOR |
| 200 | 105543 | GB | KOR | | 282 | 113458 | CN | KOR |
| 207 | 105856 | JB | KOR | | 283 | 113577 | GB | KOR |
| 210 | 105954 | GN | KOR | | 284 | 113582 | GB | KOR |
| 212 | 105997 | JN | KOR | | 285 | 115174 | GB | KOR |
| 214 | 108713 | GB | KOR | | 286 | 115180 | GB | KOR |
| 215 | 108752 | GB | KOR | | 287 | 115186 | GB | KOR |
| 218 | 108786 | GB | KOR | | 293 | 119935 | GB | KOR |
| 219 | 108852 | GW | KOR | | 294 | 119936 | GB | KOR |
| 297 | 134960 | GB | KOR | | 471 | 185713 | JB | KOR |
| 299 | 134969 | GB | KOR | | 472 | 185714 | JB | KOR |
| 300 | 134978 | GB | KOR | | 473 | 185715 | JB | KOR |

**TABLE 30.1** continued

| Sample Number | IT or Tem. IT | Region | Country of Origin | Sample Number | IT or Tem. IT | Region | Country of Origin |
|---|---|---|---|---|---|---|---|
| 301 | 135788 | GB | KOR | 4?4 | 185716 | JB | KOR |
| 302 | 136087 | GB | KOR | 4?5 | 185717 | CN | KOR |
| 305 | 138108 | GB | KOR | 477 | 185719 | CB | KOR |
| 308 | 138140 | GB | KOR | 4?3 | 185720 | CB | KOR |
| 310 | 138142 | GB | KOR | 48? | 185722 | CB | KOR |
| 311 | 138143 | GB | KOR | 48 | 185723 | CB | KOR |
| 313 | 138145 | GB | KOR | 48? | 185724 | CB | KOR |
| 372 | 148426 | GB | KOR | 49? | 191108 | GN | KOR |
| 373 | 148427 | GW | KOR | 49? | 191639 | GW | KOR |
| 374 | 148428 | GB | KOR | 49? | 194510 | GN | KOR |
| 375 | 148429 | CB | KOR | 500 | 194511 | GN | KOR |
| 377 | 155169 | GB | KOR | 502 | 194513 | JN | KOR |
| 378 | 158263 | GW | KOR | 503 | 194514 | JB | KOR |
| 380 | 160614 | JN | KOR | 506 | 195499 | GW | KOR |
| 387 | 162837 | CB | KOR | 507 | 195500 | GW | KOR |
| 389 | 162883 | JB | KOR | 536 | 208546 | GB | KOR |
| 390 | 162884 | JB | KOR | 538 | 208548 | GB | KOR |

| No. | ID | Region | Country | No. | ID | Region | Country |
| --- | --- | --- | --- | --- | --- | --- | --- |
| 392 | 175826 | GB | KOR | 544 | 208554 | JN | KOR |
| 394 | 175860 | GB | KOR | 545 | 208555 | JN | KOR |
| 395 | 175869 | GB | KOR | 548 | 208826 | JN | KOR |
| 403 | 176005 | GG | KOR | 549 | 208852 | GW | KOR |
| 404 | 178414 | JB | KOR | 552 | 209882 | GN | KOR |
| 405 | 178415 | CN | KOR | 555 | 209885 | GN | KOR |
| 406 | 178416 | CB | KOR | 556 | 210197 | GW | KOR |
| 407 | 178417 | JB | KOR | 557 | 210198 | GW | KOR |
| 421 | 180529 | JB | KOR | 561 | 212210 | JN | KOR |
| 422 | 180606 | JB | KOR | 562 | 212211 | JN | KOR |
| 423 | 180612 | GN | KOR | 563 | 212212 | JN | KOR |
| 424 | 180619 | GN | KOR | 564 | 212213 | JN | KOR |
| 425 | 180643 | GN | KOR | 567 | 214694 | GW | KOR |
| 432 | 180927 | CB | KOR | | | | |
| 433 | 180928 | CB | KOR | | | | |
| 436 | 180931 | JB | KOR | | | | |
| 437 | 181904 | JB | KOR | | | | |
| 441 | 181973 | JB | KOR | | | | |

Continued

**TABLE 30.1** continued

| Sample Number | IT or Tem. IT | Region | Country of Origin | Sample Number | IT or Tem. IT | Region | Country of Origin |
|---|---|---|---|---|---|---|---|
| 445 | 185687 | GG | KOR | | | | |
| 446 | 185688 | GN | KOR | | | | |
| 448 | 185690 | GN | KOR | | | | |
| 449 | 185691 | GN | KOR | | | | |
| 451 | 185693 | GN | KOR | | | | |
| 452 | 185694 | GN | KOR | | | | |
| 453 | 185695 | GB | KOR | | | | |
| 458 | 185700 | GB | KOR | | | | |
| 463 | 185705 | JN | KOR | | | | |
| 465 | 185707 | JN | KOR | | | | |

*CB, Chungbuk; CN, Chungnam; GB, Gyungbuk; GG, Gyunggi; GN, Gyungnam; GW, Gangwon; JB, Jeogbuk; JN, Jeonnam.

each 5 seeds of 179 accessions were germinated and cultivated in soil trays. Genomic DNA was extracted from green leaves of buckwheat seedlings. Total genomic DNA was extracted from the leaves of the seedling using a modified CTAB procedure as previously described by Kump and Javornik (1996). The DNA concentration was determined using a UV—Vis spectrophotometer (ND-1000; NanoDrop, Wilmington, DE, USA). The DNA solution was then diluted to a working concentration with distilled water and stored at $-20°C$ until use.

## Assess of Microsatellite Markers

All of the SSR markers were obtained from molecular markers developed by Ma et al., (2009) for analysis of genetic diversity and relationships in common buckwheat. Ten polymorphic SSR markers were utilized in a genetic diversity analysis of a common buckwheat population consisting of 179 accessions of diverse regions in Korea (Table 30.2). The M13F-tail PCR method was used to measure the size of PCR products, as described

**TABLE 30.2** List of Microsatellite Markers Used in This Study

| Marker | Gene Bank Accession | Primer Sequence (5'-3') | Repeat Motif |
|--------|---------------------|-------------------------|--------------|
| GB-FE-001 | EU998635 | F-TGAAACCCAACCATCAGG<br>R-CGACAGTGGCTGGAGAAC | (CAA)7 |
| GB-FE-012 | EU998636 | F-ACTGCACCCCAGAGGATT<br>R-GCTGTATCCATGCCCGTA | (CAG)5(CT)(CAG) &<br>(GAK)8 |
| GB-FE-014 | EU998637 | F-AGGAGCAGAGGTGGTGGT<br>R-CGGAGCCTCTGCAACC | (GA)10 C(GA) |
| GB-FE-035 | EU998638 | F-TGCAATGACTTGGAGGAGA<br>R-ACCACCATTCAACAAGCG | (GAY)14(GGT)(GAB)41 |
| GB-FE-043 | EU998639 | F-TTCAGCACCTGGATGGAC<br>R-TGTCCCCAATGTGAAAGG | (CCA)5 |
| GB-FE-054 | EU998640 | F-TGTTGGACTTCCTAGACCTG<br>R-CATGAAAAGGGGATGCAA | (TR)12 |
| GB-FE-055 | EU998641 | F-CTGCTTGGATCCCATTGA<br>R-AGCCTCTCGATCCCTCTG | (GAK)6 & (GAT)3 &<br>(GAT)2 |
| GB-FE-080 | EU998642 | F-CGAGGTGGGCAGTAGAGA<br>R-GAGGAGGACGAGGAGGTG | (CST)7 |
| GB-FE-169 | EU998643 | F-CAACCCTATGCAGCGTTC<br>R-GAGGGGAAGCTGCTTGTT | (ACA)6 |
| GB-FE-191 | EU998644 | F-AGT AATCAATGACCAGCACGC<br>R-CTGATGGAGGATGCCAAA | (CAT)5 |

previously (Ma et al., 2009). PCR amplification was carried out in a total volume of 20 uL, containing 2 uL of genomic DNA (10 ng/uL), 0.2 uL of the specific primer (10 pmol/uL), 0.4 uL of M13 universal primer (10 pmol/uL), 0.6 uL of normal reverse primer, 2.0 uL of 10× PCR buffer (Takara, Tokyo, Japan), 1.6 uL of dNTP (2.5 mM), and 0.2 uL of Taq polymerase (5 unit/uL; Takara). The reaction mixture was subjected to the following conditions: initial denaturation at 94°C for 3 min, followed by 30 cycles of denaturation at 94°C for 30 s, annealing at 52−55°C for 45 s, then 15 cycles at 94°C for 30 s, 53°C for 45 s, and extension at 72°C for 45 s and final extension at 72°C for 10 min. PCR was carried out in PTC-220 thermocyclers (MJ Research, Waltham, MA, USA). The PCR products were then run on an ABI PRISM 3130xl Genetic Analyzer according to the manufacturer's instructions (Applied Biosystems, USA). Fragments were sized and scored into alleles using GeneMapper v4.0 software (Applied Biosystems, USA).

## Data Analyses of Genetic Diversity and Population Structure

The total number of alleles, alleles frequency, gene diversity, and polymorphism information content (PIC) per each SSR locus were calculated with the PowerMarker version 3.25 analysis (Liu and Muse, 2005). Genetic distance between each pair of accessions were calculated from Nei's distance (Nei and Takezaki, 1983) using the program PowerMarker. Nei's distance was calculated and used the unrooted phylogeny reconstruction using neighbor-joining (NJ) method as implemented in PowerMarker version 3.25 (Liu and Muse, 2005). The tree to visualize the phylogenetic distribution of accessions was constructed using the software MEGA version 5.03 (Tamura et al., 2007) embedded in PowerMarker. The model-based program STRUCTURE (Pritchard et al., 2007) was utilized to infer population structure and assign individuals to populations based on the SSR genotypes using a burn-in of 50,000, run length of 100,000, and a model allowing for admixture and correlated allele frequencies. The number of populations ($K$) was set from 1 to 10, with 3 independent runs each. The most probable value of ($K$) corresponds to the peak in the $D(K)$, which is an ad hoc statistic $D(K)$, assisted with $L(K)$, $L'(K)$ and $L''(K)$ (Evanno et al., 2005). The $D(K)$ perceives the rate of change in log probability of the data with respect to the number of groups inferred by STRUCTURE.

## RESULTS AND DISCUSSION
### Profile of Microsatellite Markers

We assessed the genetic variability of common buckwheat landrace accessions representing diverse regional collections in Korea using SSR markers (Table 30.1). Ten microsatellite markers detected a total of 79 alleles among the 179 buckwheat accessions (Table 30.3). The number of alleles per SSR

**TABLE 30.3** Characterization of the 10 Microsatellite Loci Among Common Buckwheat Base on 179 Collected Germplasm Accessions

| Marker | $M_{AF}$ | $N_A$ | $H_E$ | $H_O$ | PIC |
|---|---|---|---|---|---|
| GB-FE-001 | 0.61 | 2.00 | 0.48 | 0.55 | 0.36 |
| GB-FE-012 | 0.63 | 8.00 | 0.55 | 0.44 | 0.51 |
| GB-FE-014 | 0.47 | 5.00 | 0.67 | 0.60 | 0.62 |
| GB-FE-035 | 0.12 | 31.00 | 0.94 | 0.25 | 0.93 |
| GB-FE-043 | 0.77 | 2.00 | 0.35 | 0.22 | 0.29 |
| GB-FE-054 | 0.37 | 9.00 | 0.74 | 0.14 | 0.70 |
| GB-FE-055 | 0.52 | 2.00 | 0.50 | 0.92 | 0.37 |
| GB-FE-080 | 0.78 | 6.00 | 0.36 | 0.36 | 0.33 |
| GB-FE-169 | 0.97 | 5.00 | 0.05 | 0.06 | 0.05 |
| GB-FE-191 | 0.42 | 9.00 | 0.67 | 0.62 | 0.61 |
| Total | 5.67 | 79.00 | 5.31 | 4.15 | 4.78 |
| Mean | 0.57 | 7.90 | 0.53 | 0.42 | 0.48 |

*MAF*, major allele frequency; *NA*, number of alleles; *HO*, observed heterozygosity; *HE*, expected heterozygosity; *PIC*, polymorphic information content.

marker locus ($N_A$) ranged from 2 (GB-FE-001, GB-FE-043 and GB-FE-055) to 31 (GB-FE-035) with an average of 7.9 alleles. The GB-FE-035 marker produced 31 alleles that were the highest number of alleles of markers and the highest PIC value 0.93. The major allele frequencies ($M_{AF}$) for the 10 polymorphic loci varied from 0.12 (GB-FE-035) to 0.97 (GB-FE-169) with an average allele frequency of 0.57. The expected heterozygosity ($H_E$) values ranged from 0.05 to 0.94 with an average of 0.53 and the observed heterozygosity ($H_O$) ranged from 0.06 to 0.92 with an average of 0.42. The overall polymorphic information contents (PIC) values ranged from 0.05 to 0.93 with an average of 0.48. We can confirm the genetic diversity among 179 common buckwheat accessions in this study. These results are compared with those detected in buckwheat using SSR markers by Iwata ct al., (2005). Our results indicated that the average $H_E$ value was lower than that of among the 19 cultivars (0.819) used by Iwata et al., (2005) in buckwheat.

## Genetic Diversity and Phylogenetic Relationships

A neighbor-joining tree of 179 landraces accessions was constructed based on Nei's genetic distance. The genetic distance matrix was generated by PowerMarker software and used to construct an unrooted neighbor-joining tree. The dendrogram revealed a complex accession distribution pattern (Fig. 30.1A). DNA polymorphism detected by 10 SSR markers allowed the

(A)  (B)

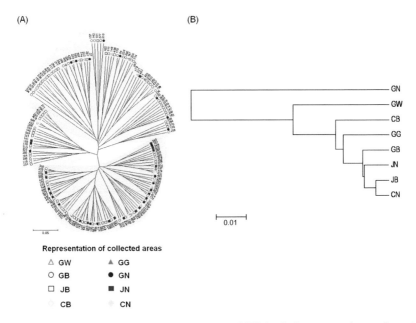

Representation of collected areas

△ GW  ▲ GG
○ GB  ● GN
□ JB  ■ JN
CB  CN

**FIGURE 30.1** Unrooted neighbor-joining trees of 179 buckwheat accessions collected from different regions in Korea based on Nei's genetic distances among 10 SSR loci (A) and the genetic relationships among different populations in different regions (B).

genetic distance to be estimated, and the UPGMA tree showed that 179 accessions of Korean buckwheat cultivars were classified in three major groups. The genetic distance among the buckwheat populations from 8 different regions was also used to construct an UPGMA tree (Fig. 30.1B). The genotypic diversity of buckwheat from 8 geographical regions is compared in Table 30.4. The genetic diversity of buckwheat populations from 8 geographical regions was characterized by an average of 4.08 alleles, ranging from 2 in GG to 6 in GB province.

The mean frequency of major alleles ($M_{AF}$) per locus was 0.613, varying from 0.542 in CB to 0.792 in GG province. The expected heterozygosity ($H_E$) values ranged from 0.310 (GG) to 0.549 (CB) with an average of 0.481 and the observed heterozygosity ($H_O$) ranged from 0.325 (CN) to 0.521 (CB) with an average of 0.415. The overall polymorphic information contents (PIC) values ranged from 0.261 (GG) to 0.497 (CB) with an average of 0.428.

The phylogenetic distribution of buckwheat accessions and populations from the 8 geographical regions indicated the complex distribution and did not cluster from the same regions. This result suggests that common buckwheat is widely dispersed with small local differentiation due to strong migration pressure into new geographical regions. Similar results were reported by other studies (Cho et al., 2011).

**TABLE 30.4** Characterization of the 10 Microsatellite Loci According to 8 Geographical Regions in Korea

| Regions | Sample Size | $M_{AF}$ | $N_A$ | $H_E$ | $H_O$ | PIC |
|---|---|---|---|---|---|---|
| GW | 19 | 0.586 | 3.80 | 0.517 | 0.408 | 0.459 |
| GG | 3 | 0.792 | 2.00 | 0.310 | 0.383 | 0.261 |
| GN | 24 | 0.609 | 4.50 | 0.491 | 0.448 | 0.438 |
| GB | 60 | 0.588 | 6.00 | 0.502 | 0.401 | 0.450 |
| JN | 14 | 0.580 | 3.90 | 0.515 | 0.449 | 0.454 |
| JB | 43 | 0.558 | 5.70 | 0.540 | 0.389 | 0.483 |
| CN | 4 | 0.650 | 2.70 | 0.422 | 0.325 | 0.378 |
| CB | 12 | 0.542 | 4.00 | 0.549 | 0.521 | 0.497 |
| Total | 179 | 4.905 | 32.60 | 3.846 | 3.324 | 3.420 |
| Average | | 0.613 | 4.08 | 0.481 | 0.415 | 0.428 |

## Population Structure

In order to check the subdivision, a model-based clustering method for multi-loci genotype data was performed to infer the population structure and assign individuals to populations using STRUCTURE. The most probable structure number of ($K$) was calculated based on Evanno et al., (2005) using and ad hoc statistics $D(K)$, assisted with $L(K)$, $L'(K)$ and $L''(K)$. The highest value of $D(K)$ for the 179 buckwheat accessions was for $K = 2$ (Fig. 30.2A). The model-based structure analysis revealed the presence of two subpopulations (Fig. 30.2B).

As shown in Table 30.5, most of the 179 buckwheat accessions, that is, 114 (63.7%) accessions, were classified into one of the two genetic groups, whereas 65 (36.3%) of the entire accessions were classified as admixed forms with varying levels of membership shared among the two genetic groups (Fig. 30.2B and Table 30.5). Group 1 consisted of 57 accessions, involving 7 GW, 10 GN, 20 GB, 2 JN, 13 JB, 1 CN, and 4 CB accessions. Group 2 (G2) consisted of 57 accessions, including 6 GW, 1 GG, 5 GN, 22 GB, 5 JN, 13 JB, 2 CN, and 3 CB accessions. The result indicated that the 179 landrace accessions of buckwheat were not distinctly grouped according to geographic distribution.

In this study, the genetic diversity of common buckwheat accessions was studied based on microsatellite markers in order to provide useful information for conservation and utilization of buckwheat genetic resources in Korea. The genetic diversity, phylogenetic relationships and population structure of the common buckwheat landraces in Korea were analyzed by the statistics

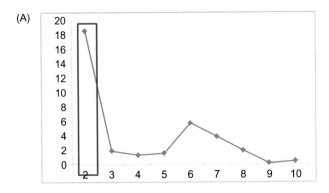

**FIGURE 30.2** Population structure of 179 buckwheat accessions based on 10 SSRs ($K = 2$). (A) (Log) Likelihood of the data ($n = 179$), $L(K)$, as a function of $K$ (the number of groups used to stratify the sample) and values of $D(K)$, with its modal value used to detect the true $K$ of the two groups ($K = 2$). (B) Model-based ancestry of 179 accessions using STRUCTURE.

**TABLE 30.5** Distribution (Inferred) of Accessions From Different Regions to Each Clusters and Admixture

| Region | Cluster 1 | Cluster 2 | Admixture | Total |
|---|---|---|---|---|
| GW | 7 | 6 | 6 | 19 |
| GG | 0 | 1 | 2 | 3 |
| GN | 10 | 5 | 9 | 24 |
| GB | 20 | 22 | 18 | 60 |
| JN | 2 | 5 | 7 | 14 |
| JB | 13 | 13 | 17 | 43 |
| CN | 1 | 2 | 1 | 4 |
| CB | 4 | 3 | 5 | 12 |
| Total | 57 | 57 | 65 | 179 |

methods. The results have shown that there are genotypic variations in common buckwheat accessions collected from 8 different regions in Korea. However, the present study showed that UPGMA tree and the division of genetic structure do not match between the model-based genetic structure and the geographical regions. In addition, the genotypes collected from the same geographical places did not form a single cluster or grouping. Similar opinions were offered by Masud et al., (1995) in pumpkin. The average number of alleles per locus among the 179 accessions of the RDA genotyped by 10 SSR markers was 7.9, slightly less than in the population studied by Iwata et al., (2005). Konishi et al., (2006) reported an average SSR PIC value of 0.79 among a worldwide core collection of common buckwheat accession, the PIC value of 0.48 in our analysis. Although common buckwheat in Korea in this study shows moderate levels of genetic diversity, the parameters are lower than the expected both from outcrossing and from 8 areas in Korea. This genetic variability is highly dependent on the number of samples and on the areas from which the samples were collected. Sinha et al., (1991) reported that selection of parents from distantly placed clusters exhibited significant high heterotic segregants and the decline of cultivated areas may be a major factor.

In conclusion, the results suggest that genetic differentiation was relatively low according to geographic region because of characters of outcrossing and self-incompatibility. Moreover, these reasons could be explained by various factors, such as migration into new geographical areas and adaptation to the climate of Korea. Murai and Ohnishi (1996) had noted a gradual decline of polymorphisms with migration out from the center of origin of the species (Yunnan or Sichchuan province). Ohnishi (1993) describes common buckwheat as a widely dispersed crop with small local differentiation. These results, including the genotype-specific alleles, genetic diversity, and population structure information, will facilitate the use of the buckwheat germplasm for crop improvement. Evaluations of the genetic diversity of Korean landraces have a very important role in the conservation program of plant genetic resources. This diversity information based on genetic variation may contribute to the evaluation of other germplasm collections and the genetic analysis of common buckwheat species in order to elucidate their evolutionary and phylogenetic relationships and to broaden the genetic base of modern buckwheat cultivars.

## ACKNOWLEDGMENT

This study was carried out with the support of the Research Program for Agricultural Science & Technology Development (Code # PJ006825), and the 2011 Post-doctoral Fellowship Program of National Academy of Agricultural Science, Rural Development Administration, Republic of Korea.

# REFERENCES

Alekseeva, E.S. 1986. Selection, cultivation and utilisation of buckwheat. In: Proceedings of the Third International Symposium on Buckwheat, Pulawy, Poland. pp.18–36.

Cho, Y.I., Park, J.H., Lee, C.W., Ra, W-H., Chung, J-W., Lee, J-R., et al., 2011. Evaluation of the genetic diversity and population structure of sesame (*Sesamum indicum* L.) using microsatellite marker. Genes&Genomics 33, 187–195.

Evanno, G., Regnaut, S., Goudet, J., 2005. Detecting the number of clusters of individuals using the software STRUCTURE: a simulation study. Mol. Ecol. 14, 2611–2620.

Goldstein, D.B., Schlötterer, C., 1999. Microsatellites. Evolution and application. Oxford University Press, New York.

Hallauer, A.R., Miranda, J.B.F., 1988. Quantitative Genetics in Maize Breeding, 2nd ed. Iowa State University Press, Ames.

Iwata, H., Imon, K., Tsumura, Y., Ohsawa, R., 2005. Genetic diversity among Japanese indigenous common buckwheat (Fagopyrum esculentum) cultivars as determined from ampliWed fragment length polymorphism and simple sequence repeat markers and quantitative agronomic traits. Genome 48, 367–377.

Javornik, B., Eggum, B.O., Kreft, I., 1981. Studies on protein fractions and protein quality of buckwheat. Genetika 13 (2), 115–121.

Kump, B., Javornik, B., 1996. Evaluation of genetic variability among common buckwheat (*Fagopyrum esculentum* Moench) populations by RAPD markers. Plant Sci. 114, 149–158.

Li, X., Yan, W., Agrama, H., Hu, B., Jia, L., Jia, M., et al., 2010. Genotypic and phenotypic characterization of genetic differentiation and diversity in the USDA rice minicore collection. Genetica 138, 1221–1230.

Litt, M., Luty, J.A., 1989. A hypervariable microsatellite revealed by in vitro amplification of a dinucleotide repeat within the cardiac muscle actin gene. Am. J. Human Genet 44, 397–401.

Liu, K., Muse, S.V., 2005. Powermarker: an integrated analysis environment for genetic marker analysis. Bioinformatics 21, 2128–2129.

Ma, K.H., Kim, N.S., Lee, G-A., Lee, S.-Y., Lee, J.K., Yi, J.Y., et al., 2009. Development of SSR markers for studies of diversity in the genus *Fagopyrum*. Theor. Appl. Genet. 119, 1247–1254.

Masud, M.A.T., Chowdhury, M.A., Hossain, M.A., Hossain, S.M.M., 1995. Multivariate analysis of pumpkin. Bangladesh J. Plant. Breed. Genet 8 (1), 45–50.

Murai, M., Ohnishi, O., 1996. Population genetics of cultivated common buckwheat, *Fagopyrum esculentum* Moench. X. DiVusion routes revealed by RAPD markers. Genes Genet Syst 71, 211–218.

Narain, P., 1979. Buckwheat cultivation in plain of UP. Indian Farming 29 (1), 3–5.

Nei, M., Takezaki, N., 1983. Estimation of genetic distances and phylogenetic trees from DNA anlysis. In: Proceedings of 5th World Cong. Genet. Appl. Livstock Prod. 21: 405–412.

Ohnishi, O., 1993. Population genetics of cultivated common buckwheat, *Fagopyrum esculentwn* Moench, VIII. Local differentiation of land races in Europe and the silk road. Jpn. J. Genet 68, 317–326.

Pritchard, J.K., Wen, X., Falush, D., 2007. Documentation for structure software: Version 2.2. Department of Human Genetics, University of Chicago; Department of Statistics, University of Oxford. Available at http://pritch.bsd.uchicago.edu/software.

Sharma, K.D., Boyes, J.D., 1961. Modified incompatibility of common buckwheat following irradiation. Can. J. Bot. 39, 1241–1246.

Sinha, P.K., Chauhan, V.S., Prasad, K., Chauhan, J.S., 1991. Genetic divergence in indigenous upland rice varieties. Indian J. Genet 51 (1), 47–50.

Smith,J.S.C., D.S. Ertl, B.A. Orman. 1995. Identification of maize varieties. In: Wrigley, C.W., (Ed.), Identification of food grain varieties. Am. Assoc. Cereal Chemists, St Paul, pp 253–264.

Tamura, K., Dudley, J., Nei, M., Kumar, S., 2007. MEGA4: molecular evolutionary genetics analysis (MEGA) software version 4.0. Mol. Biol. Evol. 24, 1596–1599.

Tautz, D., 1989. Hypervariability of simple sequences as general source for polymorphic DNA markers. Nucleic Acids Res. 17, 6463−6471.

Thuillet, A.-C., Bataillon, T., Poirier, S., Santoni, S., David, J.L., 2005. Estimation of long-term effective population sizes through the history of durum wheat using microsatellite data. Genetics 169, 1589 –1599.

Vigouroux, Y., Mitchell, S., Matsuoka, Y., Hamblin, M., Kresovich, S., Stephen, J., et al., 2005. An analysis of genetic diversity across the maize genome using microsatellites. Genetics 169, 1617−1630.

Vos, P., Horgers, R., Bleeker, M., Reijans, M., Van de Lee, T., Hornes, M., et al., 1995. AFLP: a new technique for DNA fingerprinting. Nucleic Acids Res. 23, 4407−4414.

# Chapter | Thirty One

# Applications of Plant Tissue Culture and Biotechnology in Buckwheat

## Chang Ha Park and Sang Un Park
*Chungnam National University, Daejeon, South Korea*

## INTRODUCTION

Buckwheat is an agronomic species of the Polygonaceae family that belongs to the genus *Fagopyrum*. There are two buckwheat (*Fagopyrum*) species used for food around the world (Beckman, 2005); common buckwheat (*Fagopyrum esculentum* Moench) originates from Southwest China and has gradually spread to all continents, whereas tartary buckwheat (*Fagopyrum tataricum* Gaertn.) is grown and used in the mountainous regions of Southwest China (Sichuan) and in northern India, Bhutan, and Nepal. In Europe, tartary buckwheat is currently grown as a crop only in a small part of northwest Europe (Bonafaccia and Fabjan, 2003).

Buckwheat seeds are used for human food production and as a potentially important source of rutin (Hinneburg and Neubert, 2005; Kalinova et al., 2006). Previous research findings have shown that rutin, a flavonol glycoside, has strong antioxidant effects and possesses other interesting pharmacological properties, such as antiinflammatory, anticarcinogenic, antithrombotic, cytoprotective, and vasoprotective activities (Kreft et al., 2002; Li and Zhang, 2001).

Early this century, Haberlandt (1902) suggested that individual nucleated plant cells might have the genetic capacity to be converted into complete plants either directly or through an intervening callus stage. This phenomenon has been termed "totipotency". More recently, the term "regeneration" has been broadly used in the context of tissue culture to indicate the recovery of

**333**

*Buckwheat Germplasm in the World.* DOI: https://doi.org/10.1016/B978-0-12-811006-5.00038-0
© 2018 Elsevier Inc. All rights reserved.

a whole organism from cells, tissues, organs, meristems, or zygotic embryos cultivated in vitro (Phillips and Hubstenberger, 1995).

Almost all current practical transformation systems require a plant regeneration system to establish an efficient method for gene transfer, selection, and regeneration of transgenic plants. Plant regeneration systems have been developed commercially for the micropropagation of ornamentals and disease-free stock. There are two major systems of plant regeneration, i.e., organogenesis and somatic embryogenesis (Brown and Thorpe, 1986; Thorpe, 1990).

Plant genetic transformation can be defined as the transfer of foreign genes into a plant. In plants, successful genetic transformation requires the production of normal, fertile plants which express newly inserted genes through an optimal plant regeneration system. The process of genetic transformation involves several distinct stages, namely insertion, integration, expression, and inheritance of the new foreign gene. A wide range of methods have been used for the transformation of plant cells. Major methods include *Agrobacterium*-mediated transfer, direct transfer into protoplasts, and particle-gun (microprojectile bombardment or biolistic) transformation; other techniques have also been attempted, including viral transformation, direct injection into plant cells, pollen transformation, injection into pollen tubes, and fusion with DNA-containing liposomes, with varying degrees of success. These approaches are not widely used, and in some cases, there may be doubts regarding whether transformation has actually taken place. *Agrobacterium*-mediated genetic transformation systems are the most common and inexpensive methods (Gelvin, 2003; Dunwell, 2005; Narusakaet al., 2012).

In this paper, we review various studies describing the applications of plant biotechnology, in vitro plant regeneration, hairy root cultures for bioactive compound production, and genetic transformation for crop improvement of buckwheat.

## IN VITRO PLANT REGENERATION

In 1974, the first report on the differentiation of buckwheat plants from callus culture was published (Yamane, 1974). Later, in vitro plant regeneration of buckwheat via somatic embryogenesis or shoot organogenesis was established and reported by several researchers. Protocols for plant regeneration from various explants, such as hypocotyls (Lachmann and Adachi, 1990; Lachmann, 1991; Gumerova et al., 2003, Han et al., 2011), cotyledons (Srejović and Nešković, 1981; Srejović and Nešković, 1985; Neskovic et al., 1987; Miljuš-Djukić et al., 1992; Woo et al., 2000; Park and Park, 2001), immature inflorescence (Takahata, 1988), immature embryos (Nešković et al., 1987; Rumyantseva et al., 1989), and anthers (Adachi et al., 1989; Bohanec et al., 1993), have been reported. We also developed a simple and efficient protocol for shoot organogenesis and plant regeneration from the lateral cotyledonary meristems of buckwheat (*F. esculentum* Moench) and investigated the effects of different concentrations of cytokinins (BAP, kinetin, and TDZ) on the efficiency of shoot organogenesis in buckwheat. Treatment with BAP significantly induced shoot regeneration from cotyledon cultures of buckwheat. The highest number of shoots per explant (5.7) and greatest shoot length (1.3 cm) were obtained on MS medium containing 4.0 mg/L BAP. The addition of $AgNO_3$ at an optimal concentration of 7 mg/L substantially improved the shoot regeneration frequency, which was around 30% more than that in the control; however, concentrations higher than the optimal concentration caused the shoot regeneration frequency to decrease. The regenerated shoots (about 1 cm in length) displayed a normal morphology and could be easily rooted without any plant hormone treatment. The rooted plants were hardened and transferred to soil with a 72% survival rate, where they grew normally (Lee et al., 2009).

# PLANT TRANSFORMATION AND BIOTECHNOLOGY

Before establishment of buckwheat transformation, the susceptibility of common buckwheat to tumor formation by *Agrobacterium tumefaciens* and to hairy root formation by *Agrobacterium rhizogenes* was reported (Nesovic et al., 1986). In this experiment, strain A281 exhibited remarkable virulence, stronger than those of A348, Ach5, and A6. Hairy roots developed readily on stems at the site of *A. rhizogenes* inoculation. Buckwheat is known to be highly sensitive to *Agrobacterium*, which can be used as a vector for genetic transformation. The same research group (Miljuš-Djukić et al., 1992) succeeded in *Agrobacterium*-mediated transformation of buckwheat using a reporter gene for kanamycin resistance.

Kojima et al. (2000a) developed a simple and efficient method for transformation of common buckwheat plants using *A. tumefaciens*. Apical meristems of seedlings of *F. esculentum* var. Shinano No. 1 were pricked with a needle and inoculated with *A. tumefaciens* (LBA4404, pBI121). The inoculated seedlings were grown to maturation and allowed to pollinate randomly to set the seeds (T1 plants). The transformation efficiency of the T1 plants was estimated by germination in the presence of geneticin (20 μg/mL) and by detection of the beta-glucuronidase (*GUS*) gene by polymerase chain reaction (PCR), indicating that 36% and 70% of the T1 plants were transformed, respectively. The same research group reported apical meristems of buckwheat (*F. esculentum* var. Shinano No. 1) seedlings were transformed by inoculation with *A. tumefaciens* harboring a binary vector containing cDNA of the rice MADS box gene (accession no. [DDBJ] AB003325) in either a sense or antisense orientation downstream of the CaMV35S promoter. The plants transformed (T0) with the cDNA in both orientations showed unique features; the plants transformed with the cDNA in the sense orientation were stimulated in branching, producing many branches, whereas plants transformed with the cDNA in the antisense orientation showed inhibition in both branching and growth (Kojima et al. 2000b).

Cheng et al. (2007, 2008) developed salt-tolerant transgenic buckwheat plants overexpressing AtNHX1, a vacuolar Na(+)/H(+) antiporter gene from *Arabidopsis thaliana*. These plants were able to grow, flower, and accumulate more rutin in the presence of 200 mM sodium chloride. Moreover, the content of important nutrients in buckwheat was not affected by the high salinity of the soil. These results demonstrated the potential value of these transgenic plants for agriculture use in saline soil.

## IN VITRO HAIRY ROOT CULTURE FOR BIOACTIVE COMPOUND PRODUCTION

Various strategies, such as *Agrobacterium*-mediated gene transfer, genetic variability, in vitro cell line establishment, in vitro cell suspension cultures, bioreactor cultivation, shoot cultivation, organ cultivation at the bioreactor level, and in vitro hairy root cultivation, have been demonstrated for the synthesis and production of pharmaceutically important phytochemicals (Bourgaud et al., 2001). *Agrobacterium* is a gram-negative soil-borne bacterium that is able to transfer part of its DNA, specifically the T-DNA carried on a large plasmid, to host plant cells (Tzfira et al., 2004). The hairy roots obtained by infecting plants with *A. rhizogenes* are unique with respect to their genetic and biosynthetic stability and have been used to produce selected plant metabolites with stable, high yields. Furthermore, they are appropriate for the production of valuable secondary metabolites, as they can energize growth regulators and are characterized by rapid growth (Giri and Narasu, 2000; Christey and Braun, 2004; Hu and Du, 2006; Srivastava and Srivastava, 2007).

Hairy root cultures of *F. esculentum* have been used to examine the biosynthesis of rutin and other polyphenols in vitro (Tanaka et al., 1996). However, no previous studies have examined the effects of various medium conditions on rutin production and the growth of hairy root cultures of common buckwheat. Lee et al. (2007) established a hairy root culture of buckwheat by infecting leaf explants with *A. rhizogenes* R1000 and examined the effects of different media and plant growth regulators on growth and rutin biosynthesis. They established a hairy root culture of buckwheat by infecting leaf explants with *A. rhizogenes* R1000 and tested the growth conditions and rutin production rates of these cultures. Of four tested culture media (half-strength B5, B5, half-strength MS, and MS medium), half-strength MS medium was found to induce the highest levels of growth (378 mg dry wt/ 30 mL flask) and rutin production (1.4 mg/g dry wt) of common buckwheat hairy root. In contrast, supplementation with auxins (0.1–1 mg/L indoleacetic acid [IAA], indolebutyric acid [IBA], and naphthaleneacetic acid [NAA]) increased the growth rate, but had no significant effect on rutin production in hairy root cultures. Collectively, these findings indicated that hairy root cultures of buckwheat culture could be a valuable alternative approach for rutin production.

Park et al. (2011) obtained hairy roots of *F. tataricum* after inoculating sterile young stems with *A. rhizogenes* R1000. The established roots displayed two morphological phenotypes when cultured on hormone-free medium containing Murashige-Skoog salts and vitamins (Murashige and Skoog, 1962). The thin phenotype had a higher growth rate than the thick phenotype. Furthermore, the phenolic compound content of the thin phenotype was higher than that of the thick phenotype. In terms of their total dry weight, the thin phenotype produced almost double the amount of (−)-epigallocatechin as well as more than 51.5% caffeic acid, 65% chlorogenic acid, and 40% rutin compared with the thick phenotype after 21 days of culture. Therefore, selection of the optimal morphological phenotype of hairy roots of tartary buckwheat is an important factor for improved phenolic compound production.

The efficient protocol was described for the transformation of common buckwheat (*F. esculentum* Moench.) (Kim et al., 2010) and tartary buckwheat (*F. tataricum*) (Kim et al., 2009) hairy roots that were infected by *A. rhizogenes*, a strain with the binary vector pBI121. Kanamycin-resistant buckwheat hairy roots were maintained on hormone-free medium. PCR analysis of the neomycin phosphotransferase (*NTPII*) gene confirmed transformation in kanamycin-resistant hairy root cultures. Detection of high levels of GUS transcripts and enzyme activity and analysis of GUS histochemical localization also confirmed the stable genetic transformation. Transgenic root cultures of common buckwheat and tartary buckwheat will facilitate studies of the molecular and metabolic regulation of rutin biosynthesis and evaluation of the genetic engineering potential of this species. To improve the production of rutin in common buckwheat, Park et al. (2012) overexpressed the flavonol-specific transcription factor AtMYB12 using *A. rhizogenes*-mediated transformation into hairy root culture systems. This induced the expression of flavonoid biosynthetic genes encoding phenylalanine ammonia lyase, cinnamate 4-hydroxylase, 4-coumarate:CoA ligase, chalcone synthase, chalcone isomerase, flavone 3-hydroxylase, flavonoid 30-hydroxylase, and flavonol synthase and led to accumulation of rutin in buckwheat hairy roots up to 0.9 mg/g dry wt. PAP1 expression, however, did not correlate with the production of rutin.

Some physiological studies using hairy root culture systems have been reported. An experiment was conducted to investigate the level of expression of various genes in the phenylpropanoid biosynthetic pathway to analyze in vitro production of anthocyanin and phenolic compounds from hairy root cultures derived from two cultivars of tartary buckwheat (Hokkai T8 and T10). A total of 47 metabolites were identified by gas chromatography/time-of-flight mass spectrometry and subjected to principal component analysis in order to fully distinguish between Hokkai T8 and T10 hairy roots. The expression levels of phenylpropanoid biosynthetic pathway genes were higher for almost all the genes in T10 than T8 hairy root, except for *FtF3'H-2* and *FtFLS-2*. Rutin, quercetin, gallic acid, caffeic acid, ferulic acid, 4-hydroxybenzoic acid, and two anthocyanin compounds were identified in Hokkai T8 and T10 hairy roots. The concentrations of rutin and anthocyanin in Hokkai T10 hairy roots of tartary buckwheat were several-fold higher than those obtained from Hokkai T8 hairy root (Thwe et al., 2013). Differential expression patterns of flavonoid biosynthetic pathway genes in the hairy roots of tartary buckwheat cultivars ("Hokkai T8" and "Hokkai T10") were studied over time. The gene expression levels peaked on day 5 of culture during both dark and light conditions. Notably, *FtPAL*, *Ft4CL*, *FtC4H*, *FtCHI*, *FtF3H*, *FtF3'H-1*, and *FtFLS-1* were more highly expressed in Hokkai T10 than in Hokkai T8 under dark conditions, and *FtPAL* and *FtCHI* were found to be significantly upregulated, except on day 20 of culture. Significantly higher levels of the phenolic compound rutin, along with two anthocyanins, were detected in the hairy roots of Hokkai T10 under both conditions. Furthermore, among all phenolic compounds detected, the amount of rutin in Hokkai T10 hairy roots was found to be ~5-fold (59.01 mg/g) higher than that in the control (12.45 mg/g) at the respective time periods under light and dark conditions (Thwe et al., 2014). Recently, we investigated the influence of auxins on the growth of hairy roots and accumulation of anthocyanins, including cyanidin 3-*O*-glucoside (C3gl) and cyanidin 3-*O*-rutinoside (C3r), in hairy root cultures of the tartary buckwheat cultivar Hokkai T10. C3gl and C3r contents were evaluated using high-performance liquid chromatography. Various auxins, such as 2,4-dichlorophenoxyacetic acid (2,4-D), IAA, IBA, and NAA, were added to the medium of hairy root cultures at 0.1, 0.5, and 1.0 mg/L. IAA, IBA, and 2,4-D promoted the growth of hairy roots since the dry weight of the roots was slightly higher than or comparable to that of the control. However, NAA at all concentrations suppressed the growth of hairy roots. Generally, auxin treatments resulted in higher accumulation of C3gl and C3r than that of the control, except for 0.5 mg/mL IAA and NAA. The amount of C3gl and C3r after treatment with 1.0 mg/mL IBA was highest among all treatments and was 3.24 times more than that of the control. Thus, these results suggested that auxins at appropriate concentrations may facilitate hairy root growth of tartary buckwheat and enhance the production of C3gl and C3r (Park et al., 2016).

## CONCLUSION

Plant cell and tissue cultures play an important role in the manipulation of plants for improved crop varieties. Plant regeneration systems are an essential part of micropropagation and molecular approaches leading to plant improvement in buckwheat. For plant tissues of buckwheat, establishment of optimal growing conditions in vitro is difficult. Therefore, there continues to be an urgent need for extensive work in the field of basic tissue culture protocols and plant regeneration methods for buckwheat before any practical utilization of molecular biology approaches can be achieved.

Plant transformation has become a core research tool for crop improvement and for studying gene function in plants. Various methodologies of plant transformation have been developed to increase the efficiency of transformation and to achieve stable expression of transgenes in plants. Genetic engineering using plant gene transformation systems is responsible for the production of transgenic plants with enhancement in a range of desirable traits. For the establishment of genetic engineering protocols for buckwheat, plant transformation systems are essential. Several scientists have developed transformation systems for buckwheat; however, it is still difficult to produce transgenic plants easily. Therefore, further studies are needed to establish optimal protocols for buckwheat transformation. The information presented in this chapter is expected to facilitate further studies by students and scientists in this discipline.

## REFERENCES

Adachi, T., Yamaguchi, A., Miike, Y., Hoffmann, F., 1989. Plant regeneration from protoplasts of common buckwheat (*Fagopyrum esculentum*). Plant. Cell. Rep. 8 (4), 247−250.

Beckman, C., 2005. Buckwheat: situation and outlook. Bi-Weekly Bulletin 18, 1−4.

Bohanec, B., Nešković, M., Vujičić, R., 1993. Anther culture and androgenetic plant regeneration in buckwheat (Fagopyrum esculentum Moench). Plant Cell Tissue Organ Cult. 35 (3), 259−266.

Bonafaccia, G., Fabjan, N., 2003. Nutritional comparison of tartary buckwheat with common buckwheat and minor cereals. Zbiornik Biotehniške Fakultet Univerze v Ljubljani 81 (2), 349−355.

Bourgaud, F., Gravot, A., Milesi, S., Gontier, E., 2001. Production of plant secondary metabolites: a historical perspective. Plant Sci. 161 (5), 839−851.

Brown, D.C., Thorpe, T.A., 1986. Plant regeneration by organogenesis. In: Vasil, I.K. (Ed.), Cell Culture and Somatic Cell Genetics of Plants. Academic Press, New York, pp. 49−65.

Cheng, L.H., Zhang, B., Xu, Z.Q., 2007. Genetic transformation of buckwheat (*Fagopyrum esculentum* Moench) with AtNHX1 gene and regeneration of salt-tolerant transgenic plants. Chin. J. Biotechnol. 23 (1), 51−60.

Cheng, L.H., Zhang, B., Xu, Z.Q., 2008. Salt tolerance conferred by overexpression of Arabidopsis vacuolar Na + /H + antiporter gene AtNHX1 in common buckwheat (*Fagopyrum esculentum*). Transgenic. Res. 17 (1), 121.

Christey, M.C., Braun, R.H., 2004. Production of hairy root cultures and transgenic plants by *Agrobacterium rhizogenes*-mediated transformation. In: Peña, L. (Ed.), *Transgenic Plants: Methods and Protocols. Methods in Molecular Biology*™. Humana Press, Totowa.

Dunwell, J.M., 2005. Intellectual property aspects of plant transformation. Plant Biotechnol. J. 3 (4), 371−384.

Gelvin, S.B., 2003. Agrobacterium-mediated plant transformation: the biology behind the "Gene-Jockeying" tool. Microbiol. Mol. Biol. Rev. 67 (1), 16−37.

Giri, A., Narasu, M.L., 2000. Transgenic hairy roots: recent trends and applications. Biotechnol. Adv. 18 (1), 1−22.

Gumerova, E., Galeeva, E., Chuyenkova, S., Rumyantseva, N., 2003. Somatic embryogenesis and bud formation on cultured *Fagopyrum esculentum* hypocotyls. Russian J. Plant Physiol. 50 (5), 640−645.

Haberlandt, G., 1902. Kulturversuche mit isolierten Pflanzenzellen. Sitzungsber. d. Akad. d. Wissensch. Wien, math.-naturw. Klasse 111, 1913−1921.

Han, M.H., Kamal, A.H.M., Huh, Y.S., Jeon, A., Bae, J.S., Chung, K.Y., et al., 2011. Regeneration of Plantlet Via Somatic Embryogenesis from Hypocotyls of Tartary Buckwheat (*Fagopyrum tataricum*). Austral. J. Crop Sci. 5 (7), 865.

Hinneburg, I., Neubert, R.H., 2005. Influence of extraction parameters on the phytochemical characteristics of extracts from buckwheat (*Fagopyrum esculentum*) herb. J. Agric. Food. Chem. 53 (1), 3−7.

Hu, Z.B., Du, M., 2006. Hairy root and its application in plant genetic engineering. J. Integr. Plant Biol. 48 (2), 121−127.

Kalinova, J., Triska, J., Vrchotova, N., 2006. Distribution of vitamin E, squalene, epicatechin, and rutin in common buckwheat plants (*Fagopyrum esculentum* Moench). J. Agric. Food. Chem. 54 (15), 5330−5335.

Kim, Y.K., Li, X., Xu, H., Park, N.I., Uddin, M.R., Pyon, J.Y., et al., 2009. Production of phenolic compounds in hairy root culture of tartary buckwheat (*Fagopyrum tataricum* Gaertn). J. Crop Sci. Biotechnol. 12 (1), 53−57.

Kim, Y.K., Xu, H., Park, W.T., Park, N.I., Lee, S.Y., Park, S.U., 2010. Genetic transformation of buckwheat (*Fagopyrum esculentum* M.) with *Agrobacterium rhizogenes* and production of rutin in transformed root cultures. Austral. J. Crop Sci. 4 (7), 485.

Kojima, M., Arai, Y., Iwase, N., Shirotori, K., Shioiri, H., Nozue, M., 2000a. Development of a simple and efficient method for transformation of buckwheat plants (*Fagopyrum esculentum*) using *Agrobacterium tumefaciens*. Biosci. Biotechnol. Biochem. 64 (4), 845−847.

Kojima, M., Hihara, M., Koyama, S.I., Shioiri, H., Nozue, M., Yamamoto, K., et al., 2000b. Buckwheat transformed with cDNA of a rice MADS box gene is stimulated in branching. Plant Biotechnol. 17 (1), 35−42.

Kreft, S., Štrukelj, B., Gaberščik, A., Kreft, I., 2002. Rutin in buckwheat herbs grown at different UV-B radiation levels: comparison of two UV spectrophotometric and an HPLC method. J. Exp. Bot. 53 (375), 1801−1804.

Lachmann, S., 1991. Plant cell and tissue culture in buckwheat: an approach towards genetic improvements by means of unconventional breeding techniques. *Overcoming breeding barriers by means of plant biotechnology*. Proc. Int. College Miyazaki Japan 145−154.

Lachmann, S., Adachi, T., 1990. Callus regeneration from hypocotyl protoplasts of tartary buckwheat (*Fagopyrum tataricum* Gaertn.). Fagopyrum 10, 62−64.

Lee, S.Y., Cho, S.I., Park, M.H., Kim, Y.K., Choi, J.E., Park, S.U., 2007. ). Growth and rutin production in hairy root cultures of buckwheat (*Fagopyrum esculentum* M.). Prep. Biochem. Biotechnol. 37 (3), 239−246.

Lee, S.Y., Kim, Y.K., Uddin, M.R., Park, N., Park, S.U., 2009. An efficient protocol for shoot organogenesis and plant regeneration of buckwheat (*Fagopyrum esculentum* Moench.). Romanian Biotechnol. Lett. 14 (4), 4524−4529.

Li, S.Q., Zhang, Q.H., 2001. Advances in the development of functional foods from buckwheat. Crit. Rev. Food. Sci. Nutr. 41 (6), 451−464.

Miljuš-Djukić, J., Nešković, M., Ninković, S., Crkvenjakov, R., 1992. ). Agrobacterium-mediated transformation and plant regeneration of buckwheat (*Fagopyrum esculentum* Moench.). Plant Cell Tissue Organ Cult. 29 (2), 101−108.

Murashige, T., Skoog, F., 1962. A revised medium for rapid growth and bioassays with tobacco tissue cultures. Physiol. Plant. 15 (3), 473−497.

Narusaka, Y., Narusaka, M., Yamasaki, S., Iwabuchi, M., 2012. Methods to transfer foreign genes to plants. In: Çiftçi, Y.Ö. (Ed.), Transgenic Plants-Advances and Limitations. InTech, Novi Sad, pp. 173−189.

Nesovic, M., Srejovic, V., Vujicic, R., 1986. Buckwheat (*Fagopyrum esculentum* Moench). In: Bajaj, Y.P.S. (Ed.), Biotechnology in Agriculture and Forestry 2, Crops I. Springer-Verlag, Berlin, pp. 579−593.

Nešković, M., Vujičić, R., Budimir, S., 1987. ). Somatic embryogenesis and bud formation from immature embryos of buckwheat (*Fagopyrum esculentum* Moench.). Plant Cell Rep. 6 (6), 423−426.

Park, S.U., Park, C.H., 2001. Multiple shoot organogenesis and plant regeneration from cotyledons of buckwheat. *Fagopyrum esculentum* 427−430.

Park, N.I., Xiaohua, L., Uddin, R.M., Park, S.U., 2011. Phenolic compound production by different morphological phenotypes in hairy root cultures of Fagopyrum tataricum Gaertn. Arch. Biol. Sci. 63 (1), 193−198.

Park, N.I., Li, X., Thwe, A.A., Lee, S.Y., Kim, S.G., Wu, Q., et al., 2012. Enhancement of rutin in *Fagopyrum esculentum* hairy root cultures by the Arabidopsis transcription factor AtMYB12. Biotechnol. Lett. 34 (3), 577−583.

Park, C.H., Thwe, A.A., Kim, S.J., Park, J.S., Arasu, M., Al-Dhabi, N.A., et al., 2016. Effect of auxins on anthocyanin accumulation in hairy root cultures of tartary buckwheat cultivar hokkai T10. Nat. Product Commun. 11 (9), 1283−1286.

Phillips, G.C., Hubstenberger, J.F., 1995. Micropropagation by proliferation of axillary buds. In: Gamborg, O.L., Phillips, G.C. (Eds.), Plant Cell, Tissue and Organ Culture. Springer Lab Manual. Springer-Verlag, Berlin, pp. 45−54.

Rumyantseva, N., Sergeeva, N., Khakimova, L., Salnikov, V., Gumerova, E., Lozovaya, V., 1989. Organogenesis and somatic embryogenesis in tissue culture of two buckwheat species. Fiziol Rast 36, 187−194.

Srejović, V., Nešković, M., 1981. Regeneration of plants from cotyledon fragments of buckwheat (*Fagopyrum esculentum* Moench). Zeitschrift für Pflanzenphysiologie 104 (1), 37−42.

Srejović, V., Nešković, M., 1985. Effect of gibberellic acid on organogenesis in buckwheat tissue culture. Biol. Plant. 27 (6), 432−437.

Srivastava, S., Srivastava, A.K., 2007. Hairy root culture for mass-production of high-value secondary metabolites. Crit. Rev. Biotechnol. 27 (1), 29−43.

Takahata, Y., 1988. Plant regeneration from cultured immature inflorescence of common buckwheat (*Fagopyrum esculentum* Moench) and perennial buckwheat (*F. cymosum* Meisn.). Jpn. J. Breed. 38 (4), 409−413.

Tanaka, N., Yoshimatsu, K., Shimomura, K., Ishimaru, K., 1996. Rutin and other polyphenols in *Fagopyrum esculentum* hairy roots. Nat. Med. 50 (4), 269−272.

Thorpe, T.A., 1990. The current status of plant tissue culture. In: Bhojwani, S.S. (Ed.), Plant Tissue Culture: Applications and Limitations. Elsevier, Amsterdam, pp. 1−33.

Thwe, A.A., Kim, J.K., Li, X., Kim, Y.B., Uddin, M.R., Kim, S.J., et al., 2013. Metabolomic analysis and phenylpropanoid biosynthesis in hairy root culture of tartary buckwheat cultivars. PLoS. ONE 8 (6), e65349.

Thwe, A.A., Kim, Y., Li, X., Kim, Y.B., Park, N.I., Kim, H.H., et al., 2014. Accumulation of phenylpropanoids and correlated gene expression in hairy roots of tartary buckwheat under light and dark conditions. Appl. Biochem. Biotechnol. 174 (7), 2537−2547.

Tzfira, T., Vaidya, M., Citovsky, V., 2004. Involvement of targeted proteolysis in plant genetic transformation by *Agrobacterium*. Nature 431 (7004), 87.

Woo, S.H., Nair, A., Adachi, T., Campbell, C.G., 2000. Plant regeneration from cotyledon tissues of common buckwheat (*Fagopyrum esculentum* Moench). In Vitro Cell. Dev. Biol. Plant 36 (5), 358−361.

Yamane, Y., 1974. Induced differentiation of buckwheat plants from subcultured calluses in vitro. Jpn. J. Genet. 49 (3), 139−146.

# FURTHER READING

Neskovic, M., Vinterhalter, R., Miljus-Djukic, J., Ninkovic, S., Vinterhalter, D., Jovanovic, V., et al., 1990. Susceptibility of buckwheat (*Fagopyrum esculentum* Moench.) to *Agrobacterium tumefaciens* and *A. rhizogenes*, Fagopyrum, 10. pp. 57−61.

# Index

Printed in the United States
By Bookmasters